THE UK

SCANNING DIRECTORY

Interproducts
Publishers of Specialist Radio Books
Scotland and Eire

The UK Scanning Directory

© Copyright 1993 by Interproducts

ISBN 0951978322

UK Office:
8 Abbot Street, Perth, PH2 0EB, Scotland
Tel/Fax: (0738) 441199

Introduction

In the past twelve months since the publication of the 2nd Edition of *The UK Scanning Directory*, interest in the scanning hobby has continued to mushroom, so much so that a third edition has been produced to meet the ever increasing demands of it's readers. Such is the interest in scanning, that at least one long running scanner guide, which previously provided only endless bandplans, has now ventured into the realm of quoting, with much inaccuracy, limited spot frequencies which were previously quoted correctly in *The UK Scanning Directory*. *The UK Scanning Directory* continues to be the only publication in the UK today that provides the reader with the most detailed, up-to-the-minute spot frequencies from all over the country, and one should be very cautious of other cheaper and inferior publications which claim to do the same.

Within the last year, we have all been exposed in one way or another to the revelations of Dianagate and Camillagate, and all the associated moral questions. It now seems safe to assume that whole affair was designed to undermine both the Royal Family and the scanning hobby in the UK, neither of which seems to have been acheived. One thing is certain though, the scanner hobbyists who originally were accused of eavesdropping on the Royals have been completely exhonerated, as tales continue to point to members of GCHQ in Cheltenham and NSA (National Security Agency, the American version of GCHQ) engineering the whole affair.

The legalities of scanning have also been questioned. An official DTI statement indicated that while actually listening to certain VHF and UHF may be illegal, the passing on of frequency information is not. However, there are some pit-falls. A high-level source inside the South Yorkshire Police passed on information regarding the official Police stance on the legality of scanning. He informed us that when a scanner (which is now defined by the Police as a burglar's tool along with the jemmy!) is confiscated from a suspect, it's memory channels are checked to find out if the local police frequencies are stored, which could lead to prosecution.

Finally thanks to all our readers who have taken the time to pass on information and offer their encouragement. Your calls and letters are always welcome.

April, 1993

USING THE UK SCANNING DIRECTORY

Feed-back from our readers is always welcome and plays an important part in the planning of future editions, telling us what you, the reader, really wants. Several queries have arisen regarding the layout of *The UK Scanning Directory*, and in particular as to what each column means and how frequencies are presented. In answer to all these questions, the diagram below will set all your minds at ease. It depicts a typical page of frequencies and explains what each column refers to.

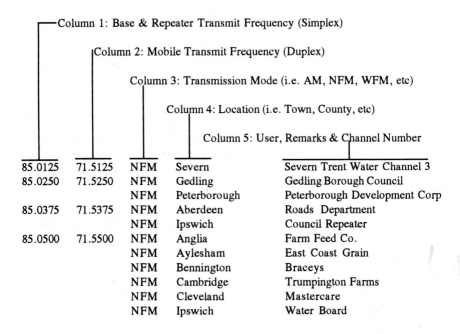

```
┌──Column 1: Base & Repeater Transmit Frequency (Simplex)
│
│   ┌Column 2: Mobile Transmit Frequency (Duplex)
│   │
│   │   Column 3: Transmission Mode (i.e. AM, NFM, WFM, etc)
│   │   │
│   │   │   Column 4: Location (i.e. Town, County, etc)
│   │   │   │
│   │   │   │   Column 5: User, Remarks & Channel Number
│   │   │   │   │
```

85.0125	71.5125	NFM	Severn	Severn Trent Water Channel 3
85.0250	71.5250	NFM	Gedling	Gedling Borough Council
		NFM	Peterborough	Peterborough Development Corp
85.0375	71.5375	NFM	Aberdeen	Roads Department
		NFM	Ipswich	Council Repeater
85.0500	71.5500	NFM	Anglia	Farm Feed Co.
		NFM	Aylesham	East Coast Grain
		NFM	Bennington	Braceys
		NFM	Cambridge	Trumpington Farms
		NFM	Cleveland	Mastercare
		NFM	Ipswich	Water Board

The gaps present in columns one and two, for example, below "85.0500 71.5500" in the extract above, are purely designed to avoid cluttering up the page too much with frequency repetitions, and indicate that the last frequency quoted is also valid on that line. Gaps appearing only in column two indicate that either the corresponding frequency is not known, in the case of a duplex channel arrangement, or that only a single frequency is used for one-way communications such as is the case with aeronautical VHF radio nagivation equipment, cordless microphones and broadcasting stations. Any further queries resulting from this edition are most welcome.

SCANNING IN THE UK

The purpose of the following short articles is to provide readers with a basic outline of the workings of just three types of different telecommunications systems. Some advanced technical details may be omitted simply because these articles are designed to be entertain and not to be used to build our own radio equipment!

CELLULAR TELEPHONES

Mobile telephones have become part of everyone's life in some way or another since the publication of the 2nd edition of the *UK Scanning Directory* as Dianagate and Camillagate hit the headlines. The truth behind these stories may never be known, but it certainly awakened the public to eavesdropping dangers of portable phones.

The advent of the portable telephone brought with it the freedom and versatility of a mobile office. The ability to contact anyone, anywhere in the world is by far one of the most useful applications of modern telecommunications engineering. Cellnet is so named as its networking mimics the cellular structure of a biological system whereby messages can be passed from cell to cell.

In the British Cellnet/Vodafone system, the entire country is split up into various regions, each of which is further split up into cells. At the centre of each cell, which typically can measure up to 50 square kilometres, is located the cellular transmitter tower known as a node which handles all calls in that area. The locations of these nodes are invariably at the highest feasible point within the cell, giving the widest and least restricted coverage possible. On page 229, a typical Cellnet node site can be seen, showing the omni-directional antenna arrangement at the top of the tower. The four flat vertical antenna systems receive and transmit signals within a 90 degree lobe centred on the node. The other vertical rod antennas visible are used to receive and transmit Cellnet data control channels which are the heart of the system.

The Cellnet voice system operates on two independent UHF frequencies; a mobile transmit frequency, which is used by the actual portable telephone to transmit voice signals to the node and a base relay frequency which the node uses to broadcast the land-line party's voice to the portable phone. Transmit and receive frequencies are allocated in the 872 & 950 MHz bands at 25 kHz intervals over two frequency segments each holding a 35 MHz band providing 2047 possible Cellular channels capable of supporting 2007 simultaneous conversations. The other 40 channels are divided equally between Cellnet and Vodaphone to be used as data control channels. Node and mobile frequencies are separated by a split of 45 MHz providing a fully duplexed channel arrangement with voice modulated in narrow band FM. To avoid interference caused by adjoined nodes transmitting signals on the same channel, each node is allocated a fixed number of frequencies in such a way that adjoining nodes are not allocated the same frequencies, thus interference is kept to a minimum.

For the network to operate so efficiently it must be controlled by fast, reliable computer hardware accompanied by versatile software. Cellnet is entirely computer controlled which provides almost instantaneous access to any Cellular telephone anywhere in the country. System management is provided by Cellnet data control channels. These channels transmit digital instructions to portable telephones on a nationwide basis and provide the information needed to initiate a link to make the call. When you pick up your normal home or office telephone and dial an 0836 or 0861 number, the number, already digitised, is flashed around the entire country being broadcast from all nodes via these individual data control channels. The portable phone, when it is switched on, will respond to this signal by transmitting a reply. If the phone is switched off, the caller will hear the recorded message to the effect "The Vodaphone you have called may be switched off. Please call again later."

If the cellphone is switched on, it will transmit a digital reply via the local data control channel to the nearest available cellular node, which will allow for the selection of suitable frequencies, cause the cellphone to ring and a link to be established between the caller (now represented by the node) and the mobile cellphone. Once the cellphone is picked up, the call is initiated and a conversation follows. While all this is going on, the node is continuously and automatically monitoring the quality of the incoming signal from the cellphone. If it is satisfactory, the node will leave the link on the previously chosen frequencies, but if the signal begins to deteriorate, the node can either select a new frequency pair that provides a better signal path or will pass the entire link onto an adjoining node. When a cellphone is mobile and is crossing over node boundaries, the frequencies used will shift to provide the best reception from the new node's bank of available frequencies. The short digital pulses that can be heard during all cellular calls is the system checking the link and instructing the mobile phone which frequency it should next hop to.

Most cellular phone users experience at one time or another, rapidly deteriorating signals that can be the cause much hilarity or serious business disagreements.

This phenomenon often occurs in open fields where the uninitiated user expects perfect signals. Unfortunately, like any system, Cellnet and Vodafone have their draw backs, notably some node sites, especially those in flat areas such as Norfolk, Suffolk and Lincolnshire, do not have the range they were designed for, resulting in blind spots that have poor reception. The only remedy appears to be, spark up your car and move a couple of miles down the road. If that doesn't work try again, and again.....and again!

Following further development in mobile telecommunications hardware, Cellnet and Vodaphone systems are going to compete with the new Rabbit Phone system, designed by the electronics giant Hutchison Telecom of Hong Kong. Rabbit Phones were set to hit the British market in early 1993 after which most cellular telecommunications would be transferred onto high speed digital data circuits, providing improved high quality signals and speech reproduction. Advanced publicity of Rabbit phones promised that it would have extraordinary sound quality and a crystal clear connection without the need for an antenna! The technology that comes with the Rabbit will most certainly lead the way to the new high quality digital cellular telephones which are set to be phased in all over Britain during the next couple of years. However the public expense thta would be required to keep up with these new technologies is of such magnitude that both Cellnet and Vodafone will remain active well into the late 1990's.

Another form of mobile communications that does not often raise much interest are commercial digital pagers such as those supplied by Mercury and British Telecom. They operate basically on the same principle as the Cellphone. Since pager transmissions are only one-way, only one frequency is required, and therefore a single nationwide frequency is allocated to pager system, usually in the 138 MHz area. When a message is sent out to a pager, it is sent via all pager nodes nationwide, however the message sent will only be displayed on the pager screen that the message is addressed to. Receipt of such messages are usually confirmed by the pager transmitting an electronic acknowledgement on a separate frequency.

COMMERCIAL PMR

The Personal Mobile Radio system (PMR) is an electronic asset that more and more companies are investing in. The ability to be able to contact cars, vans, and lorries that are away from the office provides companies within the ability to co-ordinate their external workforce remotely. Gone are the days when work for each day would have to be prepared before each vehicle left the yard in the morning. Also gone is the time wasted in workmen returning to the office from a single job only to be told that their next job is next door to the one they have just left! PMR base sets are now resident in most offices and betray their presence by the existence of a large vertical antenna conspicuously placed on the roof or wall. They are much more practical in their short range abilities to provide two-way communications, but are limited by mobile radios exceeding the cover provided by either base or repeater stations. At significantly large distances, companies such as haulage contractors still prefer the versatility of the Cellular telephone. However, to counter this trend away from PMR at longer distances, a national trunking network has been installed that allows PMR messages to be relayed around the country, effectively expanding VHF range to the length and breadth of the country.

Current commercial PMR band allocations provide for ten individual bands shown in the table below: (See listings for further details)

Band	Frequency Limits (MHz)		Step	Mode
1	84.1250 to 84.9750	Simplex	12.5 kHz	AM/NFM
2	85.0125 to 86.2975	71.5125 to 72.7975	12.5 kHz	AM/NFM
3	86.3000 to 86.7000	Simplex	12.5 kHz	AM/NFM
4	86.7125 to 87.9625	76.7125 to 77.9625	12.5 kHz	AM/NFM
5	138.0000 to 140.4875	Simplex	6.5 kHz	AM/NFM
6	165.0500 to 168.2250	169.8500 to 173.0500	12.5 kHz	AM/NFM
7	168.9500 to 169.8375	Simplex	12.5 kHz	AM/NFM
8	425.0000 to 425.5000	445.5000 to 446.0000	12.5 kHz	AM/NFM
9	425.5000 to 429.0000	440.0000 to 443.5000	12.5 kHz	AM/NFM
10	453.0000 to 454.0000	459.5000 to 460.5000	12.5 kHz	AM/NFM

Within these ten bands, users may obtain permission to operate radio equipment from the RCA (Radio Communications Agency) whereupon a suitable frequency will be allocated to the user such that no interference will be caused to other PMR users in the local area. With a basic PMR set-up, the user can operate a standard two-way communications link between base and mobile units, either on simplex or duplex bands. Simplex channels are those which carry both mobile and base sides of the conversation on the same channel, while duplex channels are those which allocate a separate channel for each side of the conversation. Such simplex arrangements are often preferred by users of handheld PMR radios rather than those who operate a base station, while base stations prefer the duplex frequency set-up.

The drawback to all the benefits of operating communications on VHF frequencies is the very limited range that these signals have, and essentially operate only on a line of sight basis. If you can picture your own town or village that may have the odd hill or two around it, just think how difficult it would be to carry on a conversation if one of the party's suddenly disappeared behind the hill. VHF signals cannot bend themselves around obstacles! The signal would be lost and the conversation could only resume when the mobile station re-appeared at the other side. It may not take him long, but these two or three minutes of lost communications could cause many problems. To solve this problem, the idea of a repeater station was developed. A repeater station simply being an intermediate station located between the base and mobile stations that can automatically relay transmissions.

In a typical duplex PMR arrangement, the base station transmits on channel A and the mobile station transmits of channel B. Therefore the mobile station will have his receiver tuned to receive transmissions from the base station on channel B and the base station will have his receiver tuned to channel A. Such a set-up is that which is standard for any duplex PMR operation. If however the mobile station is behind a hill and therefore cannot receive signals from base, then the only solution is for the signal to somehow either go around or over the hill. And in essence that is what a Repeater does, it re-routes signals around obstacles by making the communications link have two legs rather than just one. In such a case as mention above, a repeater station may be placed on the obstructing hill or on another surrounding hill that has line of sight with both stations. Once the repeater was installed, it would simultaneously listen on channel A and retransmit all it heard on channel A. Note that both base and mobile stations need not have to transmit on channel A and receive on channel B. Such repeaters are operated by sole users such as the various departments of the local district councils. A second type of repeater, known as a Community Repeater, is also available to PMR users. This is a shared repeater that can be used simultaneously by a number of users and is tone activated. These tones, often below the range of the human ear, are sent by Repeater users to activate individual channels at the repeater site. Such systems are often favoured by small local companies who do not use PMR enough to warrant the hire of their own repeater. The photograph below shows a typical VHF repeater antenna that is host to some seventeen individual antennas.

OUTSIDE BROADCASTS
When you watch television news and sports reports from outside the studio or programmes that are presented from outside studios, have you every wondered how the pictures and voices are sent back to the studio, or how the presenter knows exactly what to do, where to turn and what to say during live broadcasts? Or how the cameramen know exactly who to zoom in on or where to pan? This

---

is the realm of the media broadcaster and the technology that accompanies him. Outside broadcasts fall into two categories, radio and TV, although both are very similar in essence.

First let us consider TV outside broadcasts and as an example let's discuss a typical political conference. Often the first people at any such venue are the TV crews and their riggers who hastily erect antennas for transmission. The actual setting up of these broadcast links is not as complicated as at first it would seem. Throughout the country is a large microwave relay network that terminates at the central TV studios, such as the BBC studios at Shepherd's Bush in London or the BBC studios at Queen Margaret Drive in Glasgow. At various points along this network are access points that outside broadcast crews can feed their signals into which are then carried off to London or wherever. The network is a two-way system so that both outside broadcasters and the TV Studio can keep in constant contact by both voice and picture. When the team first arrives, priority is given to accessing the network and establishing the two-way link. While this is being set up, the broadcast manager will arrange the best locations for his cameras to capture all the action. There are often up to eight cameras at such a broadcast, each with a separate area or specific individual to capture. During this stage significant use is made of VHF radios so that quick and efficient contact can be maintained with everyone on the set.

Once antennas have been installed, links established and cameras set up, all that is left to do is test and retest the systems and await the arrival of the political guests. The control of all outside broadcasts is co-ordinated on site by the mobile control room. These vans are literally packed to the roof with all matter of electronic gadgetry that enables the broadcast manager to record, edit and dub an entire transmission that can either be relayed in real time for a live broadcast or stored on tape to be transmitted at a later date. During any broadcast, site managers operate on various VHF Talkback channels as a means of communicating information to camera crews. Studio sound is often relayed from the studio floor by low powered cordless microphones direct to dedicated receivers in the mobile control van. (These are the little box of tricks often seen clipped onto the back of a presenter belt!) Other Talkback channels combine studio sounds from microphones, instructions to cameramen and the sounds emanating from the control van to produce a single channel carrying all the information that the team requires.

Outside radio broadcasts are very similar in makeup to that of the TV outside broadcasts. They access the same microwave relay network in order send their broadcasts back to the studio. Cordless microphones again play an important role as does local talkback. An event of particular interest is the Radio One Roadshow that makes many yearly visits to towns and cities the length and breadth of the country. In both TV and radio, outside broadcasts that cover brief items of news can be recorded onto tape by a single cameraman that can be transported back to the studio for transmission at the next news broadcast. This form of news recording, known as ENG or Electronic News Gathering (much popularised on Channel 4 by the Canadian TV show of the same name), is a popular method employed by all television companies who cover brief items of news in the field. Indeed it was such an ENG crew from Sky TV News who appeared at Interproducts Company Headquarters at 8 Abbot St. back in September 1992 looking for an interview with company staff to discuss a certain very popular book!

List of Abbreviations and Terms

AFIS	Aerodrome Forecast Information Service
AFSATCOM	US Air Force Satellite Communications
AM	Amplitude Modulation
ARA	Air Refuelling Area
ATC	Air Training Corps or Air Traffic Control
ATCC	Air Traffic Control Centre
ATIS	Aerodrome Terminal Information Service
AWACS	Airborne Warning And Control System
BAe	British Aerospace
BBC	Britsih Broadcasting Corporation
BNFL	British Nuclear Fuels Ltd
BR	British Rail
BT	British Telecom
BTP	British Transport Police
CAC	Centralised Approach Control
CB	Citizen Band
CEGB	Central Electricity Generating Board
CMD	Command
Comms	Communications
CW	Continuous Wave (Morse)
DME	Distance Measuring Equipment
DTI	Department of Trade & Industry
FLTSATCOM	US Navy Fleet Satellite Communications
FM	Frequency Modulation
GCHQ	Government Communications Headquarters
IBA	Independent Broadcasting Authority
IFR	Instrument Flight Rules
ILR	Independent Local Radio
ILS	Instrument Landing System
ITN	Independent Television News
ITV	Independent Television
LWT	London Weekend Television
MoD	Ministry of Defence
Mould	MoD National Home Defence Repeater Network
MRSA	Mandatory Radar Service Area

MWL	Mid-Wales Railway Line
NATO	North Atlantic Treaty Organisation
NASA	National Aeronautics & Space Administration
NB	Narrow Band
NCB	National Coal Board
NFM	Narrow Band FM
O/B	Outside Broadcast
Ops	Operations
PAR	Precision Approach Radar
PFA	Popular Flying Association
PMR	Personal Mobile Radio
PR	Personal Radio
R	Runway (Left & Right)
RAF	Royal Air Force
RCA	Radiocommunications Agency
RN	Royal Navy
RTTY	Radio Teletype
SAR	Search And Rescue
Spec.	Specification
SRE	Surveillance Radar Element
SSB	Single Side Band
SSTV	Slow Scan Television
Std	Standard
Surv	Surveillence
TACAN	Tactical Air Navigation
TMA	Terminal Manoeuvring Area
TX	Transmission
UACC	Upper Air Control Centre
UHF	Ultra High Frequency (300 - 3000 MHz)
UKAEA	UK Atomic Energy Authority
USAFE	US Air Force Europe
USB	Upper Side Band
VFR	Visual Flight Rules
VHF	Very High Frequency (30 - 300 MHz)
VOLMET	Aviation Weather Broadcast
VOR	VHF Omni-Directional Radio Range
WFM	Wide Band FM

Base	Mobile	Mode	Location	User & Notes
24.9900 - 25.0200 MHz			Std Freq, Time Signals and Space Research SSB	
25.0200 - 25.0700 MHz			Fixed and Land Mobile USB	
25.0700 - 25.2100 MHz			Maritime Mobile USB & RTTY	
25.2100 - 25.5500 MHz			Fixed, Land and Maritime Mobile USB	
25.5500 - 25.6000 MHz			Radio Astronomy	
25.6000 - 26.1000 MHz			11m Broadcasting Band AM	
26.1000 - 26.1750 MHz			Maritime Mobiles USB	
26.1750 - 26.2350 MHz			Fixed, Land and Maritime Mobiles USB	
26.2350 - 26.8700 MHz			**Low Power Paging**	
26.5880		NFM	Nationwide	Common Paging Channel
26.5150 - 26.9550 MHz			**Illegal "Lo" CB**	
26.5150	26.5150	NFM	Nationwide	Channel 1
26.5250	26.5250	NFM	Nationwide	Channel 2
26.5350	26.5350	NFM	Nationwide	Channel 3
26.5450	26.5450	NFM	Nationwide	Channel 4
26.5550	26.5550	NFM	Nationwide	Channel 5
26.5650	26.5650	NFM	Nationwide	Channel 6
26.5750	26.5750	NFM	Nationwide	Channel 7
26.5850	26.5850	NFM	Nationwide	Channel 8
26.5950	26.5950	NFM	Nationwide	Channel 9
26.6050	26.6050	NFM	Nationwide	Channel 10
26.6150	26.6150	NFM	Nationwide	Channel 11
26.6250	26.6250	NFM	Nationwide	Channel 12
26.6350	26.6350	NFM	Nationwide	Channel 13
26.6450	26.6450	NFM	Nationwide	Channel 14
26.6550	26.6550	NFM	Nationwide	Channel 15
26.6650	26.6650	NFM	Nationwide	Channel 16
26.6750	26.6750	NFM	Nationwide	Channel 17
26.6850	26.6850	NFM	Nationwide	Channel 18
26.6950	26.6950	NFM	Nationwide	Channel 19
26.7050	26.7050	NFM	Nationwide	Channel 20
26.7150	26.7150	NFM	Nationwide	Channel 21
26.7250	26.7250	NFM	Nationwide	Channel 22
26.7350	26.7350	NFM	Nationwide	Channel 23
26.7450	26.7450	NFM	Nationwide	Channel 24
26.7550	26.7550	NFM	Nationwide	Channel 25
26.7650	26.7650	NFM	Nationwide	Channel 26
26.7750	26.7750	NFM	Nationwide	Channel 27
26.7850	26.7850	NFM	Nationwide	Channel 28
26.7950	26.7950	NFM	Nationwide	Channel 29

Base	Mobile	Mode	Location	User & Notes
26.8050	26.8050	NFM	Nationwide	Channel 30
26.8150	26.8150	NFM	Nationwide	Channel 31
26.8250	26.8250	NFM	Nationwide	Channel 32
26.8350	26.8350	NFM	Nationwide	Channel 33
26.8450	26.8450	NFM	Nationwide	Channel 34
26.8550	26.8550	NFM	Nationwide	Channel 35
26.8650	26.8650	NFM	Nationwide	Channel 36
26.8750	26.8750	NFM	Nationwide	Channel 37
26.8850	26.8850	NFM	Nationwide	Channel 38
26.8950	26.8950	NFM	Nationwide	Channel 39
26.9050	26.9050	NFM	Nationwide	Channel 40
26.9150	26.9150	NFM	Nationwide	Channel 41
26.9250	26.9250	NFM	Nationwide	Channel 42
26.9350	26.9350	NFM	Nationwide	Channel 43
26.9450	26.9450	NFM	Nationwide	Channel 44
26.9550	26.9550	NFM	Nationwide	Channel 45

26.9650 - 27.4050 MHz CEPT (UK & Europe) CB & Controlled Models

Base	Mobile	Mode	Location	User & Notes
26.9650	26.9650	NFM	Nationwide	Channel 01
26.9750	26.9750	NFM	Nationwide	Channel 02
26.9850	26.9850	NFM	Nationwide	Channel 03
26.9950	26.9950	NFM	Nationwide	'Brown' Model Channel
27.0050	27.0050	NFM	Nationwide	Channel 04
27.0150	27.0150	NFM	Nationwide	Channel 05
27.0250	27.0250	NFM	Nationwide	Channel 06
27.0350	27.0350	NFM	Nationwide	Channel 07
27.0450	27.0450	NFM	Nationwide	'Red' Model Channel
27.0550	27.0550	NFM	Nationwide	Channel 08
27.0650	27.0650	NFM	Nationwide	Channel 09
27.0750	27.0750	NFM	Nationwide	Channel 10
27.0850	27.0850	NFM	Nationwide	Channel 11
27.0950	27.0950	NFM	Nationwide	'Orange' Model Channel
27.1050	27.1050	NFM	Nationwide	Channel 12
27.1150	27.1150	NFM	Nationwide	Channel 13
27.1250	27.1250	NFM	Nationwide	Channel 14
27.1350	27.1350	NFM	Nationwide	Channel 15
27.1450	27.1450	NFM	Nationwide	'Yellow' Model Channel
27.1550	27.1550	NFM	Nationwide	Channel 16
27.1650	27.1650	NFM	Nationwide	Channel 17
27.1750	27.1750	NFM	Nationwide	Channel 18
27.1850	27.1850	NFM	Nationwide	Channel 19
27.1950	27.1950	NFM	Nationwide	'Green' Model Channel
27.2050	27.2050	NFM	Nationwide	Channel 20
27.2150	27.2150	NFM	Nationwide	Channel 21
27.2250	27.2250	NFM	Nationwide	Channel 22
27.2350	27.2350	NFM	Nationwide	Channel 23
27.2450	27.2450	NFM	Nationwide	Channel 24
27.2550	27.2550	NFM	Nationwide	Channel 25
27.2650	27.2650	NFM	Nationwide	Channel 26

Base	Mobile	Mode	Location	User & Notes
27.2750	27.2750	NFM	Nationwide	Channel 27
27.2850	27.2850	NFM	Nationwide	Channel 28
27.2950	27.2950	NFM	Nationwide	Channel 29
27.3050	27.3050	NFM	Nationwide	Channel 30
27.3150	27.3150	NFM	Nationwide	Channel 31
27.3250	27.3250	NFM	Nationwide	Channel 32
27.3350	27.3350	NFM	Nationwide	Channel 33
27.3450	27.3450	NFM	Nationwide	Channel 34
27.3550	27.3550	NFM	Nationwide	Channel 35
27.3650	27.3650	NFM	Nationwide	Channel 36
27.3750	27.3750	NFM	Nationwide	Channel 37
27.3850	27.3850	NFM	Nationwide	Channel 38
27.3950	27.3950	NFM	Nationwide	Channel 39
27.4050	27.4050	NFM	Nationwide	Channel 40

27.4050 - 27.6000 MHz   Mobiles, Weather Sondes & Low Power Alarms

27.4150 - 27.8550 MHz   Illegal "Hi" CB

Base	Mobile	Mode	Location	User & Notes
27.4150	27.4150	NFM	Nationwide	Channel 1
27.4250	27.4250	NFM	Nationwide	Channel 2
27.4350	27.4350	NFM	Nationwide	Channel 3
27.4450	27.4450	NFM	Nationwide	Channel 4
27.4550	27.4550	NFM	Nationwide	Channel 5
27.4650	27.4650	NFM	Nationwide	Channel 6
27.4750	27.4750	NFM	Nationwide	Channel 7
27.4850	27.4850	NFM	Nationwide	Channel 8
27.4950	27.4950	NFM	Nationwide	Channel 9
27.5050	27.5050	NFM	Nationwide	Channel 10
27.5150	27.5150	NFM	Nationwide	Channel 11
27.5250	27.5250	NFM	Nationwide	Channel 12
27.5350	27.5350	NFM	Nationwide	Channel 13
27.5450	27.5450	NFM	Nationwide	Channel 14
27.5550	27.5550	NFM	Nationwide	Channel 15
27.5650	27.5650	NFM	Nationwide	Channel 16
27.5750	27.5750	NFM	Nationwide	Channel 17
27.5850	27.5850	NFM	Nationwide	Channel 18
27.5950	27.5950	NFM	Nationwide	Channel 19
27.6050	27.6050	NFM	Nationwide	Channel 20
27.6150	27.6150	NFM	Nationwide	Channel 21
27.6250	27.6250	NFM	Nationwide	Channel 22
27.6350	27.6350	NFM	Nationwide	Channel 23
27.6450	27.6450	NFM	Nationwide	Channel 24
27.6550	27.6550	NFM	Nationwide	Channel 25
27.6650	27.6650	NFM	Nationwide	Channel 26
27.6750	27.6750	NFM	Nationwide	Channel 27
27.6850	27.6850	NFM	Nationwide	Channel 28
27.6950	27.6950	NFM	Nationwide	Channel 29
27.7050	27.7050	NFM	Nationwide	Channel 30
27.7150	27.7150	NFM	Nationwide	Channel 31

Base	Mobile	Mode	Location	User & Notes
27.7250	27.7250	NFM	Nationwide	Channel 32
27.7350	27.7350	NFM	Nationwide	Channel 33
27.7450	27.7450	NFM	Nationwide	Channel 34
27.7550	27.7550	NFM	Nationwide	Channel 35
27.7650	27.7650	NFM	Nationwide	Channel 36
27.7750	27.7750	NFM	Nationwide	Channel 37
27.7850	27.7850	NFM	Nationwide	Channel 38
27.7950	27.7950	NFM	Nationwide	Channel 39
27.8050	27.8050	NFM	Nationwide	Channel 40
27.8150	27.8150	NFM	Nationwide	Channel 41
27.8250	27.8250	NFM	Nationwide	Channel 42
27.8350	27.8350	NFM	Nationwide	Channel 43
27.8450	27.8450	NFM	Nationwide	Channel 44
27.8550	27.8550	NFM	Nationwide	Channel 45
27.60125 - 27.99125 MHz		**UK CB**		
27.60125	27.60125	NFM	Nationwide	Channel 01
27.61125	27.61125	NFM	Nationwide	Channel 02
27.62125	27.62125	NFM	Nationwide	Channel 03
27.63125	27.63125	NFM	Nationwide	Channel 04
27.64125	27.64125	NFM	Nationwide	Channel 05
27.65125	27.65125	NFM	Nationwide	Channel 06
27.66125	27.66125	NFM	Nationwide	Channel 07
27.67125	27.67125	NFM	Nationwide	Channel 08
27.68125	27.68125	NFM	Nationwide	Channel 09 Emergency
27.69125	27.69125	NFM	Nationwide	Channel 10
27.70125	27.70125	NFM	Nationwide	Channel 11
27.71125	27.71125	NFM	Nationwide	Channel 12
27.72125	27.72125	NFM	Nationwide	Channel 13
27.73125	27.73125	NFM	Nationwide	Channel 14 Calling
27.74125	27.74125	NFM	Nationwide	Channel 15
27.75125	27.75125	NFM	Nationwide	Channel 16
27.76125	27.76125	NFM	Nationwide	Channel 17
27.77125	27.77125	NFM	Nationwide	Channel 18
27.78125	27.78125	NFM	Nationwide	Channel 19 Calling
27.79125	27.79125	NFM	Nationwide	Channel 20
27.80125	27.80125	NFM	Nationwide	Channel 21
27.81125	27.81125	NFM	Nationwide	Channel 22
27.82125	27.82125	NFM	Nationwide	Channel 23
27.83125	27.83125	NFM	Nationwide	Channel 24
27.84125	27.84125	NFM	Nationwide	Channel 25
27.85125	27.85125	NFM	Nationwide	Channel 26
27.86125	27.86125	NFM	Nationwide	Channel 27
27.87125	27.87125	NFM	Nationwide	Channel 28
27.88125	27.88125	NFM	Nationwide	Channel 29
27.89125	27.89125	NFM	Nationwide	Channel 30
27.90125	27.90125	NFM	Nationwide	Channel 31
27.91125	27.91125	NFM	Nationwide	Channel 32
27.92125	27.92125	NFM	Nationwide	Channel 33

Base	Mobile	Mode	Location	User & Notes
27.93125	27.93125	NFM	Nationwide	Channel 34
27.94125	27.94125	NFM	Nationwide	Channel 35
27.95125	27.95125	NFM	Nationwide	Channel 36
27.96125	27.96125	NFM	Nationwide	Channel 37
27.97125	27.97125	NFM	Nationwide	Channel 38
27.98125	27.98125	NFM	Nationwide	Channel 39
27.99125	27.99125	NFM	Nationwide	Channel 40

28.0000 - 29.7000 MHz   10m Amateur Band USB

28.2000		CW	Crowborough	GB3SXE Beacon
28.2150		CW	Slough	GB3RAL Beacon
29.6000		NFM	Nationwide	FM Calling

29.7000 - 29.9700 MHz   MoD Tactical Channels 25 kHz Simplex NFM

30.0050 - 31.0250 MHz   NASA Space to Earth NFM Simplex

30.0100		NFM	Space	Downlink

30.0250 - 31.7000 MHz   USAFE Communications 25 kHz Simplex

30.0250	30.0250	NFM	RAF Fairford	Security Control
30.5000	30.5000	NFM	London	US Embassy Security
30.5500	30.5500	NFM	RAF Fairford	Base Security
30.9875	30.9875	NFM	RAF Fairford	Base Security
31.0000	31.0000	NFM	RAF Fairford	Base Security
31.1825	31.1825	NFM	RAF Fairford	Fence Security
31.2000	31.2000	NFM	RAF Fairford	Ground Maintenance
31.2500	31.2500	NFM	RAF Fairford	Base Security
31.3000	31.3000	NFM	RAF Fairford	Ground Medical
31.4000	31.4000	NFM	RAF Fairford	Tanker Ground Ops
31.5000	31.5000	NFM	Brighton	TA Barracks
		NFM	Preston	TA Barracks

31.7250 - 31.7750 MHz   Hospital Paging Emergency Speech Return

31.7250	161.0000	NFM	Nationwide	Hospital Paging
31.7500	161.0250	NFM	Nationwide	Hospital Paging
31.7750	161.0500	NFM	Nationwide	Hospital Paging

31.8000 - 34.9000 MHz   USAFE Communications 25 kHz Simplex

32.2000	32.2000	NFM	Nationwide	USAF Base Security
32.3000	32.3000	NFM	RAF Fairford	Base Security
		NFM	Brighton	TA Barracks
		NFM	Preston	TA Barracks
32.3500	32.3500	NFM	RAF Alconbury	Ground Services
		NFM	RAF Mildenhall	Security
33.1250	33.1250	NFM	Nationwide	Army Land Forces
33.2500	33.2500	NFM	RAF Lakenheath	Birdscare Ops
		NFM	RAF Mildenhall	Birdscare Ops
33.3000	33.3000	NFM	Nationwide	USAF War Training
33.5000	33.5000	NFM	RAF Lakenheath	Birdscare Ops

Base	Mobile	Mode	Location	User & Notes
		NFM	RAF Mildenhall	Birdscare Ops
33.6750	33.6750	NFM	Brighton	TA Barracks
		NFM	Preston	TA Barracks
34.1000	34.1000	NFM	Nationwide	USAF Medical Common
34.1500	34.1500	NFM	RAF Lakenheath	Crash Ops
		NFM	RAF Mildenhall	Crash Ops
34.9000	34.9000	NFM	RAF Lakenheath	Security
		NFM	RAF Mildenhall	Security
		NFM	RAF Upper Heyford	Security

34.9250 - 34.9750 MHz   Low Power Alarms for Elderly and Infirm

Base	Mobile	Mode	Location	User & Notes
34.9500		NFM	Nationwide	Alarm for Elderly & Infirm

35.0000 - 35.2500 MHz   Radio Controlled Models 10 kHz (100 mW Max)

Base	Mobile	Mode	Location	User & Notes
35.0000		NFM	Nationwide	Channel 60
35.0100		NFM	Nationwide	Channel 61
35.0200		NFM	Nationwide	Channel 62
35.0300		NFM	Nationwide	Channel 63
35.0400		NFM	Nationwide	Channel 64
35.0500		NFM	Nationwide	Channel 65
35.0600		NFM	Nationwide	Channel 66
35.0700		NFM	Nationwide	Channel 67
35.0800		NFM	Nationwide	Channel 68
35.0900		NFM	Nationwide	Channel 69
35.1000		NFM	Nationwide	Channel 70
35.1100		NFM	Nationwide	Channel 71
35.1200		NFM	Nationwide	Channel 72
35.1300		NFM	Nationwide	Channel 73
35.1400		NFM	Nationwide	Channel 74
35.1500		NFM	Nationwide	Channel 75
35.1600		NFM	Nationwide	Channel 76
35.1700		NFM	Nationwide	Channel 77
35.1800		NFM	Nationwide	Channel 78
35.1900		NFM	Nationwide	Channel 79
35.2000		NFM	Nationwide	Channel 80
35.2100		NFM	Nationwide	Channel 81
35.2200		NFM	Nationwide	Channel 82
35.2300		NFM	Nationwide	Channel 83
35.2400		NFM	Nationwide	Channel 84
35.2500		NFM	Nationwide	Channel 85

35.2500 - 37.7500 MHz   MoD Tactical Communications 25 kHz Simplex

Base	Mobile	Mode	Location	User & Notes
35.2500	35.2500	NFM	Salisbury Plain	Army Units
35.2750	35.2750	NFM	Salisbury Plain	Army Units
35.3500	35.3500	NFM	Brecon Beacons	Army Units
35.4000	35.4000	NFM	Brecon Beacons	Army Units
		NFM	RAF Upper Heyford	Ground Control
35.5750	35.5750	NFM	Salisbury Plain	Army Units
35.7750	35.7750	NFM	Salisbury Plain	Army Units

Base	Mobile	Mode	Location	User & Notes
36.0000	36.0000	NFM	Brighton	TA Barracks
		NFM	Preston	TA Barracks
36.3500	36.3500	NFM	Bovington	Army Training Camp
37.2250	37.2250	NFM	Norfolk	Stanford Army Battle Area

37.7500 - 37.850 MHz   Radio Astronomy Band

37.9000 - 40.1000 MHz   MoD Tactical Communications 25 kHz Simplex

Base	Mobile	Mode	Location	User & Notes
38.0000	38.0000	USB	Nationwide	Racal Comsec Spot Freq.
38.6250	38.6250	NFM	Oakington Camp	Army 657 Squad Ops
39.5000	39.5000	NFM	Nationwide	Army Tanks Channel
39.6000	39.6000	NFM	Ludford Cove	Army
39.6500	39.6500	NFM	Nationwide	Army Tanks Channel
39.7500	39.7500	NFM	Nationwide	Royal Signals Display
39.9000	39.9000	NFM	RAF Fairford	Base Security
40.0500	40.5000	NFM	Nationwide	Military Distress Frequency

40.6650 - 40.9550 MHz   Radio Controlled Surface Models 10 kHz
MoD Tactical Communications 25 kHz Simplex

Base	Mobile	Mode	Location	User & Notes
40.6650		NFM	Nationwide	Channel 665
40.6750		NFM	Nationwide	Channel 675
40.6850		NFM	Nationwide	Channel 685
40.6950		NFM	Nationwide	Channel 695
40.7000	40.7000	NFM	Bovington	Army Training Camp
40.7050		NFM	Nationwide	Channel 705
40.7150		NFM	Nationwide	Channel 715
40.7250		NFM	Nationwide	Channel 725
40.7350		NFM	Nationwide	Channel 735
40.7450		NFM	Nationwide	Channel 745
40.7500	40.7500	NFM	Bovington	Army Training Camp
40.7550		NFM	Nationwide	Channel 755
40.7650		NFM	Nationwide	Channel 765
40.7750		NFM	Nationwide	Channel 775
40.7850		NFM	Nationwide	Channel 785
40.7950		NFM	Nationwide	Channel 795
40.0800	40.0800	NFM	London	Sky News Link
40.8050		NFM	Nationwide	Channel 805
40.8150		NFM	Nationwide	Channel 815
40.8250		NFM	Nationwide	Channel 825
40.8350		NFM	Nationwide	Channel 835
40.8450		NFM	Nationwide	Channel 845
40.8550		NFM	Prestonwide	Channel 855
40.8650		NFM	Nationwide	Channel 865
40.8750		NFM	Nationwide	Channel 875
40.8850		NFM	Nationwide	Channel 885
40.8950		NFM	Nationwide	Channel 895
40.9050		NFM	Nationwide	Channel 905
40.9150		NFM	Nationwide	Channel 915
40.9250		NFM	Nationwide	Channel 925

Base	Mobile	Mode	Location	User & Notes
40.9350		NFM	Nationwide	Channel 935
40.9450		NFM	Nationwide	Channel 945
40.9550		NFM	Nationwide	Channel 955

41.0000 - 46.6000 MHz   MoD Tactical Communications 25 kHz Simplex

Base	Mobile	Mode	Location	User & Notes
42.1250	42.1250	NFM	Nationwide	Army War Training
43.8250	43.8250	NFM	Ouston	Army
44.0000	44.0000	NFM	Bovington	Army Training Camp
44.4500	44.4500	NFM	Nationwide	Army War Training
45.3000	45.3000	NFM	Norfolk	Stanford Army Battle Area
45.4250	45.4250	NFM	Norfolk	Stanford Army Battle Area
45.7125	45.7125	NFM	Nationwide	Army War Training
46.0000	46.0000	NFM	Nationwide	Royal Signals Display
46.1250	46.1250	NFM	Nationwide	Army War Training
46.3250	46.3250	NFM	Nationwide	Army War Training

46.0000 - 68.0000 MHz   TV Band I (Not UK, DX from Europe)

46.6100 - 46.9700 MHz   US Spec. Cordless Telephones Base (Split + 3.06 MHz)

Base	Mobile	Mode	Location	User & Notes
46.6100	49.6700	NFM	Nationwide	Channel 1
46.6300	49.8450	NFM	Nationwide	Channel 2
46.6700	49.8600	NFM	Nationwide	Channel 3
46.7100	49.7700	NFM	Nationwide	Channel 4
46.7300	49.8750	NFM	Nationwide	Channel 5
46.7700	49.8300	NFM	Nationwide	Channel 6
46.8000	46.8000	WFM	London	BBC Music Link
46.8300	49.8900	NFM	Nationwide	Channel 7
46.8700	49.9300	NFM	Nationwide	Channel 8
46.9300	49.9900	NFM	Nationwide	Channel 9
46.9700	49.9700	NFM	Nationwide	Channel 10

47.0000 - 47.4000 MHz   Future PMR Allocation, currently MoD

Base	Mobile	Mode	Location	User & Notes
47.4000			Nationwide	Car Theft Paging Alarms
47.4000	47.4000	WFM	London	BBC Music Link

47.41875 - 47.43125 MHz   Extended Range Cordless Phones

Base	Mobile	Mode	Location	User & Notes
47.41875	77.5500	NFM	Nationwide	
47.43125	77.5125	NFM	Nationwide	

47.45625 - 47.54375 MHz   DTI Approved Cordless Telephones

Base	Mobile	Mode	Location	User & Notes
47.45625	1.6420	NFM	Nationwide	Channel 1
47.46875	1.6620	NFM	Nationwide	Channel 2
47.48125	1.6820	NFM	Nationwide	Channel 3
47.49375	1.7020	NFM	Nationwide	Channel 4
47.50625	1.7220	NFM	Nationwide	Channel 5
47.51875	1.7420	NFM	Nationwide	Channel 6
47.53125	1.7620	NFM	Nationwide	Channel 7
47.54375	1.7820	NFM	Nationwide	Channel 8

Base	Mobile	Mode	Location	User & Notes
47.45625 - 47.54375 MHz			**DTI Approved Cordless Telephones**	
47.45625		NFM	Nationwide	Channel 1
47.46875	4.9850	NFM	Nationwide	Channel 2
47.48125	6.7290	NFM	Nationwide	Channel 3
47.49375	5.1150	NFM	Nationwide	Channel 4
47.50625		NFM	Nationwide	Channel 5
47.51875		NFM	Nationwide	Channel 6
47.53125		NFM	Nationwide	Channel 7
47.54375		NFM	Nationwide	Channel 8
47.5500 - 48.5500 MHz			**Broadcasting Links**	
47.6450		NFM	Nationwide	BBC Outside Broadcast
47.94375		NFM	Nationwide	ITV Engineers Channel 1
		NFM	Stockport	TV Engineers
47.95625		NFM	Nationwide	ITV Engineers Channel 2
47.96875		NFM	Nationwide	ITV Engineers Channel 3
48.05625		NFM	Isle Of Wight	IoW Feeder to 1242 kHz
48.4000 - 48.5000 MHz			**Cordless Radio Microphones 12.5 kHz**	
48.4000	48.4000	NFM	Nationwide	Channel 1
48.4125	48.4125	NFM	Nationwide	Channel 2
48.4250	48.4250	NFM	Nationwide	Channel 3
48.4375	48.4375	NFM	Nationwide	Channel 4
48.4500	48.4500	NFM	Nationwide	Channel 5
48.4625	48.4625	NFM	Nationwide	Channel 6
48.4750	48.4750	NFM	Nationwide	Channel 7
48.4875	48.4875	NFM	Nationwide	Channel 8
49.0000 - 49.4875 MHz			**One Way Non Speech Paging Systems 12.5 kHz**	
49.0000		NFM	Nationwide	Channel 1
49.0125		NFM	Nationwide	Channel 2
49.0250		NFM	Nationwide	Channel 3
49.0375		NFM	Nationwide	Channel 4
49.0500		NFM	Nationwide	Channel 5
49.0625		NFM	Nationwide	Channel 6
49.0750		NFM	Nationwide	Channel 7
49.0875		NFM	Nationwide	Channel 8
49.1000		NFM	Nationwide	Channel 9
		NFM	Nationwide	Radio Headphones
49.1125		NFM	Nationwide	Channel 10
49.1250		NFM	Nationwide	Channel 11
49.1375		NFM	Nationwide	Channel 12
49.1500		NFM	Nationwide	Channel 13
49.1625		NFM	Nationwide	Channel 14
49.1750		NFM	Nationwide	Channel 15
49.1875		NFM	Nationwide	Channel 16
49.2000		NFM	Nationwide	Channel 17
49.2125		NFM	Nationwide	Channel 18
49.2250		NFM	Nationwide	Channel 19

Base	Mobile	Mode	Location	User & Notes
49.2375		NFM	Nationwide	Channel 20
49.2500		NFM	Nationwide	Channel 21
49.2625		NFM	Nationwide	Channel 22
49.2750		NFM	Nationwide	Channel 23
49.2875		NFM	Nationwide	Channel 24
49.3000		NFM	Nationwide	Channel 25
49.3125		NFM	Nationwide	Channel 26
49.3250		NFM	Nationwide	Channel 27
49.3375		NFM	Nationwide	Channel 28
49.3500		NFM	Nationwide	Channel 29
49.3625		NFM	Nationwide	Channel 30
49.3750		NFM	Nationwide	Channel 31
49.3875		NFM	Nationwide	Channel 32
49.4000		NFM	Nationwide	Channel 33
49.4125		NFM	Nationwide	Channel 34
49.4250		NFM	Nationwide	Channel 35 Hospital Paging
49.4375		NFM	Nationwide	Channel 36 Hospital Paging
49.4500		NFM	Nationwide	Channel 37 Hospital Paging
49.4625		NFM	Nationwide	Channel 38 Hospital Paging
49.4750		NFM	Nationwide	Channel 39 Hospital Paging
49.4875		NFM	Nationwide	Channel 40

49.5000 - 49.7875 MHz   BBC Cordless Microphones NFM

49.6700 - 49.9700 MHz   US Spec. Cordless 'phones Base (Split - 3.06 MHz)

49.8200 - 49.9875 MHz   Walkie Talkies, Radio Mics and Baby Monitors

Base	Mobile	Mode	Location	User & Notes
49.8200		NFM	Nationwide	Channel 1
49.8300		NFM	Nationwide	Channel 2
		NFM	Nationwide	Microphone Channel A
		NFM	Nationwide	Maxan Walkie Talkie Ch 1
49.8400		NFM	Nationwide	Channel 3
		NFM	Nationwide	Microphone Channel UK1
49.8450		NFM	Nationwide	Microphone Channel B
		NFM	Nationwide	Maxan Walkie Talkie Ch 2
49.8500		NFM	Nationwide	Channel 4
49.8600		NFM	Nationwide	Channel 5
		NFM	Nationwide	Microphone Channel C
		NFM	Nationwide	Maxan Walkie Talkie Ch 3
49.8700		NFM	Nationwide	Channel 6
49.8750		NFM	Nationwide	Microphone Channel D
		NFM	Nationwide	Maxan Walkie Talkie Ch 4
49.8800		NFM	Nationwide	Microphone Channel UK2
49.8900		NFM	Nationwide	Channel 7
		NFM	Nationwide	Microphone Channel E
		NFM	Nationwide	Maxan Walkie Talkie Ch 5
49.9000		NFM	Nationwide	Channel 8
49.9100		NFM	Nationwide	Channel 9
49.9200		NFM	Nationwide	Channel 10

Base	Mobile	Mode	Location	User & Notes
49.9300		NFM	Nationwide	Channel 11
49.9400		NFM	Nationwide	Channel 12
49.9500		NFM	Nationwide	Channel 13
49.9600		NFM	Nationwide	Channel 14
		NFM	Nationwide	Microphone Channel UK3
49.9700		NFM	Nationwide	Channel 15
49.9800		NFM	Nationwide	Channel 16

50.0000 - 52.0000 MHz   6m UK Amateur Radio Band All Modes

Base	Mobile	Mode	Location	User & Notes
50.0000		CW	Buxton	GB3BUX Beacon
50.0200		CW	Anglesey	GB3SIX Beacon
50.0420		CW	St Austell	GB3CTC Beacon
50.0500		CW	Potters Bar	GB3NHQ Baecon
50.0600		CW	Inverness	GB3RMK Beacon
50.0620		CW	Garvagh	GB3NGI Beacon
50.0640		CW	Lerwick	GB3LER Beacon
50.0650		CW	St Helier	GB3IOJ Beacon
50.2000	50.2000	USB	Nationwide	SSB Calling
51.5100	51.5100	NFM	Nationwide	FM Calling

52.0000 - 52.3875 MHz   Broadcasting Links NFM

52.8500 - 52.9500 MHz   Cordless Radio Microphones 12.5 kHz

Base	Mobile	Mode	Location	User & Notes
52.8500		NFM	Nationwide	Channel 1
52.8625		NFM	Nationwide	Channel 2
52.8750		NFM	Nationwide	Channel 3
52.8875		NFM	Nationwide	Channel 4
52.9000		NFM	Nationwide	Channel 5
52.9125		NFM	Nationwide	Channel 6
52.9250		NFM	Nationwide	Channel 7
52.9375		NFM	Nationwide	Channel 8
53.5250		NFM	Nationwide	BBC O/B Continuity

53.7500 - 55.7500 MHz   BBC 5W Cordless Microphones

Base	Mobile	Mode	Location	User & Notes
53.7500		NFM	Nationwide	Channel 1
53.8500		NFM	Nationwide	Channel 2
53.9500		NFM	Nationwide	Channel 3
54.0500		NFM	Nationwide	Channel 4
54.1500		NFM	Nationwide	Channel 5
54.2500		NFM	Nationwide	Channel 6
54.3500		NFM	Nationwide	Channel 7
54.4500		NFM	Nationwide	Channel 8
54.5500		NFM	Nationwide	Channel 9
54.6500		NFM	Nationwide	Channel 10
54.7600		NFM	Nationwide	Channel 11
54.8500		NFM	Nationwide	Channel 12
54.9500		NFM	Nationwide	Channel 13
55.0500		NFM	Nationwide	Channel 14
55.1500		NFM	Nationwide	Channel 15

Base	Mobile	Mode	Location	User & Notes
55.2500		NFM	Nationwide	Channel 16
55.3500		NFM	Nationwide	Channel 17
55.4500		NFM	Nationwide	Channel 18
55.5500		NFM	Nationwide	Channel 19

54.0000 - 60.0000 MHz MoD Tactical Communications 25 kHz Simplex

Base	Mobile	Mode	Location	User & Notes
56.6250	56.6250	NFM	Nationwide	Royal Signals Display
56.9750		NFM	Nationwide	BBC O/B Continuity
60.2950		NFM	Nationwide	BBC O/B Continuity

60.7500 - 62.7500 MHz BBC 5W Cordless and Outside Broadcast Mics

Base	Mobile	Mode	Location	User & Notes
60.7500		NFM	Nationwide	Channel 20
60.8500		NFM	Nationwide	Channel 21
60.9500		NFM	Nationwide	Channel 22
61.0500		NFM	Nationwide	Channel 23
61.1500		NFM	Nationwide	Channel 24
61.2500		NFM	Nationwide	Channel 25
61.3500		NFM	Nationwide	Channel 26
61.4500		NFM	Nationwide	Channel 27
61.5500		NFM	Nationwide	Channel 28
61.6500		NFM	Nationwide	Channel 29
61.7600		NFM	Nationwide	Channel 30
61.8500		NFM	Nationwide	Channel 31
61.9500		NFM	Nationwide	Channel 32
62.0500		NFM	Nationwide	Channel 33
62.1500		NFM	Nationwide	Channel 34
62.2500		NFM	Nationwide	Channel 35
62.3500		NFM	Nationwide	Channel 36
62.4500		NFM	Nationwide	Channel 37
62.5500		NFM	Nationwide	Channel 38
62.9250		NFM	Nationwide	Spare Channel

64.0000 - 68.0000 MHz MoD Tactical Communications 25 kHz Simplex
BBC Outside Broadcast Microphones and Links

Base	Mobile	Mode	Location	User & Notes
64.0375		WFM	Nationwide	BBC O/B Microphones
67.0125		NFM	Nationwide	BBC O/B Control
67.0250		NFM	Nationwide	BBC O/B Control

68.0000 - 69.5000 MHz MoD, Mould & Tactical Communications 25 kHz

Base	Mobile	Mode	Location	User & Notes
68.0000		AM	S. England	MoD Engineering
68.0250		AM	Nationwide	UKAEA Transport
		AM	Cheltenham	MoD Transport
68.1000		AM	S. England	MoD Engineering
68.1250		NFM	Brecon Beacons	Army Ranges
68.1500		NFM	Nationwide	Army Pye Equipment Ch 6
68.1500	79.3500	NFM	Nationwide	Army Stores
		NFM	Maidenhead	Army Stores
		NFM	Camberley	Army Stores
68.2000	79.4000	NFM	Nationwide	Army Pye Equipment Ch 4

Base	Mobile	Mode	Location	User & Notes
68.2500		NFM	South County Down	Ulster Defence Regiment
68.3000	79.2250	NFM	Nationwide	Army Pye Equipment Ch 1
68.3250		AM	Nationwide	Mil. Airfield Gnd Services
68.3500	79.4500	NFM	Nationwide	Army Pye Equipment Ch 5
68.3875		AM	Bristol	MoD Transport
68.4000	79.2500	NFM	Nationwide	Army Pye Equipment Ch 2
68.4250		NFM	Northern Ireland	Army
68.4250	79.2750	NFM	Nationwide	Army Pye Equipment Ch 3
68.5000		NFM	RN Rosyth	Dockyard Ops
68.5000	79.8000	NFM	Nationwide	Army Pye Equipment Ch 7
68.5250	79.7000	NFM	Nationwide	Army Pye Equipment Ch 8
68.5626		AM	RAE Farnborough	Emergency Services
68.6000	79.9500	NFM	Nationwide	Army Pye Equipment Ch 9
68.6125		AM	RAE Farnborough	Tractor Control
68.6250		NFM	Larkhill	Army Ranges
68.6375		NFM	Nationwide	Army Nubian Major Ch 1
		AM	RAE Farnborough	ATC
68.6625		NFM	Nationwide	Army Nubian Major Ch 2
68.6735		NFM	Northern Ireland	Army
68.6875		NFM	Nationwide	Army Nubian Major Ch 3
		AM	Cheltenham	MoD Transport
		AM	RAE Farnborough	Ground Services Channel 1
68.6938		AM	RAE Farnborough	Repair Workshop Ch 3
68.7125		NFM	Nationwide	Army Nubian Major Ch 4
68.7375		NFM	Nationwide	Army Nubian Major Ch 5
68.7625		NFM	Nationwide	Army Land Forces
		NFM	Nationwide	Army Nubian Major Ch 6
		NFM	Salisbury Plain	Army Transport
68.7875		NFM	Nationwide	Army Stores
		AM	Cheltenham	MoD Transport
68.8688		AM	RAE Farnborough	Fire Channel 4
68.9063		AM	RAE Farnborough	Medical Channel 2
68.9875		NFM	South Wales	Mould
69.0500		NFM	Nationwide	BBC OB Control
69.1250		NFM	Northern Ireland	Army
69.1550		NFM	Northern Ireland	Army
69.2000		NFM	Northern Ireland	Army
69.2500		NFM	Northern Ireland	Army
69.3000		AM	Nationwide	Sea Cadets
69.3250		NFM	Brecon Beacons	Army
		NFM	Northern Ireland	Army
69.3500		AM	Brecon Beacons	Army Cadets Hike Control
69.3750		NFM	Nationwide	BBC OB Control
		NFM	Northern Ireland	Army
69.4000		NFM	Northern Ireland	Army
69.4750	84.5250	NFM	Nationwide	39 Inf Bgd & Sig Sqn Ch A9
69.5000	69.5000	NFM	RN Rosyth	Security

Base	Mobile	Mode	Location	User & Notes
69.5000 - 69.8000 MHz			**MoD Tactical Communications 25 kHz Simplex**	
69.5500	69.5500	NFM	Northern Ireland	Army
69.7500	69.7500	NFM	Northern Ireland	Army
69.8000	69.8000	NFM	Northern Ireland	Army
69.8250 - 69.9625 MHz			**BBC TV O/B "System One" CMCR**	
			Studio Manager Talkback & MoD Tactical Coms	
69.8225	74.7000	NFM	Nationwide	BBC Studio Manager Ch 3
69.8250		NFM	Nationwide	BBC CMCR Channel 7
		NFM	Northern Ireland	Army
69.8350	75.2688	NFM	Nationwide	BBC Studio Manager Ch 4
69.8375		NFM	Nationwide	BBC CMCR Channel 8
69.8475	74.6875	NFM	Nationwide	BBC Studio Manager Ch 5
69.8500		NFM	Nationwide	BBC CMCR Channel 9
		NFM	Northern Ireland	Army
69.8600	74.7125	NFM	Nationwide	BBC Studio Manager Ch 6
69.8625		NFM	Nationwide	BBC CMCR Channel 10
69.8725	75.2813	NFM	Nationwide	BBC Studio Manager Ch 7
69.8750		NFM	Nationwide	BBC CMCR Channel 11
69.8850		NFM	Nationwide	BBC Studio Manager Ch 8
69.8975	75.2938	NFM	Nationwide	BBC Studio Manager Ch 9
69.9000		NFM	Nationwide	BBC Ch 12 BBC Scotland
		NFM	County Armagh	Ulster Defence Regiment
69.9075	75.2875	NFM	Nationwide	BBC Studio Manager Ch 2
69.9250		NFM	Nationwide	BBC Studio Continuity
69.9500		NFM	Northern Ireland	Army
69.9625	75.2625	NFM	Nationwide	BBC Studio Manager Ch 1
69.9650 - 70.0000 MHz			**MoD Tactical Communications 25 kHz**	
70.0000 - 70.5000 MHz			**4m Amateur Radio Band**	
70.0000		CW	Buxton	GB3BUX Beacon
70.0300		CW	St Austell	GB3CTC Beacon
70.0400		CW	Chatham	GB3REB Beacon
70.0600		CW	Dundee	GB3ANG Beacon
70.1300		CW	Slane	EI4RF Beacon
70.2000	70.2000	SSB	Nationwide	SSB Calling Channel
		CW	Nationwide	CW Calling Channel
70.2600	70.2600	AM	Nationwide	AM Calling Channel
		NFM	Nationwide	FM Calling Channel
70.3000	70.3000	FAX	Nationwide	FAX Calling Channel
		RTTY	Nationwide	RTTY Calling Channel
70.3125	70.3125	PKT	Nationwide	Packet Channel
70.3250	70.3250	PKT	Nationwide	Packet Channel
70.3500	70.3500	NFM	Nationwide	Raynet Channel
70.3750	70.3750	NFM	Nationwide	Raynet Channel
70.4000	70.4000	NFM	Nationwide	Raynet Channel
70.4500	70.4500	NFM	Nationwide	FM Calling Channel
70.4875	70.4875	PKT	Nationwide	Packet Channel

Base	Mobile	Mode	Location	User & Notes	
70.5125 - 71.5000 MHz		**MHz**	**Fire Brigades (England & Wales) 12.5 kHz**		
70.5125	80.1875	NFM	Northumberland	LJ	
70.5125	80.4375	NFM	West Midlands	BW	Channel 1
70.5250	80.7375	NFM	Greater Manchester	FT	Channel 2
70.5250	80.9625	NFM	London Central	FH	Channel 1
70.5375	80.1875	NFM	Nottinghamshire	M2NZ	
70.5500		NFM	Greater Manchester	FT	Channel 1 HQ
70.5625	80.6000	NFM	Mid Glamorgan	WF	
70.5625	80.9875	NFM	Lincolnshire	NV	
70.5750	80.4625	NFM	West Midlands	FB	Channel 3
70.5875		NFM	Kent		
70.5875	80.7625	NFM	Greater Manchester	FT	Channel 3
70.5875	80.1875	NFM	Hampshire	HX	Channel 2
70.6000	81.2625	NFM	Warwickshire	M2YS	
70.6125		NFM	West Yorkshire	M2XF	
70.6125	80.1250	NFM	North Surrey	HF	
		NFM	Dyfed	WV	
70.6250		NFM	Merseyside	FO	Channel 3
70.6250	80.6125	NFM	Essex	VD	Channel 1
70.6375	80.2125	NFM	East Sussex	KD	
70.6375	80.1125	NFM	South Yorkshire	XV	
70.6500	80.9875	NFM	Wiltshire	QM	
70.6625	81.0000	NFM	Leicestershire	M2NK	
70.6750	80.5500	NFM	Lancashire	BE	Channel 1
70.6750	80.5250	NFM	South Glamorgan	WD	
70.6875	81.1250	NFM	Worcester/Hereford	YB	
70.7000	80.2000	NFM	Merseyside	M2FO	Channel 4
		NFM	Norwich	M2VF	
70.7000	81.1250	NFM	Gwent	WP	
70.7000	80.9875	NFM	West Yorkshire	XF	Channel 1
70.7125	80.8000	NFM	Derbyshire	M2ND	
70.7250		NFM	Kent		
70.7250	80.0375	NFM	Devon	QD	
70.7250	80.6750	NFM	Essex	VD	Channel 2
70.7500	80.7500	NFM	Northamptonshire	NO	
70.7625		NFM	West Yorkshire	M2XF	
70.7625	80.1500	NFM	East London	FE	Channel 3
70.7750	80.5000	NFM	Cheshire	M2CF	
		NFM	Hampshire	HX	Channel 1
70.7875	80.8000	NFM	Cornwall	QA	
70.8000	80.5125	NFM	West Sussex	KW	
70.8125	81.2125	NFM	Gwynedd	WC	
70.8250	80.3250	NFM	Devon	M2QD	
70.8250	80.7875	NFM	Greater Manchester	FT	Channel 4
70.8375		NFM	Cambridgeshire	VC	
		NFM	Kent		
70.8375	80.0375	NFM	Cumbria	BC	
70.8374	80.1250	NFM	Kent	KF	
70.8500	80.9625	NFM	Powys	WB	

Base	Mobile	Mode	Location	User & Notes	
70.8625	80.5500	NFM	Dorset	QK	
70.8750		NFM	West Yorkshire	M2XF	
70.8875	80.2125	NFM	County Durham	LF	
70.8875	80.9375	NFM	Staffordshire	YG	
70.9000	80.0375	NFM	Hertfordshire	M2VI	
70.9000	80.4000	NFM	East Anglia		
70.9000	80.6000	NFM	Lancashire	BE	Channel 2
70.9125		NFM	Essex	VD	Channel 3
		NFM	London		Data Channel
70.9375		NFM	England & Wales		Data Channel
70.9500		NFM	West Glamorgan	WZ	
70.9625	81.0875	NFM	Merseyside	FO	Channel 2
70.9626	80.1125	NFM	South London	FS	Channel 2
70.9750	80.6500	NFM	Shropshire	YU	
71.0125	80.1750	NFM	Avon	QG	
71.0375	81.0625	NFM	Merseyside	FO	
71.0750	80.1500	NFM	Humberside	XT	Channel 2
71.0750	80.6250	NFM	Gloucester	QF	
71.1000	80.6625	NFM	Oxfordshire	M2HI	
71.1000	83.6250	NFM	Humberside	XT	
71.1125		NFM	Bedfordshire	M2VM	
		NFM	Cleveland	LT	
71.1250	80.1125	NFM	Somerset	QI	
71.1375	80.4500	NFM	Buckinghamshire	HK	
71.1375	80.4375	NFM	North Yorkshire	LY	
71.1500	80.5125	NFM	West Midlands	FBW	Channel 2
71.1625	80.8750	NFM	Clwyd	M2WK	
71.1750		NFM	North Yorkshire	LY	
71.1750	80.2125	NFM	North London	FN	Channel 4
71.2000	80.2250	NFM	Berkshire	HD	
71.2750	81.0625	NFM	Isle of Wight	HP	
71.2750	81.0875	NFM	Suffolk	VN	
71.3000		NFM	Tyne & Wear	LP	Channel 1
71.3125		NFM	England & Wales		Data Channel
71.3375		NFM	England & Wales		Data Channel
		NFM	London	FH	Channel 5
		NFM	Tyne & Wear	LP	Channel 2
71.3875		NFM	Gloucestershire	QC	
		NFM	Lancashire	BE	Channel 3
71.4000	81.2500	NFM	Nationwide	RW	Radio Eng's
71.4250		NFM	Cambridgeshire	VC	

71.5125 - 72.7875 MHz PMR Low Band Mobiles 12.5 kHz (Split + 10.5 MHz)

72.8000 - 73.9250 MHz MoD Tactical Communications 25 kHz Duplex

Base	Mobile	Mode	Location	User & Notes
72.8000		NFM	Brecon Beacons	Army Range Ops
72.8125		NFM	Southampton	Royal Navy Loading
72.8250		NFM	RAF Fairford	Base Ops
72.9875	72.9875	NFM	HMS Drake	Naval Provost

Base	Mobile	Mode	Location	User & Notes
73.0000		NFM	Brecon Beacons	Army Range Ops
73.0000	78.9000	NFM	Nationwide	Army Pye Equipment Ch 7
73.0250	78.6500	NFM	Nationwide	Army Pye Equipment Ch 8
73.1000		NFM	Aldershot	Military Police
73.1250	78.4000	NFM	Nationwide	Army Pye Equipment Ch 1
73.1500	78.4250	NFM	Nationwide	Army Pye Equipment Ch 2
73.2000	78.4500	NFM	Nationwide	Army Pye Equipment Ch 3
73.2125		NFM	Southampton	Royal Navy Transport
73.2250	78.4750	NFM	Nationwide	Army Pye Equipment Ch 4
73.2500	78.5000	NFM	Nationwide	Army Pye Equipment Ch 5
73.3250	78.8000	NFM	Nationwide	Army Pye Equipment Ch 6
73.3375		NFM	RN Portsdown	Base Ops
73.3500		NFM	RAF High Wycombe	Security
		NFM	Nationwide	MoD Security Police
73.4000		NFM	Plymouth	MoD Operations
73.4500		NFM	RN Portsdown	Security
73.4750		NFM	RAF Molesworth	Security Channel 1
		NFM	RAF Greenham Common	Security
73.4750	73.4750	NFM	Nationwide	MoD Police
73.5000	78.6500	NFM	Nationwide	Army Pye Equipment Ch 1
73.5250	78.7250	NFM	Nationwide	Army Pye Equipment Ch 2
73.5375		NFM	Plymouth	MoD Operations
73.5500	78.7750	NFM	Nationwide	Army Pye Equipment Ch 9
73.6125		NFM	Plymouth	MoD Police
73.6500		NFM	London	MoD Police
73.6250		NFM	RAF Molesworth	Security Channel 2
73.6500		NFM	RAF Molesworth	Security Channel 3
73.6750		NFM	RAF Molesworth	Security Channel 4
73.7000	84.1250	NFM	Nationwide	61 Sig Sqn Channel 1
73.7250	84.1500	NFM	Nationwide	61 Sig Sqn Channel 2
73.7500	84.1750	NFM	Nationwide	61 Sig Sqn Channel 3
73.7750	84.2500	NFM	Nationwide	61 Sig Sqn Channel 4
73.8000		NFM	Brecon Beacons	Army Range Control
73.8000	84.2750	NFM	Nationwide	61 Sig Sqn Channel 5
73.8250	84.3250	NFM	Nationwide	61 Sig Sqn Channel 6
73.8500		NFM	RAF Molesworth	Security Channel 5
73.8500	84.3050	NFM	Nationwide	61 Sig Sqn Channel 7
73.8750	84.3750	NFM	Nationwide	61 Sig Sqn Channel 8
73.9000	84.4250	NFM	Nationwide	61 Sig Sqn Channel 9
73.9250	84.4750	NFM	Nationwide	61 Sig Sqn Channel 10

73.9250 - 74.1000 MHz MoD Mould Repeaters 12.5 kHz Duplex

Base	Mobile	Mode	Location	User & Notes
74.0125		NFM	Northamptonshire	Mould
74.0250		NFM	Hampshire	Mould
		NFM	Lincolnshire	Mould
		NFM	Midlands	Mould
74.0375		NFM	Shropshire	Mould
		NFM	Wiltshire	Mould
74.0500		NFM	Brecon Beacons	Mould

Base	Mobile	Mode	Location	User & Notes
		NFM	Midlands	Mould
		NFM	Northamptonshire	Mould
74.0625		NFM	Manchester	Mould
		NFM	Midlands	Mould
		NFM	Northamptonshire	Mould
		NFM	Yorkshire	Mould
74.0750		NFM	Devon	Mould
		NFM	South Wales	Mould
74.0875		NFM	Manchester	Mould
		NFM	Perthshire	Mould
		NFM	South Wales	Mould

74.1000 - 74.7875 MHz   MoD Mould & Tactical Channels 12.5 kHz

Base	Mobile	Mode	Location	User & Notes
74.1000	79.2000	NFM	RAF Honington	RAF Police Channel 3
74.1125		NFM	Colchester	Barracks
		NFM	Manchester	Mould
		NFM	Midlands	Mould
		NFM	South Wales	Mould Channel 1 Primary
74.1250		NFM	Brecon Beacons	Mould
		NFM	Midlands	Mould
74.1375		NFM	Brecon Beacons	Mould
74.1500		NFM	Midlands	Mould
		NFM	Shropshire	Mould
74.1625		NFM	Manchester	Mould
		NFM	Midlands	Mould
74.1875		NFM	Yorkshire	Mould
74.2000	79.3000	NFM	Nationwide	Military Transport Security
74.2125	79.2125	NFM	East Anglia	Mould
		NFM	South Wales	Mould
74.2250		NFM	Lincolnshire	Mould
		NFM	Midlands	Mould
		NFM	Nationwide	Royal Ordnance Corps
		NFM	Nottinghamshire	Mould
		NFM	Upper Heyford	Mould
74.2375		NFM	Lincolnshire	Mould
		NFM	Perthshire	Mould
		NFM	Shropshire	Mould
74.2500		NFM	RAF Bentwaters	USAF Police
74.2500	79.2875	NFM	Colchester	Barracks
74.2500	79.3500	NFM	Anglia	Mould
74.2625		NFM	South Wales	Mould
74.3125		NFM	South Wales	Mould
		NFM	Shropshire	Mould
74.3375		NFM	Manchester	Mould
74.3500	79.4500	NFM	Brecon Beacons	Mould
74.3625		NFM	Chester	Mould
		NFM	Hampshire	Mould
74.3750		NFM	Brecon Beacons	Mould
74.3875	79.4125	NFM	Devon	Mould

Base	Mobile	Mode	Location	User & Notes
		NFM	South Wales	Mould
74.4000		NFM	Brecon Beacons	Mould
		NFM	Cambridge Airport	Marshalls Ltd Fire Channel
74.4125		NFM	Brecon Beacons	Mould
		NFM	Shropshire	Mould
		NFM	Yorkshire	Mould
74.4375	79.6625	NFM	East Anglia	Mould
		NFM	Midlands	Mould
74.4500	79.4500	NFM	Brecon Beacons	Mould
74.4625		NFM	South Wales	Mould
74.4675	79.7125	NFM	South Wales	Mould
74.4875	79.9265	NFM	East Anglia	Mould
		NFM	Hampshire	Mould
		NFM	Wiltshire	Mould
74.5125		NFM	Brecon Beacons	Mould
		NFM	Midlands	Mould
74.5250		NFM	Brecon Beacons	Mould
74.5375		NFM	Brecon Beacons	Mould
		NFM	Lincolnshire	Mould
		NFM	Midlands	Mould
74.5500	79.5500	NFM	Nationwide	RAF Police
74.5625		NFM	Brecon Beacons	Mould
74.5750		NFM	Brecon Beacons	Mould
		NFM	Manchester	Mould
		NFM	Shropshire	Mould
		NFM	Yorkshire	Mould
74.5875		NFM	Devon	Mould
		NFM	Hampshire	Mould
		NFM	South Wales	Mould
		NFM	Wiltshire	Mould
74.6000		NFM	Midlands	Mould
74.6125		NFM	Brecon Beacons	Mould
		NFM	Midlands	Mould
74.6250		NFM	Brecon Beacons	Mould
		NFM	Devon	Mould
74.6500	79.6125	NFM	Hampshire	Mould
74.6625		NFM	Wales	Mould
74.6750	79.7125	NFM	South Wales	Mould
74.6875	79.6875	NFM	Cambridgeshire	Mould Channel 15
74.7000		NFM	Perthshire	Mould
		NFM	South Wales	Mould
74.7125		NFM	Oxfordshire	Mould
74.7250		NFM	Brecon Beacons	Mould
		NFM	Manchester	Mould
		NFM	Shropshire	Mould
74.7375		NFM	Hampshire	Mould
		NFM	London	Mould
74.7500		NFM	Devon	Mould
		NFM	Manchester	Mould

Base	Mobile	Mode	Location	User & Notes
		NFM	Shropshire	Mould
		NFM	Wiltshire	Mould
74.7625		NFM	South Wales	Mould
74.7875		NFM	South Wales	Mould
75.0000		AM	Nationwide	Runway Marker Beacons

75.2500 - 75.3000 MHz   BBC O/B Talkback & MoD Tactical Comms

Base	Mobile	Mode	Location	User & Notes
75.2500		AM	Dyfed	Pendine Range
		NFM	Midlands	USAF Police
75.2625	69.9625	NFM	Nationwide	BBC Talkback Channel 1
75.2688	69.8350	NFM	Nationwide	BBC Talkback Channel 4
75.2813	69.8725	NFM	Nationwide	BBC Talkback Channel 7
75.2875	69.9075	NFM	Nationwide	BBC Talkback Channel 2
75.2938	69.8975	NFM	Nationwide	BBC Talkback Channel 9

75.3000 - 76.7000 MHz   MoD Police, Mould & USAFE Security 12.5 kHz

Base	Mobile	Mode	Location	User & Notes
75.3000	75.3000	AM	Dyfed	Pendine Range
		NFM	Porton Down	Security
		NFM	East Anglia	USAF Police
75.3250	75.3250	NFM	Nationwide	USAF Ground Common
		NFM	Plymouth	Royal Marines
75.3375	75.3375	NFM	RAF Bentwaters	Ground
73.3500	73.3500	NFM	Midlands	USAF Police
75.4000	75.4000	AM	Dyfed	Pendine Range
		NFM	East Anglia	USAF Police
75.4250	75.4250	NFM	RAF Upper Heyford	Ground
75.4500	75.4500	NFM	Nationwide	USAF Base Security
		NFM	RAF Upper Heyford	Security
75.4750		NFM	Brecon Beacons	Mould
75.5000	75.5000	NFM	RAF Upper Heyford	Security
75.5250	75.5250	NFM	RAF Woodbridge	Security
75.5500	75.5500	NFM	RAF Alconbury	Security
		NFM	RAF Upper Heyford	Security
75.5750	75.5750	NFM	Nationwide	USAF Ground Common
75.5875	75.5875	NFM	RAF Greenham Common	Base Commander
		NFM	RAF Upper Heyford	Security
75.6750	75.6750	NFM	RAF Lakenheath	Security
		NFM	RAF Mildenhall	Security
75.7125		NFM	Hampshire	Mould
		NFM	Oxfordshire	Mould
75.7375		NFM	London	Mould
75.7500		NFM	Devon	Mould
		NFM	South Wales	Mould
		NFM	Wiltshire	Mould
75.7625	75.7625	NFM	Salisbury Plain	Close Support Group
75.7875		NFM	Lothian	Mould
		NFM	South Wales	Mould
75.8000	75.8000	NFM	RAF Upper Heyford	Security
75.8125		NFM	Brecon Beacons	Army

Base	Mobile	Mode	Location	User & Notes
75.8250	75.8250	NFM	Nationwide	USAF Special Agents
75.8625		NFM	Oxfordshire	Mould
75.8750		NFM	Devon	Mould
		NFM	Dorset	Territorial Army Mould
		NFM	London	Mould
		NFM	South Wales	Mould
75.9000		NFM	South Wales	Mould
75.9125		NFM	South Wales	Mould
75.9375		NFM	Eastern Scotland	Mould
		NFM	Hampshire	Mould
		NFM	South Wales	Mould
75.9625		NFM	Hampshire	Mould
76.0000	76.0000	NFM	Nationwide	USAF Police
76.0125		NFM	Hampshire	Mould
		NFM	Midlands	Mould
		NFM	South Wales	Mould
76.0500	76.0500	NFM	RAF Lakenheath	Security Channel 6
76.0625		NFM	Devon	Mould
		NFM	South Wales	Mould
76.1125		NFM	Hampshire	Mould
		NFM	South Wales	Mould
76.1250		NFM	Midlands	Mould
		NFM	South Wales	Mould
76.1625		NFM	Hampshire	Mould
76.2250		NFM	Devon	Mould
		NFM	South Wales	Mould
76.2250	76.2250	NFM	Nationwide	USAF Base to Mobile
76.2625		NFM	South Wales	Army
76.3250		NFM	Devon	Mould
		NFM	Hampshire	Mould
		NFM	Lothian	Mould
		NFM	London	Mould
76.3250	76.3250	NFM	RAF Lakenheath	Security
		NFM	RAF Upper Heyford	Maintenance Channel 4
		NFM	Ruislip	MoD Police
76.3500	76.3500	NFM	RAF Alconbury	Security
76.3625		NFM	Midlands	Mould
		NFM	Northamptonshire	Mould
76.3875		NFM	Hampshire	Mould
76.4000	76.4000	NFM	RAF Greenham Common	Security Fence Control
76.4375		NFM	Central London	Mould
		NFM	Midlands	Mould
		NFM	Oxfordshire	Mould
76.4500	76.4500	NFM	RAF Fairford	Security
76.5000	76.5000	NFM	Ruislip	MoD Police
76.5250	76.5250	NFM	Camberley	WRAC Gate Security
		NFM	RAF Lakenheath	Security Channel 1
76.5500	76.5500	NFM	RAF Alconbury	Security
		NFM	RAF Greenham Common	Security Channel 3

Base	Mobile	Mode	Location	User & Notes
		NFM	RAF Upper Heyford	Security
76.5750	76.5750	NFM	RAF Lakenheath	Security Channel 2
76.6000	76.6000	NFM	Cumbria	Range Control
76.6250	76.6250	NFM	RAF Greenham Common	Security Channel 1
76.6500	76.6500	NFM	RAF Alconbury	Security
		NFM	RAF Greenham Common	Security Channel 2
		NFM	RAF Upper Heyford	Security
76.6750	76.6750	NFM	RAF Greenham Common	Security Channel 4
		NFM	RAF Lakenheath	Security Channel 3

76.7000 - 76.9500 MHz   HM Customs, BT, & USAFE Comms 12.5 kHz

Base	Mobile	Mode	Location	User & Notes
76.7000		NFM	Cumbria	Range Control
		NFM	Nationwide	British Telecom Channel A
		NFM	Nationwide	Customs and Excise Ch 4
		NFM	Nationwide	USAF Police
76.7000	86.7125	NFM	Nationwide	Customs and Excise Ch 1
76.7125		NFM	Nationwide	Customs and Excise Ch 5
76.7250		NFM	Nationwide	Customs and Excise Ch 6
76.7250	86.7125	NFM	Nationwide	Customs and Excise Ch 2
76.7375		NFM	Nationwide	British Telecom Channel B
		NFM	Nationwide	Customs and Excise
76.7375	86.7375	NFM	Nationwide	Customs and Excise Ch 9
76.7500		NFM	Nationwide	British Telecom Channel C
		NFM	Nationwide	Customs and Excise
76.7625		NFM	Nationwide	British Telecom Channel D
76.8000		NFM	RAF Bentwaters	Base Security
76.8250	86.8250	NFM	Nationwide	Thames TV Talkback
76.9000		NFM	Midlands	USAF Base Gate Security
76.9250		NFM	Midlands	USAF Base Gate Security
76.9375		NFM	Nationwide	Customs and Excise Ch 7
76.9375	86.9375	NFM	Nationwide	Customs and Excise Ch 3
76.9500		NFM	Nationwide	Customs and Excise Ch 8
76.9500	86.9500	NFM	Nationwide	Customs and Excise Ch 10

76.9625 - 77.5000 MHz   PMR Low Band Mobiles 12.5 kHz Duplex

77.8000 - 79.0000 MHz   MoD & USAFE, BBC O/B Networks 12.5 kHz

Base	Mobile	Mode	Location	User & Notes
77.8750	77.8750	NFM	Nationwide	ITN OB Film Mobiles
78.0000	78.0000	NFM	Hampshire	Army Ops
		AM	RAF Finningley	ATC Air Show Stewards
78.0250	78.0250	NFM	RAF Greenham Common	USAF Security Foot Patrols
78.1000	78.1000	AM	Nationwide	RAF ATC Channel V1
		NFM	RAF Bentwaters	Command Post
78.1375	78.1375	NFM	Blandford	Royal Signals Security
78.1500	78.1500	NFM	Nationwide	BBC O/B Talkback Ch 1
78.1625	78.1625	NFM	Plymouth	MoD Police
78.1875	78.1875	NFM	Nationwide	BBC O/B Camera Ch 1
78.2000	78.2000	NFM	Nationwide	BBC O/B Camera Ch 2
78.2125	78.2125	NFM	Nationwide	BBC O/B Engineering Ch 3

Base	Mobile	Mode	Location	User & Notes
78.2250	78.2250	NFM	Nationwide	BBC O/B Engineering Ch 4
		NFM	RAF Greenham Common	Security
78.2275	78.2275	NFM	Nationwide	BBC O/B Rigging Channel
78.2375	78.2375	NFM	Nationwide	BBC O/B Engineering Ch 5
		NFM	Nationwide	BBC Microwave Link Setup
78.2400	78.2400	NFM	Nationwide	BBC O/B Rigging Channel
78.2500	78.2500	NFM	Nationwide	BBC O/B Lighting Ch 6
78.2750	78.2750	NFM	Colchester	Barracks Channel 1
		NFM	Hampshire	Army Ops
		NFM	Waterbeach Est.	Security
78.2875	78.2875	NFM	Colchester	Barracks Channel 4
78.3000	78.3000	NFM	Colchester	Barracks Channel 2
		AM	Nationwide	Combined Cadet Force
		NFM	Norfolk	Stanford Army Battle Area
		NFM	Waterbeach Est.	Security
78.3125	78.3125	NFM	Salisbury Garrison	Military Police
78.3250	78.3250	NFM	Colchester	Barracks Channel 3
		NFM	Hampshire	Military Police
		NFM	London	Military Police
		NFM	Norfolk	Stanford Army Battle Area
		NFM	Waterbeach Est.	Security
78.3375		NFM	London	MoD Police
78.3875		NFM	London	MoD Police
		NFM	RN Portsmouth	RN Police
78.4000	73.1250	NFM	Nationwide	Military Equipment Chan 1
		NFM	London	MoD Police
78.4250	73.1500	NFM	Nationwide	Military Equipment Chan 2
78.4500	73.2000	NFM	Nationwide	Military Equipment Chan 3
78.4750	73.2250	NFM	Nationwide	Military Equipment Chan 4
78.4875		NFM	Colchester	Military Police
78.5000		NFM	Nationwide	Military Equipment Chan 5
		NFM	RAF Bentwaters	USAF Police
		NFM	RAF Woodbridge	USAF Police
78.5250		NFM	Northern Ireland	Ulster Defence Reg. Police
		NFM	RAF Bentwaters	USAF Police
		NFM	RAF Greenham Common	USAF Police
		NFM	RAF Woodbridge	USAF Police
		NFM	Welford	USAF Bomb Disposal Units
78.5500		NFM	Shropshire	Army Fire Channel
78.5750		NFM	RNAS Culdrose	Base Ops
		NFM	RNAS Yeovilton	Base Ops
78.5875		NFM	RAF Woodbridge	USAF Police
78.6000		NFM	RAF Wittering	Channel 52
78.6250		NFM	RAF Bentwaters	USAF Police
78.6500	73.5000	NFM	Nationwide	Military Equipment Chan 1
78.6750		NFM	RAF Wittering	Channel 53
78.6750	73.2500	NFM	Nationwide	Military Equipment Chan 8
78.6875		NFM	RN Portsmouth	Security
78.6975	78.6975	NFM	Nationwide	RAF Police

Base	Mobile	Mode	Location	User & Notes
78.7000		NFM	RAF Fairford	Base Ops
78.7250		NFM	RAF Bentwaters	USAF Police
		NFM	RAF Woodbridge	USAF Police
78.7750		NFM	London	Military Police
		NFM	Ternhill	Army Fire Channel
78.7750	73.5500	NFM	Nationwide	Military Equipment Chan 9
78.8000		NFM	Hampshire	Army
		NFM	RNAS Portland	Tower
		NFM	RSRE Malvern	Base Security
		NFM	RAF Bentwaters	Ground Maintenance
78.8000	73.3500	NFM	Nationwide	Military Equipment Chan 6
78.8125		NFM	Newbury	Army
78.8250		NFM	Donington Salop	Army Ordinance Depot Fire
		NFM	London	MoD Police
78.8500		NFM	Brecon Beacons	Army
		NFM	RAF Wittering	Channel 54
78.9000		NFM	RN Lee-on-Solent	Tower Rescue
		NFM	RNAS Portland	Operations
78.9000	73.0000	NFM	Nationwide	Military Equipment Chan 7
78.9500		NFM	RNAS Culdrose	Base Ops
78.9750		NFM	GCHQ Cheltenham	Security and Transport

79.0000 - 80.0000 MHz   MoD and RAF Ground Services 12.5 kHz

Base	Mobile	Mode	Location	User & Notes
79.0000		NFM	RAF Odiham	Ops
		NFM	RAF Uxbridge	Ops
		NFM	RNAS Yeovilton	Ground
79.0500		NFM	RAF Honington	Ops
		NFM	RAF Lyneham	Crew Buses
		NFM	RAF Northolt	Paintbox Control
79.0750		NFM	RAF Cosford	ATC
		NFM	RAF Northolt	Tower - Crash Tenders
79.1000		NFM	Nationwide	RAF Police Channel 1
		NFM	RAF Honington	ATC Channel 1
79.1250		NFM	RAF Brize Norton	Ground Services
		NFM	RAF St Athan	Loadmaster and Ground
79.1500		NFM	Nationwide	RAF Police Channel 2
		NFM	RAF Abingdon	Tower Channel 2
79.1750		NFM	RAF Lyneham	Tower - Ground
		NFM	RAF Northolt	Forward Control
		NFM	RAF Wattisham	74 Sqn Ops
79.1875		NFM	RAF Northolt	Link Repeater
79.2000		NFM	Nationwide	RAF Police Channel 3
		NFM	RAF Honington	Fire and Refuel Channel 3
		NFM	RAF Wattisham	ATC Channel 3
79.2125	74.1125	NFM	East Anglia	Mould
		NFM	Manchester	Mould
		NFM	South Wales	Mould
79.2250		NFM	RAF Abingdon	RAF Police
		NFM	RAF Wittering	Channel 3/13

Base	Mobile	Mode	Location	User & Notes
79.2250	68.3000	NFM	Nationwide	Military Channel 1
79.2500		NFM	Nationwide	RAF Police Channel 4
		NFM	RAF Honington	Fire/CrashSapho Control
		NFM	RAF Wattisham	MaintenanceGadfly
79.2500	68.4000	NFM	Nationwide	Military Channel 2
79.2550		NFM	RAF Wittering	Channel 4/14
79.2750.		NFM	RAF Lyneham	Ground Control
		NFM	RAF Northolt	Ground Control
		NFM	RAF St Athan	Transport and Ground
		NFM	RAF Wittering	Channel 5/15
79.2750	68.4250	NFM	Nationwide	Military Channel 3
79.3000		NFM	Nationwide	RAF Police Channel 5
		NFM	RAF Odiham	Tower
		NFM	RAF Wittering	Channel 60/70
79.3250		NFM	RAF Cosford	Ground Control
		NFM	RAE Farnborough	Ground Control
		NFM	RAF Honington	Ground Control
		NFM	RAF Lossiemouth	Tower
79.3250	79.3250	NFM	Nationwide	RAF Police
79.3500		NFM	Nationwide	RAF Police Channel 6
		NFM	RAF Kemble	Tower Channel 1
79.3750		NFM	RAF Honington	RAF Police
		NFM	RN Gosport	Tanzy Control
79.4000		NFM	Nationwide	Military Channel 4
		NFM	Nationwide	RAF Police Channel 7
		NFM	RAF Tain	Range Control
		NFM	RAF Wittering	Channel 7/17
79.4125	74.3875	NFM	South Wales	Mould
79.4250		NFM	RAF Brize Norton	Tower - Ground
79.4500		NFM	Nationwide	RAF Police Channel 8
		NFM	RAF Cosford	Charlie Control
		NFM	RAF Wittering	Channel 8/18
79.4750		NFM	LyddLydd	Army Camp Security
		NFM	RAF Northolt	Papa Control
79.4750	79.4750	NFM	Nationwide	RAF Police
79.4875		NFM	Southend Island	MoD Repeater
		NFM	S Ireland	RAF Repeater
79.5000		NFM	Nationwide	RAF Police Channel 9
		NFM	RAF Kemble	Tower Channel 2
		NFM	RAF Wittering	Channel 9/19
		NFM	LyddLydd	Army Camp Security
79.5250		NFM	RAF Honington	RAF Police
		NFM	RAF Wattisham	Network Repeater
		NFM	RAF Wittering	Channel 10/20
79.5500		NFM	Nationwide	RAF Police Channel 10
		NFM	RAF Wattisham	56 Sqn Basil Control
79.5620		NFM	Ternhill	Army Staff Cars
79.5750		NFM	RAF Wattisham	RAF Police Silicon Control
79.6000		NFM	Nationwide	RAF Police Channel 11

Base	Mobile	Mode	Location	User & Notes
		NFM	RAF Brize Norton	Brize Ops
79.6125	79.6125	NFM	Cambridgeshire	Mould Channel 4
79.6500		NFM	Nationwide	RAF Police Channel 12
		NFM	RAF Benson	Tower
		NFM	RAF Brize Norton	Tower - Ground
79.6750		NFM	RAF Brize Norton	Loadmaster
79.7000		NFM	RAF Honington	RAF Police
79.7000	68.1500	NFM	Nationwide	Military Channel 6
		NFM	Nationwide	BBC Engineering Channel
79.7125		NFM	Colchester	Barracks
		NFM	Norfolk	Stanford Army Battle Area
79.7125	74.4675	NFM	South Wales	Mould
79.7625	74.2625	NFM	South Wales	Mould
79.7750		NFM	Colchester	Barracks
		NFM	RAF Binbrook	Fire
		NFM	Nationwide	RAF Fire Channel
79.8000		NFM	Nationwide	Military Channel 7
79.8250		NFM	RAF Wittering	Channel 22/24
79.8250	84.8250	NFM	RAF Wittering	RAF Police Channel 2/12
79.8500		NFM	RAF Abingdon	Salvage Transport
79.9000	68.5250	NFM	Nationwide	Military Channel 8
79.9500	68.6000	NFM	Nationwide	Military Channel 9

80.0000 - 82.5000 MHz Fire Brigades (England & Wales)
(Split - 10.5 MHz) **Royal Observer Corps 12.5 kHz**

Base	Mobile	Mode	Location	User & Notes
80.0125		NFM	London	Fire Tender-Tender Ch 21
80.0375		NFM	Barrow	Fire Brigade Channel
80.0750		NFM	London	Fire Tender-Tender Ch 22
81.1250		NFM	Nationwide	Fire Brigade Channel
81.2625		NFM	Tayside	ROC Link
81.3125		NFM	Tayside	ROC Link
81.3500		NFM	Tayside	ROC Link
81.5750		NFM	Nationwide	Police Radio Engineering
81.7000	68.2000	NFM	SE England	Customs & Excise
81.7750	68.2250	NFM	SE England	Customs & Excise
81.7875	68.2375	NFM	SE England	Customs & Excise
81.9250	68.4250	NFM	E Birmingham	Taxi Service
81.9500		NFM	London	Fire Brigade Ops Rooms
81.9625	68.4625	AM	Plymouth	Plymouth Taxis

82.5000 - 84.0000 MHz Limited Police Mobile 12.5 kHz NFM

84.1250 - 84.9750 MHz MoD Police Communications 25 kHz

Base	Mobile	Mode	Location	User & Notes
84.2250	84.2250	NFM	Nationwide	Royal Air Force
84.3000	84.3000	NFM	Nationwide	RAF Mountain Rescue
84.3500	84.3500	NFM	Nationwide	Military Police
84.3625	84.3625	NFM	Nationwide	Military Police
84.3750	84.3750	NFM	Nationwide	Military Police
84.3875	84.3875	NFM	Nationwide	Military Police

Base	Mobile	Mode	Location	User & Notes
84.4000	84.4000	NFM	Nationwide	Military Police
84.4125	84.4125	NFM	Nationwide	Military Police
84.4250	84.4250	NFM	Nationwide	Military Police
84.4375	84.4375	NFM	Nationwide	Military Police
84.4500	84.4500	NFM	Nationwide	Military Police
84.4625	84.4625	NFM	Nationwide	Military Police
84.4750	84.4750	NFM	Dyfed	Castlemartin RAC Range
		NFM	Nationwide	Military Police
84.4875	84.4875	NFM	Nationwide	Military Police
84.5000	84.5000	NFM	Nationwide	Military Police Escorts
84.5125	84.5125	NFM	Nationwide	Military Police
84.5250	84.5250	NFM	Aldershot	Military Police
84.5375	84.5375	NFM	Nationwide	Military Police
84.5500	84.5500	NFM	London	Military Police
84.5625	84.5625	NFM	Nationwide	Military Police
84.5750	84.5750	NFM	Nationwide	Military Police
84.5875	84.5875	NFM	Nationwide	Military Police
84.6000	84.6000	NFM	Nationwide	Military Mountain Rescue
		NFM	Southampton	Royal Navy
84.7750	84.7750	NFM	Southampton	Royal Navy
84.8250	84.8250	NFM	Nationwide	Royal Air Force
84.9250	84.9250	NFM	Southampton	Royal Navy
84.9750	84.9750	NFM	Lulworth	Range Control

85.0125 - 86.2975 MHz   PMR Low Band Base Repeaters 12.5 kHz

Base	Mobile	Mode	Location	User & Notes
85.0125	71.5125	NFM	Severn	Severn Trent Water Ch 3
85.0250	71.5250	NFM	Gedling	Gedling Borough Council
		NFM	Peterborough	Peterborough Develop. Corp
85.0375	71.5375	NFM	Aberdeen	Roads Department
		NFM	Ipswich	Council Repeater
85.0500	71.5500	NFM	Anglia	Farm Feed Co.
		NFM	Aylesham	East Coast Grain
		NFM	Bennington	Braceys
		NFM	Cambridge	Trumpington Farms
		NFM	Cleveland	Mastercare
		NFM	Ipswich	Water Board
		NFM	Jersey	Abbey
		NFM	London	Baron Transport
		NFM	Reepham	Salle Farm Co.
85.0625	71.5625	NFM	Anglia	James Abbotts Ltd
		NFM	Cleveland	Mastercare
		NFM	Jersey	C.I. Bakery
		NFM	Jersey	Amy, Mark Ltd
		NFM	London	Diamond Cars
		NFM	London	Lee Vans
		NFM	Midlands	Mastercare
		NFM	Witham	Anglia Land Drainage
85.0750	71.5750	NFM	Great Yarmouth	Wolsey Taxis
		NFM	Ispwich	Water Board

Base	Mobile	Mode	Location	User & Notes
		NFM	Lakenheath	H. Palmer
		NFM	London	David Marshall
		NFM	Wrexham	Trafford Estate
85.0875	71.5875	NFM	Baldock	Winifred Express
		NFM	Cambridge	Regency Cars
		NFM	Colchester	Wooldridge
		NFM	Newmarket	Six Mile Bottom Estate
85.1000	71.6000	NFM	Brecon	Welsh Water Channel 8
		NFM	Ipswich	Water Board
85.1125	71.6125	NFM	Abergavenny	Welsh Water Channel 9
		NFM	Colchester	Roadworks Depot
		NFM	Ispwich	Roadworks Depot
		NFM	Norwich	Anglian Water
		NFM	Thames Valley	Thames Valley Water
		NFM	Yorkshire	British Pipeline
85.1250	71.6250	NFM	Nationwide	British Pipeline
85.1375	71.6375	NFM	Nationwide	British Telecom Channel 6
		NFM	Edinburgh	British Telecom
85.1500	71.6500	NFM	Nationwide	British Telecom Channel 2
85.1625	71.6625	NFM	Nationwide	British Telecom Channel 4
85.1750	71.6750	NFM	Jersey	Jersey Electric Co. Ch 1
		NFM	Nationwide	British Telecom Channel 1
		NFM	Perth	British Telecom
85.1875	71.6875	NFM	Nationwide	British Telecom Channel 5
		NFM	Perth	BT Data
		NFM	Edinburgh	British Telecom Voice Link
85.2000	71.7000	NFM	Nationwide	British Telecom Channel 3
85.2125	71.7125	NFM	Folkestone	Water Board
		NFM	Ipswich	Water Board Channel 4
		NFM	Lea Valley	Southern Water Channel 2
		NFM	Montgomery	Severn Trent Water
		NFM	South Wales	South Wales Water
85.2250	71.7250	NFM	Lea Valley	Southern Water Channel 4
		NFM	London	Chelsea Council
		NFM	London	Westminster Council
		NFM	Thames Valley	Thames Valley Water
85.2375	71.7375	NFM	Aberdeen	Water Board
		NFM	Gwynedd	Welsh Water
		NFM	Suffolk	Suffolk Water Channel 10
		NFM	Thames Valley	Thames Valley Water
		NFM	Wessex	Wessex Water
85.2500	71.7500	NFM	Anglia	Anglia Water Channel 14
		NFM	Hampshire	Hants Water
		NFM	Jersey	Jersey Milk
		NFM	Leeds	Automobile Association
		NFM	Surrey	Southern Water
		NFM	West Yorkshire	Automobile Association
85.2625	71.7625	NFM	Kent	Kent Water
		NFM	Severn	Severn Trent Water Chan 1

Base	Mobile	Mode	Location	User & Notes
		NFM	Thames Valley	Thames Valley Water Ch 9
		NFM	West Hampshire	Hants Water
85.2750	71.7750	NFM	Ipswich	Anglia Water Channel 1
		NFM	Jersey	Mascot Motors
85.2875	71.7875	NFM	Anglia	Anglia Water Channel 13
		NFM	North Kent	Kent Water Channel 1
		NFM	Perthshire	Water Board Channel 6
		AM	Cornwall	South West Water
		NFM	Sussex	Sussex Water Channel 5
		NFM	Yorkshire	Yorkshire Water
85.3000	71.8000	NFM	Aberdeen	Council Dog Catcher
		NFM	Breckland	Council HQ
		NFM	Ipswich	Community Repeater
		NFM	Thames Valley	Thames Valley Water Ch 5
85.3125	71.8125	NFM	Huntingdon	Anglia Water Channel 11
		NFM	Humberside	Council
		NFM	Perthshire	Water Board
		NFM	West Sussex	Sussex Water
85.3250	71.8250	NFM	Aberdeen	British Gas
		NFM	Edinburgh	Water Board
		NFM	Bournemouth	Dorset Water
		NFM	Folkestone	Community Repeater
		NFM	Gloucester	Gloucester Water Channel 7
		NFM	Ispwich	Anglia Water Channel 15
85.3375	71.8375	NFM	Aberdeen	Water Board
		NFM	Humberside	Humberside Water
		NFM	Ipswich	Anglia Water Channel 2
		NFM	Kidderminster	Worcestershire Water
		AM	Plymouth	South West Water
		NFM	South Wales	Severn Trent Water
85.3500	71.8500	NFM	Guernsey	Civil Defence
		NFM	Kent	Kent Water
85.3625	71.8625	NFM	Huntingdon	Anglia Water Channel 7
		NFM	Wessex	Wessex Water
85.3750	71.8750	NFM	Jersey	Motor Transport
85.3875	71.8875	NFM	Cambridge	Anglia Water Channel 8
		NFM	Perth	Tayside Construction
85.4000	71.9000	NFM	Hounslow	CTCSS
		NFM	Jersey	Resources Recovery
		NFM	Kent	Kent Water Channel 6
		NFM	Sheffield	CTCSS
85.4125	71.9125	NFM	Ipswich	Anglia Water Channel 12
		NFM	Midlands	Midland Water
		NFM	West Sussex	Sussex Water
		NFM	Wye	Welsh Water Channel 33
85.4250	71.9250	NFM	Guernsey	Fire Service
		NFM	Ipswich	Community Repeater
		NFM	Kent	Council CTCSS
		NFM	Perthshire	Water Board

Base	Mobile	Mode	Location	User & Notes
85.4375	71.9375	NFM	Alford	Roads Department
		NFM	Cornwall	Cornwall Water
		NFM	Cotswolds	Cotswolds Water
		NFM	Eastbourne	Eastbourne Water
		NFM	Gower	Welsh Water
		NFM	Lea Valley	Anglia Water
		NFM	Surrey	Surrey Water
85.4500	71.9500	NFM	Pitcaple	Roads Department
		NFM	Anglia	Anglia Water
		NFM	Gwynedd	Welsh Water
		NFM	Kent	Kent Water Channel 2
85.4625	71.9625	NFM	Anglia	Anglia Water
		NFM	Severn	Severn Trent Water Ch 2
		NFM	West Sussex	Sussex Water
85.4750	71.9750	NFM	Bristol	Automobile Association
		NFM	Hastings	Hastings Water
		NFM	Jersey	Falles
		NFM	Kings Lynn	Dow Chemicals
85.4875	71.9875	NFM	Nationwide	Automobile Assoc. Ch 6
		NFM	London	Automobile Association
85.5000	72.0000	NFM	Anglia	Pye's Managing Dir. Chan 3
		NFM	Nationwide	Automobile Assoc. Ch 2
		NFM	London	Automobile Association
85.5125	72.0125	NFM	Anglia	Pye's Managing Dir. Chan 5
		NFM	Nationwide	Automobile Assoc. Chan 4
		NFM	London	Automobile Association
		NFM	Perth	Automobile Association
85.5250	72.0250	NFM	Anglia	Pye's Managing Dir. Chan 2
		NFM	Nationwide	Automobile Assoc. Chan 1
		NFM	Aberdeen	Automobile Association
		NFM	London	Automobile Association
85.5375	72.0375	NFM	Nationwide	Automobile Assoc. Chan 5
		NFM	London	Automobile Association
		NFM	Perth	Automobile Association
85.5500	72.0500	NFM	Anglia	Pye's Managing Dir. Chan 4
		NFM	Nationwide	Automobile Assoc. Chan 3
		NFM	Aberdeen	Automobile Association
		NFM	London	Automobile Association
85.5625	72.0625	NFM	Guernsey	Water Board
		NFM	Nationwide	Automobile Assoc. Chan 7
		NFM	Edinburgh	Automobile Association
		NFM	London	Automobile Association
85.5750	72.0750	NFM	Jersey	Jersey Electric Chan 2
		NFM	Thames	Thames Water Chan 3
85.5875	72.0875	NFM	Nationwide	Automobile Assoc. Chan 8
		NFM	London	Automobile Association
85.6000	72.1000	NFM	Anglia	Royal Automobile Club
85.6125	72.1125	NFM	Dorset	Poole Adventure Centre
		NFM	Humberside	Community Repeater

Base	Mobile	Mode	Location	User & Notes
		NFM	London	Concord Ltd
85.6250	72.1250	NFM	Abbotts	RiptonFellows Estate
		NFM	Anglia	Automobile Association
		NFM	Colchester	J. Collie Ltd
		NFM	London	Battersea Cars
		NFM	London	Globe Bikes
		NFM	London	Haden Carriers
85.6375	72.1375	NFM	Aberdeen	Taxi
		NFM	'Guernsey	Fruit Export
		NFM	Solent	Solent Waters Rescue
85.6500	72.1500	NFM	Anglia	Royal Automobile Club
		NFM	London	Chequers Transport
		NFM	London	Riva Communications Ltd
		NFM	St Ives	Tyrell Contractors
85.6625	72.1625	NFM	Anglia	M. Crouch Ltd
		NFM	London	Anderson Young Ltd
85.6750	72.1750	NFM	Aberdeen	Taxi
		NFM	Anglia	Hughes TV Servicing
		NFM	Colchester	Eastern Tractors
		NFM	Jersey	LuxiCabs
		NFM	London	Petchey & Velite Cars
		NFM	Thetford	Lloyd & Marriot Vets
85.6875	72.1875	NFM	Anglia	Royal Automobile Club
		NFM	London	Belsize Ltd Channel 1
85.7000	72.2000	NFM	Aberdeen	Shanks Transport
85.7125	72.2125	NFM	London	Summit Cars
		NFM	M25	AA Recovery
		NFM	Nationwide	Curry's Master Care
85.7250	72.2250	NFM	Anglia(A12)	AA Recovery
		NFM	Bedford	Carlow Radio
		NFM	Brighton	Rediffusion
		NFM	Shrewsbury	Taxis
		NFM	London	Ascot & Bracknell
85.7375	72.2375	NFM	Colchester	Fieldspray Ltd
		NFM	Edinburgh	Taxi
		NFM	Jersey	Dyno-rod
		NFM	London	Swift & Safe
85.7500	72.2500	NFM	A55	Automobile Association
		NFM	Guernsey	Warry's Bakery
		NFM	Hitchin	Rorall Taxis
		NFM	Lakenheath	Trevor Cobbold
		NFM	London	American Cars
		NFM	London	Putney Cars
85.7625	72.2625	NFM	Aberdeen	Port Maintenence
		NFM	Anglia	Automobile Association
		NFM	Cambridge	Plant Growing Institute
		NFM	London	Galaxy Cars
		NFM	Long	StratonC.P.S Fuels
85.7750	72.2750	NFM	Anglia	Automobile Association

Base	Mobile	Mode	Location	User & Notes
		NFM	Guernsey	Le Pelley Taxi
		NFM	London	Allways Ltd
		NFM	Norwich	R.C. Snelling
		NFM	Oxford	LuxiCabs
		NFM	Sheffield	Hargreaves Clearwaste Co
		NFM	Stoke on Trent	Lucky Seven Taxis
85.7875	72.2875	NFM	Anglia	Pye's Managing Dir. Chan 5
		NFM	Edinburgh	Taxi
		NFM	Nationwide	Automobile Assoc. Chan 9
		NFM	London	Automobile Association
85.8000	72.3000	NFM	Aberdeen	City Council
		NFM	Plymouth	City Council Cleansing Dept
		NFM	Nationwide	Express Security Vans
		NFM	Nationwide	Parceline Ltd
		NFM	Surrey	Automobile Association
85.8125	72.3125	NFM	Aberdeen	Taxi
		NFM	Nationwide	DTI Channel L0065
85.8250	72.3250	NFM	Cornwall	Pye Transport
		NFM	Guernsey	Huelin
		NFM	Letchworth	Joe's Taxis
		NFM	London	Kwik Cars
		NFM	Norwich	Beeline Taxis
		NFM	Soham	P. Lyon
85.8375	72.3375	NFM	Guernsey	Falles Hire Cars
		NFM	Littleport	J.C. Rains Ltd
		NFM	London	Belsize Ltd
		NFM	London	Courier 83 Ltd
		NFM	Luton	James Early Ltd
85.8500	72.3500	NFM	Nationwide	Pye Telecom Channel 2
		NFM	Nationwide	Philips Transport Scheme
85.8625	72.3625	NFM	London	Fisher Sylvester Ltd
		NFM	London	G & R Tyres
		NFM	London	Pronto Cars
85.8750	72.3750	NFM	Nationwide	DTI 28 Day Hire
		NFM	Nationwide	Pye Telecom Channel 1
85.8875	72.3875	NFM	Aylsham	Aylsham Produce
		NFM	Diss	East Coast Grain
		NFM	London	Teleportation Ltd
		NFM	Norfolk	East Coast Grain
		NFM	Suffolk	Aylsham Produce
85.9000	72.4000	NFM	Cambridge	John's of Cambridge
		NFM	Diss	G.W. Padley
		NFM	Edinburgh	TV Repairs
		NFM	Jersey	Rank Taxis
		NFM	London	Sensechoice
		NFM	RN Portland	Naval Provest
		NFM	Spalding	Glen Heat and Irrigation
85.9125	72.4125	NFM	Aberdeen	Breakdown Services
		NFM	Edinburgh	TV Repairs

Base	Mobile	Mode	Location	User & Notes
		NFM	Ipswich	Taxi
		NFM	London	Anglo Spanish
		NFM	London	Arrival Couriers
		NFM	London	Central Motors
		NFM	London	K Cars
85.9250	72.4250	NFM	Jersey	LuxiCabs
		NFM	London	Laurie Buxton
		NFM	London	Town & Country
85.9375	72.4375	NFM	Kent	Porlant Car Hire
		NFM	London	Action Cars
85.9500	72.4500	NFM	Guernsey	Crossways Agricultural
		NFM	London	Action Cars
		NFM	London	Galaxy Bikes
		NFM	London	Super Express
85.9625	72.4625	NFM	Great Massingham	Gilman Ltd
		NFM	London	Globe Cars
		NFM	Norfolk	Don Robin Farms
85.9750	72.4750	NFM	Anglia	Stanway Taxis
		NFM	Guernsey	Stan Brouard Ltd
		NFM	Nationwide	Tarmac Roadstone
		NFM	Sutton	Darby Plant
		NFM	Woodbridge	Wm Kerr Farms
85.9875	72.4875	NFM	London	Avery Cars
		NFM	London	City & Suburban
		NFM	London	Commutercars
		NFM	London	Parkward Ltd
		NFM	London	Southampton Way Cars
		NFM	St Neots	Eyrsbury Plant Hire
86.0000	72.5000	NFM	Aberdeen	Rig Maintenence
		NFM	London	Kilburn Cars
		NFM	London	Windmill Cars
		NFM	Nationwide	Community Repeater
86.0125	72.5125	NFM	Jersey	Blue Coaches
		NFM	Leeds	Automobile Association
86.0250	72.5250	NFM	Bedfordshire	Bedford Social Services
		NFM	Humberside	Community Repeater
		AM	Scotland	Scottish Ambulance Ch L82
		AM	Spay Valley	Scottish Ambulance
86.0375	72.5375	AM	Scotland	Scottish Ambulance Ch L83
		AM	Glasgow	Scottish Ambulance
		NFM	Guernsey	Public Works
		NFM	Nationwide	St Johns Private Ambulance
86.0500	72.5500	AM	Scotland	Scottish Ambulance Ch L84
		AM	Dundee	Scottish Ambulance
		NFM	Lincolnshire	Community Repeater
		NFM	London	Medicall
86.0625	72.5625	NFM	Gt. Ashfield	G. Miles
		NFM	Norfolk	Sandringham Estate
		AM	Scotland	Scottish Ambulance Ch L85

Base	Mobile	Mode	Location	User & Notes
86.0750	72.5750	AM	Scotland	Scottish Ambulance Ch L86
		AM	Aberdeen	Scottish Ambulance
		AM	Edinburgh	Scottish Ambulance
		NFM	Nationwide	Community Repeater
86.0875	72.5875	AM	Scotland	Scottish Ambulance Ch L87
		AM	Braemar	Scottish Ambulance
		AM	Edinburgh	Scottish Ambulance
		NFM	Framlingham	Cabban, Breeze & Douglas
86.1000	72.6000	AM	Scotland	Scottish Ambulance Ch L88
		AM	Motherwell	Scottish Ambulance
		NFM	Hadleigh	Lemon & Sutherland
86.1125	72.6125	AM	Scotland	Scottish Ambulance Ch L89
		AM	Dunfermline	Scottish Ambulance
		NFM	Lincolnshire	Community Repeater
86.1250	72.6250	NFM	Cardiff	F.W. Morgan Builders
		AM	Scotland	Scottish Ambulance Ch L90
		AM	Edinburgh	Scottish Ambulance
		NFM	Norfolk	E.P.H. Radio Repeater
86.1375	72.6375	NFM	Anglia	Storno Radio Telephone Co
86.1500	72.6500	AM	Scotland	Scottish Ambulance Ch L91
		AM	Perth	Scottish Ambulance
		NFM	Jersey	Gruchy Vets
86.1625	72.6625	AM	Scotland	Scottish Ambulance Ch L92
		AM	Glasgow	Scottish Ambulance
		NFM	Kent	County Council
		NFM	Lancashire	Regional Health Ambulance
		NFM	Nationwide	Radiofone Channel 1
		NFM	Nationwide	Car Radio Telephone Hire
86.1750	72.6750	NFM	Anglia	Ipswich Transport Ltd
86.1875	72.6875	AM	Scotland	Scottish Ambulance Ch L93
		AM	Falkirk	Scottish Ambulance
		NFM	Nationwide	Automobile Assoc. Ch 10
86.2000	72.7000	NFM	Nationwide	Automobile Assoc. Ch 11
86.2125	72.7125	NFM	London	Enterprise Ltd
		NFM	Norfolk	William Cory Heating
86.2250	72.7250	NFM	Anglia	Ipswich Transport Ltd
		NFM	Perth	PortaCabin
86.2500	72.7500	NFM	Nationwide	Hotpoint Channel 1
		NFM	Aberdeen	Snowploughs
86.2625	72.7625	AM	Aberdeen	Council
		NFM	Humbershire	Council
		NFM	Nationwide	Hotpoint Channel 2
86.2750	72.7750	NFM	Nationwide	Cryston Communications
86.2875	72.7875	NFM	Ispwich	RSPCA
		NFM	Great Yarmouth	Container Depot

86.3000 - 86.7000 MHz PMR Low Band 12.5 kHz

86.3000	86.3000	NFM	Jersey	Jersey Hospital
		NFM	Nationwide	IBA Aerial Riggers

Base	Mobile	Mode	Location	User & Notes
86.3125	86.3125	AM	Nationwide	Mountain Rescue Channel 1
86.3250	86.3250	AM	Jersey	Tourism Lifeguards
		AM	Nationwide	Mountain Rescue Reserve
		NFM	Nationwide	Park Ranger Service Ch 2
		NFM	Nationwide	St Johns Ambulance Ch 1
86.3375	86.3375	NFM	Norwich	Sir Robin Lee
86.3500	86.3500	NFM	Jersey	TV Aerial Erectors
		NFM	Jersey	D.E. Payn Electrics
		AM	Nationwide	Mountain Rescue Channel 3
		AM	Nationwide	Red Cross Emergencies
		NFM	Nationwide	St Johns Ambulance Ch 3
86.3625	86.3625	NFM	Nationwide	Boy Scouts Channel 1
86.3750	86.3750	NFM	Nationwide	REACT CB Emergency
86.4000	86.4000	NFM	Felixstowe	Docks
86.4125	86.4125	AM	Nationwide	Mountain Rescue Channel 2
		NFM	Nationwide	St Johns Ambulance Ch 2
86.4250	86.4250	NFM	Jersey	St Johns Ambulance
		NFM	Nationwide	Forestry Commission Ch 3
86.4375	86.4375	NFM	Nationwide	RAC Rally Medical/Safety
86.4500	86.4500	NFM	Nationwide	Forestry Commission Ch 2
		NFM	Nationwide	Wimpey Construction Ch 1
		AM	Severn Valley	Severn Trent Water
86.4625	86.4625	NFM	Nationwide	Local Authority Common
86.4750	86.4750	NFM	Jersey	Ambulance
		NFM	Linton	T.B. Fairy
		NFM	Nationwide	British Rail Incidents
		NFM	Nationwide	Forestry Commission Ch 1
86.5000	86.5000	NFM	Luton Airport	McAlpine Aviation
		NFM	Nationwide	IBA Aerial Riggers
		NFM	Nationwide	BNFL Nuclear Incident Ch 1
		NFM	Suffolk	Suffolk County Council
86.5250	86.5250	NFM	Nationwide	BBC OB Riggers
		NFM	Nationwide	BNFL Nuclear Incident Ch 2
86.5375	86.5375	NFM	Derbyshire	Middleton Top Rangers
86.5500	86.5500	NFM	Nationwide	BNFL Nuclear Incident Ch 3
86.5750	86.5750	NFM	Nationwide	CEGB Mine Rescue
86.5625	86.5625	NFM	Nationwide	DHL International
		NFM	Aberdeen	RGIT
86.6250	86.6250	NFM	Jersey	St Johns Ambulance
		NFM	Nationwide	Boy Scouts Channel 2
		NFM	Nationwide	Wimpey Construction Ch 2
86.6375	86.6375	NFM	Bedford	RSPB
		NFM	Jersey	Telefitters
		NFM	Nationwide	Wimpey Construction Ch 3
86.6750	86.6750	NFM	Nationwide	UKAEA Radiation Survey
86.7000	86.7000	NFM	Dungeness	Power Station
		NFM	Jersey	Emergency Ambulances
		NFM	Nationwide	AA Emergency
		NFM	Nationwide	UKAEA Health Physics

Base	Mobile	Mode	Location	User & Notes
86.7125 - 87.9625 MHz PMR Low Band 12.5 kHz Duplex			
86.7125	76.7125	NFM	Nationwide	Customs & Excise Ch 1
86.7250	76.7250	NFM	Nationwide	Customs & Excise Ch 2
86.7375	76.7375	NFM	Nationwide	Customs & Excise Ch 9
86.7625	76.2625	NFM	Perth	Community Repeater
86.8000	76.3000	NFM	Nationwide	Vibroplant
86.8250	76.8250	NFM	Thames	ITN Channel 4
86.8625	76.8250	NFM	Plymouth	Plymouth Vets
86.9500	76.4500	NFM	Dover	HM Customs
		NFM	Dundee	Tayside Water Board
86.9625	76.8625	NFM	Nationwide	RSPCA Channel 1
		NFM	North Wales	Forestry Commission Ch 4
86.9375	76.9375	NFM	Nationwide	Customs & Excise Ch 3
86.9500	76.9500	NFM	Nationwide	Customs & Excise Ch 4
86.9875	76.9875	NFM	London	Guarda Security
87.0000	77.0000	NFM	Nationwide	RAC Channel 2
		NFM	Edinburgh	Data Link
		NFM	Wymondham	Ayton Asphalt
87.0125	77.0125	NFM	Nationwide	RAC Channel 4
87.0250	77.0250	NFM	Nationwide	RAC Channel 1
		NFM	Perth	Data Link
87.0375	77.0375	NFM	Nationwide	RAC Channel 5
87.0500	77.0500	NFM	Nationwide	RAC Channel 3
		NFM	Leeds	Armor Guard
87.0625	77.0625	NFM	Bishops Stortford	Rougewell Ltd
		NFM	South Wales	Forestry Commission Ch 2
87.0750	77.0750	NFM	Ispwich	Council
		NFM	Jersey	Taxi Rank
87.0875	77.0875	NFM	Jersey	Jersey Evening Post
		NFM	London	Savoy Rolls Royce
		NFM	London	Stanstead Containers Ltd
		NFM	Nationwide	RSPCA Channel 2
87.1000	77.1000	NFM	London	B.J. Transport
		NFM	Leicester	ABC Taxis
87.1250	77.1250	NFM	SW/SE Wales	Forestry Commission Ch 3
		NFM	Whitemoor Yard	British Rail Channel 1
87.1375	77.1375	NFM	Hampshire	Council
87.1500	77.1500	NFM	Nationwide	Radiofone Channel 1
87.1625	77.1625	NFM	Aberdeen	Council
		NFM	Buckinghamshire	Council
87.1875	77.1875	NFM	Cambridgeshire	Council
		NFM	Lancashire	Snowploughs & Gritters
		NFM	Radnor	Powys Council Highways
87.2125	77.2125	NFM	Nationwide	Council Common
87.2250	77.2250	NFM	London	Met Police Clamping
		NFM	Oxford	City Council
		NFM	Whitemoor Yard	British Rail Channel 2
87.2375	77.2375	NFM	Oxfordshire	Council
		NFM	Perth	Tayside Regional Council

Base	Mobile	Mode	Location	User & Notes
87.2500	77.2500	NFM	Birmingham	City Council Engineers
		NFM	Hertfordshire	Council
		NFM	Brecon	Powys County Highways
87.2625	77.2625	NFM	England	Council Common
		NFM	S. Oxfordshire	County Council
87.2750	77.2750	NFM	West Yorkshire	Bus Inspectors
87.3000	77.3000	NFM	Jersey	Waterworks
		NFM	Derbyshire	Derbyshire Council Roads
87.3125	77.3125	NFM	Perth	Tayside Regional Council
		NFM	Windsor	Royal Parks
87.3250	77.3250	NFM	Derbyshire	Council
		NFM	South Yorkshire	Council
87.3750	77.3750	NFM	Jersey	Dr. Scott Warren
		NFM	Lincolnshire	Community Repeater
87.3875	77.3875	NFM	Ipswich	Community Repeater
87.4000	77.4000	NFM	Aberdeen	Aberdeen Skip Hire
		NFM	Nationwide	Doctor Scheme
87.4125	77.4125	NFM	Scotland	Council Highways Common
87.4375	77.4375	NFM	Bedford	County Surveyors Channel 1
		NFM	Scotland	Council Highways Common
87.4500		NFM	London	Information Service
87.4500	77.4500	NFM	Cambridgeshire	Highways Channel 3
87.4625	77.4625	NFM	England	Council Common
87.5000	77.5000	NFM	Guernsey	Telecom
87.5125	77.5125	AM	Aberdeen	Regional Council Roads
87.5250	77.5250	NFM	Bedford	County Surveyors Channel 2
		NFM	Northern Ireland	Ambulance
87.5375	77.5375	NFM	England	Council Common
87.5500	77.5500	NFM	England	Council Common
		NFM	Northern Ireland	Ambulance
87.5625	77.5625	NFM	England	Council Common
		NFM	Oxfordshire	County Council
87.5750	77.5750	NFM	England	Council Common
		NFM	Northern Ireland	Ambulance
87.5875	77.5875	NFM	England	Council Common
87.6000	77.6000	NFM	Perthshire	County Council
87.6125	77.6125	NFM	England	Council Common
87.6250	77.6250	NFM	England	Council Common
		NFM	Northern Ireland	Ambulance
87.6375	77.6375	NFM	England	Council Common
87.6500	77.6500	NFM	England	Council Common
		NFM	Northern Ireland	Ambulance
87.6625	77.6625	NFM	Devon	County Council
87.6750	77.6750	NFM	Essex	Council Channel 4
		NFM	Northern Ireland	Ambulance
87.6875	77.6875	NFM	Essex	Council Channel 1
		NFM	London	Apollo Carriers
87.7000	77.7000	NFM	London	Middlesex Cars
		NFM	London	Rimington Minicabs

Base	Mobile	Mode	Location	User & Notes
		NFM	Wymondham	Ayton Asphalt
87.7125	77.7125	NFM	London	Reliable Cars
87.7250	77.7250	NFM	Birmingham	Council
		NFM	Guernsey	Gas
		NFM	London	Camberwell Cars
87.7375	77.7375	NFM	Ispwich	Council
		NFM	Lakenheath	Hammond Taxis
		NFM	Nottinghamshire	DG Taxis
87.7500	77.7500	NFM	Croydon	Metro Cars
87.7625	77.7625	NFM	Nationwide	Forestry Commission Ch 2
87.7875	77.7875	NFM	Great Yarmouth	PLG Perfect Haulage Ltd
87.8125	77.8125	NFM	Ipswich	Community Repeater
		NFM	Kings Lynn	Anglia Canners Ltd
87.8250	77.8250	NFM	Nationwide	Forestry Commission Ch 1
87.8375	77.8375	NFM	London	Concord Ltd
		NFM	London	Scott Cars
87.8750	77.8750	NFM	Jersey	Evening Post Newspaper
87.9000	77.9000	NFM	London	Intercity Couriers
87.9125	77.9125	NFM	London	Cheeta Cars
87.9250	77.9250	NFM	Manchester	Manchester Guardian
		NFM	Nationwide	Star Sat Racing Results
87.9625	77.9625	NFM	Nationwide	Forestry Commission Ch 3
		NFM	London	Helafields

88.0000	**-90.2000**	**MHz**	**Nationwide BBC Radio 2**	
88.1000		WFM	Ballachulish Area	
		WFM	Border Counties	
		WFM	Devon & East Cornwall	
		WFM	Festiniog	
		WFM	Guildford & Farnham Area	
		WFM	Llandloes Area	
		WFM	Mallaig Area	
		WFM	Manningtree Area	
		WFM	North Cumbria Area	
		WFM	South West Devon & Plymouth	
		WFM	West Islay	
88.2000		WFM	Betws-Y-Coed Area	
		WFM	Elgin Area	
		WFM	Knock More Area	
		WFM	Nailsworth Area	
88.3000		WFM	Forfar Area	
		WFM	Kilmarnock Area	
		WFM	Leicestershire	
		WFM	Leven & Renton Area	
		WFM	Lochgilphead Area	
		WFM	Midland Counties	
		WFM	Ness of Lewis	
		WFM	Shetland Isl	
		WFM	Staffordshire & Shorpshire Area	

Base	Mobile	Mode	Location	User & Notes
		WFM	Trowbridge Area	
		WFM	Wensleydale Area	
		WFM	Windermere Area	
		WFM	Melody FM North Normandy	
88.4000		WFM	Ammanford Area	
		WFM	Campbeltown Area	
		WFM	Ebbw Vale Area	
		WFM	Folkestone Area	
		WFM	Hebden Bridge Area	
		WFM	Isle Of Man	
		WFM	Peebles Area	
		WFM	South London	
		WFM	Walsden Area	
		WFM	Wharfdale Area	
88.5000		WFM	Barnstaple Area	
		WFM	Blaenavon Area	
		WFM	Cowal Peninsular & Rothesay Area	
		WFM	Central South of England	
		WFM	North Eastern Counties	
		WFM	West Skye	
88.6000		WFM	BBC Radio Sheffield	
		WFM	Bedford Area	
		WFM	Hereford & Welsh Borders	
		WFM	Inveraray Area	
		WFM	North West Lancashire	
		WFM	Ross On Wye Area	
		WFM	Warrenpoint Area	
88.7000		WFM	Abergavenny Area	
		WFM	Aberystwyth Area & Cardigan Bay	
		WFM	Ayr Area	
		WFM	Creetown Area	
		WFM	Grampian Area	
		WFM	Kirkconnel Area	
		WFM	Londonderry Area	
		WFM	Okehampton Area	
88.8000		WFM	BBC Radio Jersey	
		WFM	Cambridgeshire	
		WFM	Chippenham Area	
		WFM	Gower Peninisular	
		WFM	Lincolnshire	
		WFM	Isles of Scilly	
88.9000		WFM	Ballycastle Area	
		WFM	Brecon	
		WFM	Carmarthen Area	
		WFM	Cambridge Area	
		WFM	Keighley Area	
		WFM	Llangollen	
		WFM	Northampton Area	
		WFM	Oban Area	

Base	Mobile	Mode	Location	User & Notes
		WFM	Todmorden Area	
		WFM	Ystalyfera Area	
89.0000		WFM	Abertillery	
		WFM	Bath City Area	
		WFM	Chesterfield	
		WFM	Exeter Area	
		WFM	Kendal Area	
		WFM	North Gloucestershire	
		WFM	Manx Radio - Isle of Man	
		WFM	Perth Area	
		WFM	South Islay	
89.1000		WFM	Border Counties	
		WFM	Greater London	
		WFM	Home Counties	
		WFM	Humberside	
		WFM	Kingussie Area	
		WFM	Larne Area	
		WFM	Llandrindod Wells Area	
		WFM	Llanfyllin	
		WFM	Rosneath Area	
		WFM	Ullapool & Lewis Area	
89.2000		WFM	Aberdare Area	
		WFM	Pitlochry Area	
		WFM	Pontypool Area	
		WFM	Stroud Area	
		WFM	France Culture Cherbourg	
89.3000		WFM	Bristol City Area	
		WFM	Cheshire Area	
		WFM	Derbyshire	
		WFM	Fort William Area	
		WFM	Greater Manchester	
		WFM	Lancashire	
		WFM	Merseyside	
		WFM	Nottinghamshire	
		WFM	Orkney Isl	
		WFM	South West Wales	
		WFM	Wrexham & Deeside Area	
		WFM	Yorkshire	
89.4000		WFM	Enniskillen Area	
		WFM	Kincardine Area	
		WFM	Machynlleth Area	
		WFM	Ventnor	
		WFM	France MusiqueBrest	
89.5000		WFM	East Skye	
		WFM	Grantham Area	
		WFM	Greenholm & Darvel Area	
		WFM	Innerleithin Area	
		WFM	Pennar	
		WFM	Oxfordshire & Wiltshire	

Base	Mobile	Mode	Location	User & Notes
		WFM	South Knapdale	
		WFM	Swansea Area	
		WFM	Tobermoray	
89.6000		WFM	Channel Islands	
		WFM	Inverness Area	
		WFM	Limavady	
		WFM	Ludlow Area	
		WFM	North East Scotland	
		WFM	North Wales Coast & Anglesey	
		WFM	Welshpool Area	
		WFM	Whitby Area	
89.7000		WFM	Cornholme Area	
		WFM	Kinlochleven Area	
		WFM	Lampeter Area	
		WFM	Llanbydder Area	
		WFM	Newhaven Area	
		WFM	Norfolk	
		WFM	Saddleworth Area	
		WFM	South East London	
		WFM	Stranraer Area	
		WFM	Suffolk	
		WFM	Wearsdale Area	
		WFM	West Cornwall	
89.8000		WFM	Grantown Area	
		WFM	Matlock Area	
		WFM	Newcastle Area	
		WFM	Pudsey Area	
		WFM	Salisbury Area	
		WFM	Sheffield Area	
		WFM	Stornaway	
		WFM	West of Oswestry & Welshpool	
89.9000		WFM	Forth Valley	
		WFM	Girvan Area	
		WFM	Gloucester & Somerset Area	
		WFM	Haslingden Area	
		WFM	High Wycombe Area	
		WFM	Lowland Scotland	
		WFM	Scottish Central Lowlands	
		WFM	Somerest & North Devon	
		WFM	South & East Wales	
		WFM	Scarborough Area	
		WFM	France Musique Rennes	
90.0000		WFM	Buxton Area	
		WFM	Morecambe Bay Area	
		WFM	South Cumbria	
		WFM	South East Kent	
		WFM	Weymonth Area	
90.1000		WFM	Belfast & East Counties	
		WFM	Berwick & Borders	

		WFM	Brighton Area	
		WFM	Deeside Area	
		WFM	Dolgellau Area	
		WFM	Llandtfriog Area	
		WFM	Newcastle Emlyn Area	
		WFM	Newton Area	
		WFM	Peterborough	
		WFM	Wick Area	

90.2000 -92.4000 MHz   **Nationwide BBC Radio 3**

Base		Mode	Location
90.3000		WFM	Ballachulish Area
		WFM	Border Counties
		WFM	Devon & East Cornwall
		WFM	Festiniog Area
		WFM	Grantham Area
		WFM	Guildford & Farnham Area
		WFM	Llandloes Area
		WFM	Mallaig Area
		WFM	Manningtree Area
		WFM	North Cumbria
		WFM	South West Devon & Plymouth
		WFM	West Skye
90.4000		WFM	Betws-Y-Coed Area
		WFM	Elgin Area
		WFM	Knock More Area
		WFM	Nailsworth Area
90.5000		WFM	Forfar Area
		WFM	Kilmarnock Area
		WFM	Leicestershire
		WFM	Leven & Renton Area
		WFM	Lockgilphead Area
		WFM	Midland Counties
		WFM	Shetland Isl
		WFM	Staffordshire & Shorpshire Area
		WFM	Stornaway
		WFM	Trowbridge Area
		WFM	Wensleydale Area
		WFM	Windermere Area
90.6000		WFM	Ammanford Area
		WFM	Campbeltown Area
		WFM	Ebbw Vale Area
		WFM	Folkestone Area
		WFM	Hebden Bridge Area
		WFM	Isle of Man
		WFM	Peebles Area
		WFM	South London
		WFM	Walsden Area
		WFM	Wharfdale Area
90.7000		WFM	Barnstaple Area

Base	Mobile	Mode	Location	User & Notes
		WFM	Blaenavon Area	
		WFM	Central South of England	
		WFM	Cowal Peninsular & Rothesay Area	
		WFM	North Eastern Counties	
		WFM	Scarborough Area	
		WFM	West Skye	
90.8000		WFM	Bedford Area	
		WFM	Hereford & Welsh Borders	
		WFM	Kinross Area	
		WFM	Inveraray Area	
		WFM	North West Lancashire	
		WFM	Ross On Wye Area	
		WFM	Warrenpoint Area	
90.9000		WFM	Abergavenny Area	
		WFM	Aberystwyth Area & Cardigan Bay	
		WFM	Ayr Area	
		WFM	Cambridgeshire	
		WFM	Creetown Area	
		WFM	Grampian Area	
		WFM	Kirkconnel Area	
		WFM	Lincolnshire	
		WFM	Londonderry Area	
		WFM	Newark Area	
		WFM	Okehampton Area	
		WFM	Stranraer Area	
91.0000		WFM	Ballycastle Area	
		WFM	Chippenham Area	
		WFM	Gower Peninsular	
		WFM	Isles of Scilly	
91.1000		WFM	Brecon	
		WFM	Cambridge Area	
		WFM	Carmarthen Area	
		WFM	Channel Islands	
		WFM	Girvan Area	
		WFM	Keighley Area	
		WFM	Llangollen Area	
		WFM	Northampton Area	
		WFM	Oban Area	
		WFM	Todmorden Area	
		WFM	Ystalyfera Area	
91.2000		WFM	Abertillery Area	
		WFM	Bath City Area	
		WFM	Chesterfield Area	
		WFM	Exeter Area	
		WFM	Kendal Area	
		WFM	Kingussie Area	
		WFM	North Gloucestershire	
		WFM	Perth Area	
		WFM	South Islay	

Base	Mobile	Mode	Location	User & Notes
91.3000		WFM	Border Counties	
		WFM	Greater London	
		WFM	Kingussie Area	
		WFM	Larne Area	
		WFM	Llandrindod Wells Area	
		WFM	Llanfyllin Area	
		WFM	Home Counties	
		WFM	Ullapool & Lewis Area	
91.4000		WFM	Aberdare Area	
		WFM	Pitlochry Area	
		WFM	Pontypool Area	
		WFM	Rosneath Area	
		WFM	Stroud Area	
91.5000		WFM	Bristol City Area	
		WFM	Cheshire Area	
		WFM	Derbyshire	
		WFM	Fort William Area	
		WFM	Greater Manchester	
		WFM	Humberside	
		WFM	Orkney Isl	
		WFM	South Kanpdale	
		WFM	Lancashire	
		WFM	Merseyside	
		WFM	Nottinghamshire	
		WFM	South West Wales	
		WFM	Wrexham & Deeside Area	
		WFM	Yorkshire	
		WFM	France Culture Caen	
91.6000		WFM	Enniskillen Area	
		WFM	Kincardine Area	
		WFM	Machynlleth Area	
91.7000		WFM	East Skye	
		WFM	Grantham Area	
		WFM	Greenholme & Darvel Area	
		WFM	Innerleithin Area	
		WFM	Isle Of Mull	
		WFM	Oxfordshire & Wiltshire	
		WFM	Pennar	
		WFM	Swansea Area	
		WFM	Tobermoray	
		WFM	Ventnor	
91.8000		WFM	Inverness Area	
		WFM	Limavady Area	
		WFM	Ludlow Area	
		WFM	North East Scotland	
		WFM	North Wales Coast & Anglesey	
		WFM	Welshpool Area	
		WFM	Whitby Area	
91.9000		WFM	Cornholme Area	

Base	Mobile	Mode	Location	User & Notes
		WFM	Kinlochleven Area	
		WFM	Lampeter Area	
		WFM	Llanybydder Area	
		WFM	Newhaven Area	
		WFM	Norfolk	
		WFM	Saddleworth Area	
		WFM	South East London	
		WFM	Suffolk	
		WFM	Wearsdale Area	
		WFM	West Cornwall	
92.0000		WFM	Granton Area	
		WFM	Matlock Area	
		WFM	Newcastle Area	
		WFM	Pudsey Area	
		WFM	Salisbury Area	
		WFM	Stornaway	
		WFM	West of Oswestry & Welshpool	
92.1000		WFM	Forth Valley	
		WFM	Gloucester & Somerset Area	
		WFM	Haslingden Area	
		WFM	High Wycome Area	
		WFM	Lowland Scotland	
		WFM	Scarborough Area	
		WFM	Scottish Central Lowlands	
		WFM	Sheffield Area	
		WFM	Somerset & North Devon	
		WFM	South & East Wales	
92.2000		WFM	Berwick & Borders	
		WFM	Buxton Area	
		WFM	Morecambe Bay Area	
		WFM	Peterborough	
		WFM	South Cumbria	
		WFM	Weymouth Area	
92.3000		WFM	Belfast & East Counties	
		WFM	Brighton Area	
		WFM	Deeside Area	
		WFM	Dolgellau Area	
		WFM	Llandyfriog Area	
		WFM	Newcastle Emlyn Area	
		WFM	Newton Area	
		WFM	Peterborough Area	
		WFM	Wick Area	
		WFM	France MusiqueCherbourg	
92.4000		WFM	BBC Radio Leeds	
		WFM	South East Kent	

92.4000 -94.6000 MHz   Nationwide BBC Radio 4

Base	Mobile	Mode	Location	User & Notes
92.5000		WFM	BBC Radio Scotland Ballachulish Area	
		WFM	BBC Radio Scotland Mallaig Area	

Base	Mobile	Mode	Location	User & Notes
		WFM	BBC Radio Scotland W Islay	
		WFM	Border Counties	
		WFM	Devon & East Cornwall	
		WFM	Grantham Area	
		WFM	Guildford & Farnham Area	
		WFM	Manningtree Area	
		WFM	North Cumbria	
		WFM	South West Devon & Plymouth	
92.6000		WFM	BBC Radio Scotland Elgin Area	
		WFM	BBC Radio Scotland Knock More Area	
		WFM	Nailsworth Area	
92.7000		WFM	BBC Radio Scotland Forfar Area	
		WFM	BBC Radio Scotland Kilmarnock Area	
		WFM	BBC Radio Scotland Leven & Renton Area	
		WFM	BBC Radio Scotland Lochgilphead Area	
		WFM	BBC Radio Scotland Ness of Lewis	
		WFM	BBC Radio Scotland Shetland Isl	
		WFM	Leicestershire	
		WFM	Midland Counties	
		WFM	Staffordshire & Shorpshire Area	
		WFM	Trowbridge Area	
		WFM	Wensleydale Area	
		WFM	Windermere Area	
92.8000		WFM	BBC Radio Scotland Campbeltown Area	
		WFM	BBC Radio Scotland Peebles Area	
		WFM	Hebden Bridge Area	
		WFM	Isle Of Man	
		WFM	South London	
		WFM	Walsden Area	
		WFM	Wharfdale Area	
92.9000		WFM	BBC Radio Scotland Cowal Peninsular	
		WFM	BBC Radio Scotland West Skye	
		WFM	Central South of England	
		WFM	North Eastern Counties	
93.0000		WFM	BBC Radio Scotland Inveraray Area	
		WFM	Bedford Area	
		WFM	Hereford & Welsh Borders	
		WFM	North West Lancashire	
		WFM	Radio Ulster Warrenpoint Area	
		WFM	Ross On Wye Area	
93.1000		WFM	Ayr Area	
		WFM	BBC Radio Scotland Ayr Area	
		WFM	BBC Radio Scotland Grampian Area	
		WFM	BBC Radio Scotland Kirkconnel Area	
		WFM	Cambridgeshire	
		WFM	Folkestone Area	
		WFM	Lincolnshire	
		WFM	Newark Area	
		WFM	Okehampton Area	

Base	Mobile	Mode	Location	User & Notes
		WFM	Radio Foyle - Londonderry	
		WFM	Radio Ulster - Londonderry Area	
93.2000		WFM	Ballycastle Area	
		WFM	BBC Radio Guernsey	
		WFM	Chippenham Area	
		WFM	Isles of Scilly	
93.3000		WFM	BBC Radio Scotland Girvan Area	
		WFM	BBC Radio Scotland Kinross Area	
		WFM	BBC Radio Scotland Oban Area	
		WFM	Cambridge Area	
		WFM	Keighley Area	
		WFM	Northampton Area	
93.4000		WFM	Abertillery	
		WFM	BBC Radio Scotland South Islay	
		WFM	Bath City Area	
		WFM	BBC Radio Scotland Perth Area	
		WFM	Chesterfield	
		WFM	Exeter Area	
		WFM	Kendal Area	
		WFM	North Gloucestershire	
93.5000		WFM	BBC Radio Scotland Border Counties	
		WFM	BBC Radio Scotland Kingussie Area	
		WFM	BBC Radio Scotland Ullapool & Lewis Area	
		WFM	Bristol City Area	
		WFM	Greater London	
		WFM	Home Counties	
		WFM	Radio Ulster Larne Area	
		WFM	France Inter Rennes	
93.6000		WFM	BBC Radio Scotland Pitlochry	
		WFM	BBC Radio Scotland Rosneath Area	
		WFM	Stroud Area	
93.7000		WFM	BBC Radio Scotland Fort William Area	
		WFM	BBC Radio Scotland Orkney Isl	
		WFM	BBC Radio Scotland South Knapdale	
		WFM	Cheshire Area	
		WFM	Derbyshire	
		WFM	Greater Manchester	
		WFM	Humberside	
		WFM	Lancashire	
		WFM	Merseyside	
		WFM	Nottinghamshire	
		WFM	Wrexham & Deeside Area	
		WFM	Yorkshire	
93.8000		WFM	BBC Radio Scotland Kincardine Area	
		WFM	Radio Ulster Enniskillen Area	
		WFM	Ventnor	
93.9000		WFM	BBC Radio Scotland East Skye	
		WFM	BBC Radio Scotland Greenholm & Darvel	
		WFM	BBC Radio Scotland Innerleithen Area	

Base	Mobile	Mode	Location	User & Notes
		WFM	BBC Radio Scotland Tobermoray & Isle Of Mull	
		WFM	Grantham Area	
		WFM	Oxfordshire & Wiltshire	
		WFM	Pennar	
		WFM	Todmorden Area	
94.0000		WFM	BBC Radio Scotland Inverness	
		WFM	BBC Radio Scotland North East Scotland	
		WFM	Limavady Area	
		WFM	Ludlow Area	
		WFM	Whitby Area	
		WFM	Black County Sounds, W. Midlands	
94.1000		WFM	BBC Radio Scotland Kinlochleven Area	
		WFM	BBC Radio Scotland Stranraer	
		WFM	Cornholme Area	
		WFM	Newhaven Area	
		WFM	Norfolk	
		WFM	Saddleworth Area	
		WFM	South East London	
		WFM	Suffolk	
		WFM	Wearsdale Area	
		WFM	West Cornwall	
		WFM	France InterCherbourg	
94.2000		WFM	BBC Radio Scotland Stornaway	
		WFM	BBC Radio Derby	
		WFM	Matlock Area	
		WFM	Newcastle Area	
		WFM	Pudsey Area	
		WFM	Salisbury Area	
94.3000		WFM	BBC Radio Scotland Forth Valley	
		WFM	BBC Radio Scotland Lowland Scotland	
		WFM	Gloucester & Somerset Area	
		WFM	Haslingden Area	
		WFM	High Wycombe Area	
		WFM	Scarborough Area	
		WFM	Sheffield	
		WFM	Somerset & North Devon	
		WFM	South & East Wales	
94.4000		WFM	Buxton Area	
		WFM	Morecambe Bay Area	
		WFM	South Cumbria	
		WFM	South East Kent	
		WFM	Weymouth Area	
94.5000		WFM	BBC Radio Scotland Deeside Area	
		WFM	BBC Radio Scotland Wick Area	
		WFM	Berwick & Borders	
		WFM	Brighton Area	
		WFM	Peterborough Area	
		WFM	Radio Ulster Belfast & East Counties	

Base	Mobile	Mode	Location	User & Notes
94.6000	**-97.6000**	**MHz**	**Nationwide Independent Radio**	
94.6000		WFM	BBC Radio Scotland	Granton
		WFM	BBC Radio Stoke On Trent	Cheshire
		WFM	BBC Radio Surrey & Berkshire	Henley
94.7000		WFM	BBC Radio Hereford & Worc's	Hereford
		WFM	BBC Radio Solway	Solway
94.8000		WFM	BBC Radio 4	Jersey
94.9000		WFM	BBC CWR	Coventry
		WFM	BBC Radio Bristol	Bristol City
		WFM	BBC Radio Lincolnshire	Lincs
		WFM	BBC Radio Scotland	Perth
		WFM	BBC Radio 4	Londonderry
		WFM	Greater London Radio	London
95.0000		WFM	BBC Radio Cleveland	Teesside
		WFM	BBC Radio Gloucestershire	Stroud
		WFM	BBC Radio Leicestershire	Leicester
		WFM	BBC Radio Shropshire	Ludlow
		WFM	BBC Radio Sussex	Newhaven
95.1000		WFM	BBC Radio Gwent	South Wales
		WFM	BBC Radio Sussex	Horsham
		WFM	BBC Radio Norfolk	Norfolk
		WFM	Greater Manchester Radio	G Manchester
		WFM	Radio Ulster	Ballycastle
95.2000		WFM	BBC Radio Cornwall	E Cornwall
		WFM	BBC Radio Cumbria	Kendal
		WFM	BBC Radio Oxford	Oxfordshire
95.3000		WFM	BBC Radio Derby	Matlock
		WFM	BBC Radio Essex	Southend
		WFM	BBC Radio Leeds	Wharfdale
		WFM	BBC Radio Sussex	Brighton
95.4000		WFM	BBC Radio Newcastle	Tyneside
		WFM	BBC Radio Surrey & Berkshire	Windsor
		WFM	Radio Ulster	Limavady
95.5000		WFM	BBC Radio Bedfordshire	Bedford
		WFM	BBC Radio Bristol	Taunton
		WFM	BBC Radio Lancashire	E Lancs
		WFM	BBC Radio Nottinghamshire	Mansfield
		WFM	BBC Radio York	Scarborough
		WFM	France Inter	Brest
95.6000		WFM	BBC Radio Cumbria	Cumbria
		WFM	BBC Radio West Midlands	W Midlands
		WFM	France Musique	Caen
95.7000		WFM	BBC Radio Cambridge	Peterborough
		WFM	BBC Radio Dorset	Dorset
95.8000		WFM	BBC Radio Devon	Exeter
		WFM	BBC Radio Merseyside	Merseyside
		WFM	Capital 95.8 FM	London
95.9000		WFM	BBC Radio Gwent	South Wales
		WFM	BBC Radio Humberside	Yorkshire

The UK Scanning Directory

Base	Mobile	Mode	Location	User & Notes
		WFM	BBC Radio Newcastle	Borders
		WFM	BBC Radio 4	SW Wales
		WFM	Invicta FM	Thanet
96.0000		WFM	BBC Radio Cambridge	Cambridgeshire
		WFM	BBC Radio Cornwall	Scilles
		WFM	BBC Radio Devon	Okehampton
		WFM	BBC Radio Shropshire	Shropshire
		WFM	Radio Ulster	Belfast
96.1000		WFM	BBC Radio Solent	S Hants
		WFM	Hallam FM	Rotherham
		WFM	Invicta FM	Ashford
		WFM	BBC Radio Furness	SW Cumbria
96.2000		WFM	Trent FM	Notts & Derby
96.3000		WFM	Essex Radio	Southend
		WFM	GWR	Avon
		WFM	Radio Aire	Leeds
96.4000		WFM	BRMB Radio	Birmingham
		WFM	County Sound	Guildford
		WFM	Devonair Radio	Torbay
		WFM	Downtown Radio	Limavady
		WFM	Premier Radio	NW Surrey
		WFM	Radio Tay	Perth
		WFM	Saxon Radio	Bury St Ed.
		WFM	Signal Cheshire	S Cheshire
		WFM	Swansea Sound	Swansea
96.5000		WFM	GWR	Marlborough
		WFM	Orchard FM	Taunton
96.6000		WFM	Downtown Radio	Enniskillen
		WFM	Northants 96	Northampton
		WFM	Radio In Tavistock	Tavistock
		WFM	TFM Radio	Teesside
96.7000		WFM	BBC Radio Kent	North Kent
		WFM	Centresound	Stirling
		WFM	City FM	Merseyside
		WFM	Light FM	N Hants
		WFM	West Sound	Ayrshire
96.8000		WFM	Radio Borders	Selkirk
96.9000		WFM	Chiltern Radio	Bedford
		WFM	Choice FM	Brixton
		WFM	Manx Radio	Isle of Man
		WFM	North Sound	Aberdeen
		WFM	Signal Stafford	S Staffs
		WFM	Southern Sound	Newhaven
		WFM	Viking FM	Humberside
97.0000		WFM	Devonair Radio	Exeter
		WFM	Invicta FM	Dover
		WFM	Mercia Sound	Coventry
		WFM	Plymouth Sound	Plymouth
		WFM	Radio 210	Thames Valley

Base	Mobile	Mode	Location	User & Notes
97.1000		WFM	BBC Radio 1	Jersey
		WFM	Delta Radio	Haslemere
		WFM	Metro FM	Tyneside
		WFM	Orchard FM	Yeovil
		WFM	Radio Orwell	Ipswich
97.2000		WFM	Beacon West Midlands	Wolverhampt.
		WFM	For The People	Bristol
		WFM	GWR	Swindon
		WFM	South West Sound	SW Scotland
97.3000		WFM	Crown FM	London
		WFM	Radio Forth RFM	Edinburgh
97.4000		WFM	Chiltern Radio	Luton
		WFM	Cool FM	Belfast
		WFM	Fox FM	Banbury
		WFM	Hallam FM	Sheffield
		WFM	Moray Firth Radio	Inverness
		WFM	Pennine FM	Bradford
		WFM	Radio Force 7	St. Malo
		WFM	Red Dragon Radio	South Wales
		WFM	Red Rose Radio	Lancashire
97.5000		WFM	Ocean Sound	E Solent
		WFM	Radio Borders	Berwick
		WFM	Radio Mercury	Horsham
		WFM	Southern Sound	Hastings
		WFM	West Sound	Girvan
97.6000		WFM	Chiltern Radio	Luton
		WFM	Radio Wyvern	Hereford

97.6000 -99.8000 MHz   Nationwide BBC Radio 1

Base	Mobile	Mode	Location	User & Notes
97.7000		WFM	BBC Radio 4 Ballachulish Area	
		WFM	BBC Radio 4 Mallaig Area	
		WFM	Border Counties	
		WFM	Devon & East Cornwall	
		WFM	East Cornwall	
		WFM	Grantham Area	
		WFM	Guildford & Farnham Area	
		WFM	North Cumbria	
		WFM	South West Devon & Plymouth	
97.8000		WFM	Midland Counties Area	
		WFM	Nailsworth Area	
		WFM	Staffordshire & Shropshire	
		WFM	France CultureBrest	
97.9000		WFM	BBC Radio 4 Lochgilphead Area	
		WFM	BBC Radio 4 Ness of Lewis	
		WFM	Forfar Area	
		WFM	Leicestershire	
		WFM	Perth Area	
		WFM	Trowbridge Area	
98.0000		WFM	South London	

Base	Mobile	Mode	Location	User & Notes
98.1000		WFM	BBC Radio 4 West Skye Area	
		WFM	North Eastern Counties	
98.2000		WFM	BBC Radio 4 Inveraray Area	
		WFM	BBC Radio 4 Kinross Area	
		WFM	Bedford Area	
		WFM	Central South of England	
		WFM	North West Lancashire	
98.3000		WFM	Aberystwyth Area & Cardigan Bay	
		WFM	Cambridgeshire	
		WFM	Folkestone Area	
		WFM	Grampian Area	
		WFM	Newark Area	
		WFM	Lincolnshire	
		WFM	France Culture Rennes	
98.4000		WFM	Chippenham Area	
98.5000		WFM	BBC Radio 4 Oban Area	
98.6000		WFM	Abertillery	
		WFM	Bath City Area	
		WFM	Perth	
98.7000		WFM	BBC Radio 4 Ullapool & Lewis Area	
		WFM	Border Counties Area	
98.8000		WFM	Greater London	
		WFM	Home Counties	
		WFM	Rosneath Area	
		WFM	Stroud Area	
		WFM	Wrexham & Deeside Area	
		WFM	Radio Profile Normandy	
98.9000		WFM	BBC Radio 4 Fort William Area	
		WFM	BBC Radio 4 South Knapdale Area	
		WFM	Cheshire Area	
		WFM	Derbyshire	
		WFM	Greater Manchester	
		WFM	Humberside	
		WFM	Lancashire	
		WFM	Merseyside	
		WFM	Nottinghamshire	
		WFM	South East Wales	
		WFM	Yorkshire	
99.1000		WFM	BBC Radio 4 East Skye Area	
		WFM	BBC Radio 4 Tobermoray & Isle of Mull	
		WFM	Greenholm & Darvel Area	
		WFM	Oxfordshire & Wiltshire	
99.2000		WFM	Inverness	
		WFM	North East Scotland	
		WFM	Pennar	
99.3000		WFM	BBC Radio 4 Kinlochleven Area	
		WFM	Saddleworth Area	
		WFM	South East London	
		WFM	Suffolk & Norfolk Area	

Base	Mobile	Mode	Location	User & Notes
99.4000		WFM	BBC Radio 4 Stornaway Area	
		WFM	North Wales Coast & Anglesey	
		WFM	Pudsey Area	
		WFM	Salisbury Area	
99.5000		WFM	Central Lowlands	
		WFM	Forth Valley	
		WFM	Gloucester Area	
		WFM	Lowland Scotland	
		WFM	Scarborough Area	
		WFM	Somerset & North Devon	
		WFM	South & East Wales	
99.6000		WFM	Buxton Area	
		WFM	Morecambe Bay Area	
		WFM	South Cumbria	
		WFM	Weymouth Area	
		WFM	France InterCaen	
99.7000		WFM	Brighton Area	
		WFM	Peterborough Area	

99.8000 - 103.8000 MHz Nationwide Independent Radio

Base	Mobile	Mode	Location	User & Notes
100.0000		WFM	Kiss FM	Greater London
102.0000		WFM	Sunset Radio	Manchester
102.2000		WFM	GWRW	Wiltshire
		WFM	London Jazz Radio	Greater London
102.3000		WFM	Radio Tay	Dundee
		WFM	Two Counties Radio	E Dorest
102.4000		WFM	Buzz FM	Birmingham
		WFM	Downtown Radio	Londonderry
		WFM	Radio Broadland	Norfolk Broads
		WFM	Severn Sound	Cheltenham
		WFM	Southern Sound	N Sussex
102.5000		WFM	Clyde FM	Glasgow
		WFM	Fox FM	Oxford
		WFM	Pennine FM	Halifax
102.6000		WFM	Essex Radio	Chelmsford
		WFM	Orchard FM	Mendips
		WFM	Radio Harmony	Coventry
		WFM	Signal Radio	N Staffs
102.7000		WFM	Hereward Radio	eterborough
		WFM	Radio Mercury	Reigate
102.8000		WFM	Invicta FM	SE Kent
		WFM	Radio Wyvern	Worcester
		WFM	Trent FM 945	Derby
102.9000		WFM	Hallam FM	Barnsley
		WFM	Mercia Sound	Leamington
		WFM	Radio 210	Andover
103.0000		WFM	CN FM 103	Newmarket
		WFM	Devonair Radio	W Dorset
		WFM	GWR	Bath City

Base	Mobile	Mode	Location	User & Notes
		WFM	Key 103	Manchester
		WFM	Radio Borders	Peebles
103.1000		WFM	Beacon Shropshire	Shropshire
		WFM	Invicta FM	Medway
		WFM	France Armorique	Rennes
103.2000		WFM	Bradford City Radio	Bradford
		WFM	Power FM	Solent
		WFM	Red Dragon Radio	Cardiff
		WFM	Sound FM	Leicester
103.3000		WFM	Horizon Radio	Milton Keynes
		WFM	London Greek Radio	Haringey
103.4000		WFM	BBC Radio Devon	Devon
		WFM	Hallam FM	Doncaster
		WFM	Marcher Sound MFM	Wrexham
		WFM	Radio Borders	Eyemouth
		WFM	Wear FM	Sunderland
103.5000		WFM	BBC Radio Essex	Chelmsford
		WFM	East End Radio	Glasgow
		WFM	Southern Sound	Brighton
		WFM	Wiltshire Sound	Salisbury
103.6000		WFM	BBC Radio Northampton	Leicester
		WFM	BBC Radio 4	Anglesey
		WFM	BBC Radio 4	W Islay
		WFM	Wiltshire Sound	Swindon
103.7000		WFM	BBC CWR	Birmingham
		WFM	BBC Radio York	Yorkshire
103.8000		WFM	BBC Radio Bedfordshire	Luton
		WFM	BBC Radio Nottinghamshire	Notts
		WFM	Independent Radio Thamesmead	Thamesmead
103.9000		WFM	BBC Radio Cornwall	W Cornwall
		WFM	BBC Radio Lancashire	Central Lancs
		WFM	BBC Radio Suffolk	S Suffolk
104.0000		WFM	BBC Radio Sussex	N Sussex
		WFM	BBC Radio 4	Perth Area
104.1000		WFM	BBC Radio Sheffield	S Yorks
		WFM	BBC Radio Surrey & Berkshire	W Berks
104.2000		WFM	BBC Radio Cumbria	Windermere
		WFM	BBC Radio Kent	Dover
		WFM	BBC Radio Northampton	Northants
104.3000		WFM	BBC Radio York	Northallerton
		WFM	Wiltshire Sound	W Wiltshire
104.4000		WFM	BBC Radio Newcastle	Gateshead
		WFM	BBC Radio Norfolk	W Norfolk
		WFM	BBC Radio Surrey & Berkshire	Reading
104.5000		WFM	BBC Radio Bedfordshire	Milton Keynes
		WFM	BBC Radio Derby	Derbyshire
		WFM	BBC Radio Lancashire	Lancaster
		WFM	BBC Radio Sussex	Mid Sussex
104.6000		WFM	BBC Radio Bristol	Bath City

Base	Mobile	Mode	Location	User & Notes	
		WFM	BBC Radio Suffolk		N Suffolk
		WFM	BBC Radio Surrey		Guildford
104.7000		WFM	BBC Radio Gloucestershire		Severn Valley
104.9000		WFM	BBC Radio Essex		Manningtree
		WFM	BBC Radio Leicestershire		Coalville
		WFM	BBC Radio Surrey & Berkshire		High Wycombe
		WFM	BBC Radio 4		South Islay
		WFM	KFM Radio		Stockport
		WFM	Melody Radio		Greater London
		WFM	Radio Force 7		Granville

105.00000 - 107.99875 MHz Local Authority PMR, British Rail & Alarms

Base	Mobile	Mode	Location	User & Notes
105.36250		NFM	Nationwide	British Rail Rail Phones
106.28125		NFM	Nationwide	Gas Board
107.05625		NFM	Nationwide	British Coal Mine Rescue
107.707375		NFM	Nationwide	Alarm for Elderly & Infirm
107.80825		NFM	Nationwide	Alarm for Elderly & Infirm
107.81975		NFM	Nationwide	Local Authorities
107.83375		NFM	Nationwide	Local Authorities
107.90625		NFM	Nationwide	Local Authorities
107.94375		NFM	Nationwide	N.C.T.

108.0000 - 112.0000 MHz TACAN and DME Idents, ILS Localisers and VOR Aero Navigation Beacons

Base	Mobile	Mode	Location	User & Notes	
108.0000		AM	RAF Greenham Common	GCN	TACAN
108.1000		AM	Belfast	I-BFH	DME
		AM	Dundee	DDE	DME
		AM	Guernsey	I-UY	ILS Runway 09
		AM	Guernsey	I-GH	ILS Runway 27
		AM	RAF Abingdon	AB	ILS Runway 36
		AM	RAF Chivenor	CV	ILS Runway 28
		AM	RAF Cottesmore	CTM	TACAN
		AM	RAF Mildenhall	I-MIL	ILS Runway 11
		AM	RAF Mildenhall	I-MLD	ILS Runway 29
108.1500		AM	Blackpool	I-BPL	ILS Runway 28
		AM	Lydd	I-LYX	DME/ILS R22
108.2000		AM	RAF Boscombe Down	BDN	TACAN
108.3000		AM	Bedford	BQ	ILS Runway 27
		AM	RAF Lakenheath	I-LKH	ILS Runway 24
108.4000		AM	RAF Valley	VYL	TACAN
108.5000		AM	RAF Benson	BO	ILS Runway 19
108.6000		AM	Kirkwall	KWL	VOR/DME
		AM	RAF Bentwaters	BTW	TACAN
108.7000		AM	Newton Point	NTP	TACAN
		AM	RAF Alconbury	ALC	ILS Runway 30
		AM	RAF Leuchars	LU	ILS Runway 27
		AM	RAF St Mawgan	SM	ILS Runway 31
		AM	RAF Shawbury	SY	ILS Runway 19
108.7500		AM	Humberside	I-HS	DME/ILS R21

Base	Mobile	Mode	Location	User & Notes	
108.8000		AM	Weathersfield	WET	TACAN
108.9000		AM	Cranfield	I-CR	ILS Runway 22
		AM	Edinburgh	I-VG	DME/ILS R07
		AM	Edinburgh	I-TH	DME/ILS R25
		AM	Kerry	IKR	DME/ILS R25
		AM	RAF Woodbridge	WDB	DME/ILS R27
		AM	Ventnor	VNR	TACAN
108.9500		AM	Woodford	I-WU	DME/ILS R25
109.0000		AM	RAF Alconbury	ALC	TACAN
109.0500		AM	Yeovil	YVL	DME
109.1500		AM	Luton	I-LTN	DME/ILS R08
		AM	Luton	I-LJ	DME/ILS R26
109.2000		AM	Inverness	INS	VOR/DME
		AM	Swansea	SWZ	DME
109.3000		AM	Glasgow	I-OO	ILS Runway 23
		AM	RAF Church Fenton	CF	ILS Runway 24
		AM	RAF Wattisham	WTM	TACAN
		AM	RAF Wyton	WT	ILS Runway 27
109.3500		AM	Biggin Hill	I-BGH	DME
109.4000		AM	Barrow	WL	DME
		AM	Guernsey	GUR	VOR/ATIS
109.5000		AM	London/Heathrow	I-BB	ILS R09R
		AM	London/Heathrow	I-LL	ILS R27L
		AM	Manchester	I-MM	DME/ILS R06
		AM	Manchester	I-NN	DME/ILS R24
		AM	Plymouth	I-PLY	DME/ILS R31
		AM	Shannon	SA	ILS Runway 24
109.6000		AM	RAF Linton-On-Ouse	LOZ	TACAN
		AM	RAF Odiham	ODH	TACAN
109.7000		AM	Belfast/Aldergrove	I-AG	ILS Runway 25
		AM	Dinard	DR	ILS Runway 36
		AM	RAF Cranwell	CW	ILS Runway 27
		AM	RAF Kinloss	KS	ILS Runway 26
		AM	RAF Lyneham	LA	ILS Runway 25
		AM	RAF Valley	VY	ILS Runway 14
109.7500		AM	Coventry	I-CT	ILS Runway 23
109.8000		AM	RAF Kinloss	KSS	TACAN
		AM	RAF Lyneham	LYE	TACAN
109.8500		AM	Fair Oaks	FRK	DME
109.9000		AM	Aberdeen/Dyce	I-AX	ILS Runway 16
		AM	Aberdeen/Dyce	I-ABD	ILS Runway 34
		AM	Cherbourg	MP	ILS Runway 29
		AM	Cork	ICA	ILS Runway 17
		AM	Cork	ICN	ILS Runway 35
		AM	East Midlands	I-EMW	ILS Runway 09
		AM	East Midlands	I-EME	ILS Runway 27
		AM	Exeter	I-XR	ILS Runway 26
		AM	RAF Bentwaters	I-BTW	ILS Runway 25
		AM	Stornoway	I-SV	ILS Runway 18

Base	Mobile	Mode	Location	User & Notes	
		AM	Warton	WQ	DME/ILS R26
110.1000		AM	Birmingham	I-BIR	ILS Runway 15
		AM	Birmingham	I-BM	ILS Runway 33
		AM	Glasgow	I-UU	ILS Runway 05
		AM	RAF Marham	MR	ILS Runway 24
		AM	Rennes	RS	ILS Runway 29
110.1500		AM	Bristol	I-BON	ILS Runway 09
		AM	Bristol	I-BTS	ILS Runway 27
110.2000		AM	RAF Lakenheath	LKH	TACAN
110.3000		AM	Jersey	I-DD	ILS Runway 27
		AM	London/Heathrow	I-AA	DME/ILS R09L
		AM	London Heathrow	I-RR	ILS R29R
		AM	Prestwick	I-KK	ILS Runway 31
		AM	Prestwick	I-PP	ILS Runway 13
		AM	RAF Cottesmore	CM	ILS Runway 23
		AM	RAF Leeming	LI	ILS Runway 16
110.4000		AM	Perth	PTH	VOR
110.5000		AM	Bournemouth	I-BMH	ILS Runway 08
		AM	Bournemouth	I-BH	ILS Runway 26
		AM	London/Stansted	I-SED	DME/ILS R05
		AM	London/Stansted	I-SX	DME/ILS R23
		AM	RAF Leuchars	LUK	TACAN
		AM	RAF Scampton	SAP	ILS Runway 05
		AM	RAF Scampton	SAM	ILS Runway 23
110.5500		AM	Filton	I-BRF	ILS Runway 10
		AM	Filton	I-FB	ILS Runway 28
110.7000		AM	Cardiff	I-CDF	ILS Runway 12
		AM	Cardiff	I-CWA	ILS Runway 30
		AM	Carlisle	CO	DME
		AM	Connaught	I-CK	ILS Runway 27
		AM	London/Heathrow	I-CC	ILS Runway 23
		AM	RAF Coningsby	CY	ILS Runway 26
		AM	RAF Linton-On-Ouse	LO	ILS Runway 22
110.9000		AM	Belfast/Aldergrove	I-FT	ILS Runway 17
		AM	Jersey	I-JJ	ILS Runway 09
		AM	Leeds & Bradford	I-LF	ILS Runway 32
		AM	Leeds & Bradford	I-LBF	ILS Runway 14
		AM	London/Gatwick	I-GG	ILS R08R
		AM	London/Gatwick	I-WW	ILS R26L
		AM	Norwich	I-NH	ILS Runway 27
		AM	Ronaldsway	I-RY	DME/ILS R27
111.0000		AM	RN Yeovilton	VLN	TACAN
111.1000		AM	RAF Coningsby	CGY	TACAN
		AM	RAF Fairford	I-FFA	ILS Runway 27
		AM	RAF Fairford	I-FFD	ILS Runway 09
		AM	RAF Lossiemouth	LM	ILS Runway 23
		AM	RAF Waddington	WA	ILS Runway 21
		AM	RAF Wattisham	WT	ILS Runway 23
111.3000		AM	Hatfield	I-HD	ILS Runway 24

Base	Mobile	Mode	Location		User & Notes	
		AM	Perth	I-PRF	ILS Runway 21
		AM	Tees-side	I-TD	ILS Runway 23
111.4000		AM	RAF Binbrook	BNK	TACAN
111.5000		AM	Newcastle	I-NC	DME/ILS R07
		AM	Newcastle	I-NWC	DME/ILS R25
		AM	RAF Coltishall	CS	ILS Runway 22
		AM	RAF Fairford	FFA	TACAN
		AM	RAF Finningley	FY	ILS Runway 20
		AM	RAF Upper Heyford	I-UH	ILS Runway 27
111.6000		AM	RAF Chivenor	CVR	TACAN
111.7000		AM	RAF Boscombe Down	BD	ILS Runway 24
111.7500		AM	Liverpool	LVR	DME/ILS R09
		AM	Liverpool	I-LQ	DME/ILS R27
111.9000		AM	RAF Brize Norton	BZN	TACAN
		AM	RAF Brize Norton	BZA	ILS Runway 08
		AM	RAF Brize Norton	BZB	ILS Runway 26
		AM	RAF Honington	HT	ILS Runway 27

112.0000 - 117.9750 MHz TACAN and DME Idents, ATIS and VOR Aero Navigation Beacons

Base	Mobile	Mode	Location		User & Notes	
112.1000		AM	Pole Hill	POL	VOR/DME
112.2000		AM	Isle Of Man	IOM	VOR/DME
		AM	Jersey	JSR	VOR/ATIS
112.3000		AM	Heathrow Info		Arrival Info.
112.5000		AM	Cherbourg	CBG	VOR
		AM	St Abbs	SAB	VOR/DME
112.6000		AM	RAF St Mawgan	SMG	TACAN
112.7000		AM	Berry Head	BHD	VOR/DME
112.8000		AM	Gamston	GAM	VOR/DME
		AM	Rennes	RNE	VOR
113.1000		AM	Strumble	STU	VOR/DME
113.2000		AM	St Anthony	YAY	VOR/DME
		AM	Warton	WTN	TACAN
113.3000		AM	Shannon	SNN	DVOR/DME
113.3500		AM	Southampton	SAM	VOR/ATIS
113.5500		AM	Manchester	MCT	VOR/DME
113.6000		AM	London	LON	VOR/DME
		AM	Wick	WIK	VOR
		AM	Wick	WIZ	TACAN
113.6500		AM	Honiley	HON	VOR/DME
113.7000		AM	RAF Upper Heyford	UPH	TACAN
113.7500		AM	Bovingdon	BNN	VOR/DME
		AM	Heathrow		ATIS
113.8000		AM	Talla	TLA	VOR/DME
113.8500		AM	Southend		ATIS
113.9000		AM	Ottringham	OTR	VOR/DME
114.0000		AM	Midhurst	MID	VOR/DME
114.0500		AM	Lydd	LYD	VOR
114.1000		AM	Wallasey	WAL	VOR/DME

Base	Mobile	Mode	Location	User & Notes	
114.2000		AM	Lands End	LND	VOR/DME
114.2500		AM	Newcastle	NEW	VOR/ATIS
114.3000		AM	Aberdeen/Dyce	ADN	VOR/ATIS
		AM	Dinard	DIN	VOR
114.3500		AM	Compton	CPT	VOR/DME
114.4000		AM	Benbecula	BEN	VOR
		AM	Benbecula	BEZ	TACAN
114.5500		AM	Clacton	CLN	VOR/DME
114.6000		AM	Cork	CRK	DVOR/DME
114.7500		AM	Goodwood	GWC	VOR
114.8000		AM	RAF Sculthorpe	SKT	TACAN
114.9000		AM	Vallafield	VFD	TACAN
114.9500		AM	Dover	DVR	VOR/DME
115.1000		AM	Stornoway	STN	VOR
		AM	Stornoway	STZ	TACAN
115.2000		AM	Dean Cross	DCS	VOR/DME
115.2500		AM	Biggin Hill		ATIS
115.3000		AM	Ockham	OCK	VOR/DME
115.4000		AM	Glasgow	GOW	VOR/ATIS
		AM	Heathrow		ATIS
115.5500		AM	Gloucestershire	GOS	DME
115.6000		AM	Lambourne	LAM	VOR/DME
115.7000		AM	Trent	TNT	VOR/DME
115.9000		AM	RAF Mildenhall	MLD	TACAN
116.0000		AM	RAF Machrihanish	MAC	DVOR
		AM	RAF Machrihanish	MAZ	TACAN
116.1000		AM	RAF Church Fenton	CHF	TACAN
116.2000		AM	Blackbushe	BLC	DME
116.2500		AM	Barkway	BKY	VOR/DME
116.4000		AM	Daventry	DTY	VOR/DME
116.5000		AM	RAF Coltishall	CSL	TACAN
		AM	Cranfield	CFD	VOR
116.6000		AM	RAF Brawdy	BDY	TACAN
116.7500		AM	Cambridge	CAB	DME
117.0000		AM	Seaford	SFD	VOR/DME
117.1000		AM	Burnham	BUR	VOR
		AM	RAF Woodbridge	WDB	TACAN
117.2000		AM	Belfast	BEL	VOR/DME
117.3000		AM	Detling	DET	VOR/DME
117.3500		AM	Sumburgh	SUM	VOR/DME
117.4000		AM	Connaught	CON	DVOR/DME
		AM	RAF Cranwell	CWZ	TACAN
117.4500		AM	Brecon	BCN	VOR/DME
117.5000		AM	Brookmans Park	BPK	VOR/DME
		AM	Turnberry	TRN	VOR/DME
117.6000		AM	RAF Wittering	WIT	TACAN
117.7000		AM	Oxford	OX	DME
		AM	Tiree	TIR	VOR/DME
117.9000		AM	Mayfield	MAY	VOR/DME

Base	Mobile	Mode	Location	User & Notes
118.0000 - 136.9750 MHz International Civil Aviation Band 50 kHz				
118.0000	118.0000	AM	Nationwide	Air-Air Display Coord
		AM	Nationwide	Marlboro Aerobatic Display
118.0250	118.0250	AM	Leeds/Bradford	ATIS
118.0500	118.0500	AM	Birmingham	Radar
118.0750	118.0750	AM	London Docklands	Tower
118.1000	118.1000	AM	Aberdeen/Dyce	Tower
		AM	Granville	Tower
		AM	Liverpool	Tower
		AM	Penzance Heliport	Tower
118.1500	118.1500	AM	Prestwick	Radar
		AM	Stansted	Tower
118.2000	118.2000	AM	Ronaldsway	Radar
		AM	Southampton	Tower
118.2250	118.2250	AM	Brough	Approach & SRE
118.2500	118.2500	AM	Sumburgh	Tower
118.3000	118.3000	AM	Belfast/Aldergrove	Tower
		AM	Birmingham	Tower
		AM	Kirkwall	Tower
118.3250	118.3250	AM	Ispwich	Tower
118.3500	118.3500	AM	Brest	ACC
		AM	Burnaston	Tower
118.3750	118.3750	AM	MoD Bedford	Radar
		AM	RAF Alconbury	Talkdown
118.4000	118.4000	AM	Blackpool	Tower
118.4250	118.4250	AM	RAF Lyneham	Approach
		AM	RN Corsham Heliport	Tower
118.4500	118.4500	AM	Liverpool	Radar
118.5000	118.5000	AM	Dublin	Radar
		AM	Heathrow	Tower
		AM	Newcastle	Radar
		AM	Tees-side	Approach
118.5250	118.5250	AM	RAF Manston	Radar
118.5500	118.5500	AM	Humberside	Tower
		AM	Jersey	Radar
118.6000	118.6000	AM	Dublin	Tower
		AM	Gatwick	Radar
118.6250	118.6250	AM	Brough	Approach
		AM	Manchester	Tower
118.6500	118.6500	AM	Bournemouth/Hurn	Approach
118.7000	118.7000	AM	Edinburgh	Tower
		AM	Heathrow	Tower
		AM	Shannon	Tower
118.8000	118.8000	AM	Barra	Tower
		AM	Cork	Radar
		AM	Glasgow	Tower
118.8250	118.8250	AM	Dunsfold	Radar
118.8500	118.8500	AM	Tees-side	Approach
118.8750	118.8750	AM	Oxford/Kidlington	Tower

Base	Mobile	Mode	Location	User & Notes
118.9000	118.9000	AM	Guernsey	Radar
		AM	Norwich	Tower
		AM	RAF Lossiemouth	Tower
		AM	Ronaldsway	Tower
118.9500	118.9500	AM	Gatwick	Radar
119.0000	119.0000	AM	RAF Bentwaters	Tower
		AM	RAF Brize Norton	Approach
		AM	RAF Cranwell	Approach
		AM	RAF Fairford	Approach
		AM	RAF Woodbridge	Approach
119.0500	119.0500	AM	Exeter	Radar
119.1000	119.1000	AM	Glasgow	Approach
119.1500	119.1500	AM	RAF Fairford	Tower
		AM	RAF Woodbridge	Tower
119.2000	119.2000	AM	Benbecula	Approach/Tower
		AM	Heathrow	Approach
119.2500	119.2500	AM	Coventry	Approach
		AM	Sumburgh	Approach
119.3000	119.3000	AM	Cork	Tower
		AM	Glasgow	Radar
		AM	Hatfield	Radar
119.3500	119.3500	AM	Norwich	Approach
		AM	RAF Kinloss	Approach
		AM	RAF Lossiemouth	Approach
119.4000	119.4000	AM	Manchester	Approach
		AM	St. Brieuc	Tower
119.4250	119.4250	AM	London Docklands	Tower
119.4500	119.4500	AM	Jersey	Tower
		AM	Prestwick	Approach
119.4750	119.4750	AM	Cardiff	ATIS
119.5000	119.5000	AM	Exeter	Approach
		AM	Heathrow	Approach
119.5500	119.5500	AM	Dublin	Radar
119.6000	119.6000	AM	Gatwick	Approach
119.6250	119.6250	AM	Bournemouth/Hurn	Approach
119.6500	119.6500	AM	East Midlands	Approach
119.7000	119.7000	AM	Chichester/Goodwood	Tower
		AM	Dinard	Radar
		AM	Newcastle	Tower
		AM	Swansea	Tower
		AM	Wick	Tower
119.8000	119.8000	AM	Exeter	Tower
		AM	Gatwick	Police Helicopter Ops
		AM	Perth	Tower
		AM	Tees-side	Tower
119.8250	119.8250	AM	Dunsfold	Radar
119.8500	119.8500	AM	Altcar Heliport	Army Helicopter Ops
		AM	Burtonwood	US Army Helicopter Ops
		AM	Liverpool	Approach

Base	Mobile	Mode	Location	User & Notes
119.8750	119.8750	AM	Prestwick	Scottish ATC
119.9000	119.9000	AM	Cork	Approach
		AM	Heathrow	Radar
119.9500	119.9500	AM	Blackpool	Radar
		AM	Guernsey	Tower
119.9750	119.9750	AM	Luton	Tower
120.0000	120.0000	AM	Belfast/Aldergrove	Approach
		AM	Bognor Regis	Tower
		AM	Ford	Tower
		AM	Malvern	Navy Helicopter Ops
120.0500	120.0500	AM	Cardiff	Radar
120.1250	120.1250	AM	East Midlands	Radar
120.1500	120.1500	AM	Dinard	Approach
120.2000	120.2000	AM	Shannon	Approach
120.2250	120.2250	AM	Southampton	Radar
120.2500	120.2500	AM	Dinard	Tower
		AM	Panshanger	Tower
120.2750	120.2750	AM	Redhill	Tower/AFIS
120.3000	120.3000	AM	Jersey	Radar
		AM	Leeds/Bradford	Tower
120.3500	120.3500	AM	Deauville	Approach
		AM	RAF Finningley	Approach
120.4000	120.4000	AM	Aberdeen/Dyce	Approach
		AM	Heathrow	Approach
120.4500	120.4500	AM	Great Yarmouth	Tower
		AM	Jersey	Air Traffic Control
		AM	North Denes	Tower
		AM	North Sea	Oil Rig Heliport Common
120.5000	120.5000	AM	Rennes	Tower
120.5500	120.5500	AM	Bristol/Lulsgate	Tower
		AM	Prestwick	Approach
120.5750	120.5750	AM	Luton	ATIS
120.6000	120.6000	AM	Cumbernauld	Tower
120.6500	120.6500	AM	Goodwood	Tower
		AM	Hamble	Tower
		AM	Woodvale	Tower
120.7000	120.7000	AM	Lydd	Approach
120.7500	120.7500	AM	Swansea	Radar
120.8000	120.8000	AM	Nationwide	Battle Of Britain Flight
		AM	RAF Coningsby	Approach
		AM	RAF Leuchars	Ground
120.8250	120.8250	AM	Holland	Dutch Military ATC
120.8500	120.8500	AM	Ronaldsway	Approach
120.9000	120.9000	AM	RAF Abingdon	Approach
		AM	Belfast/Aldergrove	Radar
		AM	RAF Benson	Approach
120.9500	120.9500	AM	Paris	ACC/UACC
121.0000	121.0000	AM	Heathrow	Tower
		AM	RAF Woodvale	Zone

Base	Mobile	Mode	Location	User & Notes
121.0250	121.0250	AM	RAF West Drayton	London ATCC
121.0500	121.0500	AM	Leeds/Bradford	Radar
121.0750	121.0750	AM	Duxford	Air Display Channel
		AM	Silverstone	Tower
121.1000	121.1000	AM	Dublin	Approach
121.2000	121.2000	AM	Cardiff	Tower
		AM	Edinburgh	Approach
121.2500	121.2500	AM	Aberdeen/Dyce	Radar
121.3000	121.3000	AM	Glasgow	Radar
121.3500	121.3500	AM	Manchester	Radar
121.4000	121.4000	AM	Leavesden	Radar
		AM	Shannon	Approach
121.5000	121.5000	AM	Nationwide	Distress & Emergency
121.6000	121.6000	AM	Nationwide	Airfield Fire & Rescue
		AM	Aberdeen/Dyce	Fire Channel
		AM	Kerry	Ground
121.7000	121.7000	AM	Aberdeen/Dyce	Ground
		AM	Bournemouth/Hurn	Ground
		AM	Cork	Tower
		AM	Coventry	Ground
		AM	Glasgow	Ground
		AM	Heathrow	Tower
		AM	Manchester	Ground
121.7500	121.7500	AM	Belfast/Aldergrove	Ground
		AM	Edinburgh	Ground
		AM	Gatwick	Ground
		AM	Luton	Ground
		AM	Oxford/Kidlington	Ground
		NFM	Space	Soyuz Module Down Link
121.7750	121.7750	AM	Booker	Ground
		AM	London Docklands	Tower
121.8000	121.8000	AM	Birmingham	Ground
		AM	Cork	Ground
		AM	Dublin	Ground/Tower
		AM	Gatwick	Ground
		AM	Guernsey	Ground
		AM	Prestwick	Tower
		AM	Shannon	Ground
		AM	Southend	ATIS
121.8500	121.8500	AM	Aberdeen/Dyce	ATIS
		AM	Heathrow	Departure
		AM	Manchester	Ground
		AM	Wroughton	PFA Delivery
121.8750	121.8750	AM	Biggin Hill	ATIS
		AM	RAF Cranwell	ATIS
121.9000	121.9000	AM	Connaught	Ground
		AM	East Midlands	Ground
		AM	Heathrow	Ground
		AM	Jersey	Ground

Base	Mobile	Mode	Location	User & Notes
121.9250	121.9250	AM	Wroughton	PFA Ground
121.9500	121.9500	AM	Bournemouth/Hurn	ATIS
		AM	Gatwick	Ground
		AM	Halfpenny Green	Ground
		AM	Oxford	ATIS
122.0000	122.0000	AM	Baldonnel	Approach
		AM	Coventry	Radar
		AM	Headcorn	Tower
		AM	North Sea	BP Buchan Field
		AM	North Sea	BP Cyprus Field
		AM	North Sea	BP Forties Field
		AM	North Sea	BP Gyda Field
		AM	RAF Mona	ATIS
122.0500	122.0500	AM	Aberdeen/Dyce	British Airways
		AM	Nuthampstead	Tower
		AM	Nationwide	Brymon Airways Comp Ch
		AM	North Sea	Britoil Thistle Field
		AM	North Sea	Chevron Ninian Field
		AM	North Sea	Conoco Murchison Field
		AM	North Sea	Hamilton Argyll Field
		AM	North Sea	Ninian Field
		AM	North Sea	Shell/Esso Auk Field
		AM	North Sea	Shell/Esso Fulmar Field
		AM	North Sea	Shell/Esso Kittiwake Field
		AM	Nottinghamshire	Hutchins Crop Sprayers
122.0750	122.0750	AM	Duxford	Tower
122.1000	122.1000	AM	Falmouth	Radar
		AM	Middle Wallop	Tower
		AM	Nationwide	Military Tower Common
		AM	Northern UK	Fisheries Protection Service
		AM	RAF Abingdon	Radar
		AM	RAF Alconbury	Tower
		AM	RAF Benson	Tower
		AM	RAF Bentwaters	Tower
		AM	RAF Brawdy	Tower
		AM	RAF Chivenor	Tower
		AM	RAF Church Fentor	Tower
		AM	RAF Coltishall	Tower
		AM	RAF Coningsby	Tower
		AM	RAF Cosford	Tower
		AM	RAF Cottesmore	Tower
		AM	RAF Cranwell	Tower
		AM	RNAS Culdrose	Tower
		AM	RAF Dishforth	Tower
		AM	RAF Finningley	Tower
		AM	RAF Greenham Common	Tower
		AM	RAF Honnington	Tower
		AM	RAF Kinloss	Tower
		AM	RAF Lakenheath	Tower

Base	Mobile	Mode	Location	User & Notes
		AM	RAF Leeming	Tower
		AM	RAF Leuchars	Tower
		AM	RAF Linton-on-Ouse	Tower
		AM	RAF Lossiemouth	Tower
		AM	RAF Lyneham	Tower
		AM	RAF Manston	Tower
		AM	RAF Marham	Tower
		AM	RAF Newton	Tower
		AM	RAF Northolt	Tower
		AM	RAF Odiham	Tower
		AM	RNAS Portland	Tower
		AM	RAF St Anthan	Tower
		AM	RAF St Mawgan	Tower
		AM	RAF Scampton	Tower
		AM	RAF Sculthorpe	Tower
		AM	RAF Shawbury	Tower
		AM	RAF Swinderby	Tower
		AM	RAF Topcliffe	Tower
		AM	RAF Upper Heyford	Tower
		AM	RAF Valley	Tower
		AM	RAF Waddington	Tower
		AM	RAF Wattisham	Tower
		AM	RAF Wittering	Tower
		AM	RAF Woodbridge	Tower
		AM	RAF Wyton	Tower
		AM	RNAS Yeovilton	Tower
		AM	Sligo	AFIS
		AM	Wembury Range	Range Control
122.1250	122.1250	AM	Leicester	Tower
122.1500	122.1500	AM	Flotta	Tower
		AM	Langford Lodge	Tower
		AM	Leavesden	Tower
		AM	MoD Aberporth	AFIS
122.1750	122.1750	AM	North Sea	Mobil Beryl Field
122.1750	122.1750	AM	Turweston	Tower
122.2000	122.2000	AM	Cambridge	Tower
		AM	Crosland Moor	Tower
		AM	Haverfordwest	Tower
		AM	Huddersfield	Tower
		AM	Tatenhill	Tower
122.2500	122.2500	AM	Caernarfon	Tower
		AM	North Sea	Shell/Esso Brent Field
		AM	Rochester	Tower
122.3000	122.3000	AM	Blackbush	AFIS
		AM	Eglington	Tower
		AM	Lands End	Tower
		AM	Perth	Approach
		AM	Peterborough	Tower
122.3250	122.3250	AM	Clacton	Tower

Base	Mobile	Mode	Location	User & Notes
		AM	North Sea	Hamilton Esmond Field
		AM	North Sea	Hamilton Forbes Field
		AM	North Sea	Hamilton Gordon Field
122.3500	122.3500	AM	Audley End	Tower
		AM	Brooklands	Tower
		AM	Cardiff	Operations
		AM	Cuxwold	Cuxwold Ground
		AM	Grimsby	Tower
		AM	East Midlands	Air Bridge Carriers Ops
		AM	Edinburgh	Execair Operations
		AM	Heathrow	Gulf Air Terminal 3
		AM	Hethel	Tower
		AM	Hitchen	Tower
		AM	Luton	Reed Aviation
		AM	North Sea	Total Alwyn Field
		AM	Rush Green	Tower
		AM	Tees-side	Air Cam
122.3750	122.3750	AM	North Sea	BP Magnus Field
		AM	Morecambe Bay	BP Morecambe Field
		AM	Peterhead/Longside	Tower
		AM	Plockton	Tower
		AM	Strubby	Tower
122.4000	122.4000	AM	Dounreay	Tower
		AM	Elstree	Tower
		AM	Little Snoring	Tower
		AM	Scatsa	Radar
		AM	Westen	AFIS
122.4250	122.4250	AM	Earls Colne	Tower
122.4500	122.4500	AM	Chichester	Military Police Heli Ops
		AM	Belfast/Aldergrove	Approach
		AM	Goodwood	Approach
		AM	North Sea	Occidental Claymore Field
		AM	North Sea	Texaco Tartan Field
		AM	North Sea	Piper
		AM	North Sea	Claymore & Tartan
		AM	Sleap	Tower
		AM	Swanton Morley	Tower
		AM	Wickenby	Tower
122.5000	122.5000	AM	Bitteswell	Approach
		AM	Brussels	ACC
		AM	Cranmore	Tower
		AM	Galway	Tower
		AM	Inishmaan	Tower
		AM	MoD Llanbedr	Approach
		AM	RAE Farnborough	Tower
		AM	Weston-super-Mare	Tower
		AM	Woodford	Tower
122.5250	122.5250	AM	North Sea	Hamilton Pipe Field
122.5500	122.5500	AM	Dunsfold	Approach

Base	Mobile	Mode	Location	User & Notes
		AM	Holme-on-Spalding-Moor	Approach
		AM	RAF Mildenhall	Tower
		AM	RAF West Freugh	Tower
122.6000	122.6000	AM	Abbeyshrule	Air-Ground
		AM	Castlebar	Tower
		AM	Deptford Down	Tower
		AM	Dingwall	Tower
		AM	Faranfoe	Tower
		AM	Inverness	Approach
		AM	Kerry	Tower
		AM	Lerwick	Tower
		AM	Plymouth	Tower
		AM	Seething	Tower
		AM	Sherburn	Tower
		AM	White Waltham	Tower
122.6250	122.6250	AM	North Sea	Conoco Viking Field
122.7000	122.7000	AM	Barton	Tower
		AM	Bodmin	Tower
		AM	Compton Abbas	Tower
		AM	Manchester	Tower
		AM	Northampton	Tower
		AM	Silverstone	Tower
		AM	Sywell	AFIS
		AM	Tiree	Tower/AFIS
		AM	Usworth	Tower
122.7250	122.7250	AM	Filton	Approach
		AM	RAF Lyneham	Approach
122.7500	122.7500	AM	Cowden	Range Control
		AM	Tain	Range Control
122.7750	122.7750	AM	Crowfield	Tower
122.8000	122.8000	AM	Baldonnel	Approach
		AM	Brest	VFR Control
		AM	North Sea	Unionoil Heather Field
		AM	Nottingham	Tower
		AM	Stapleford	Tower
122.8250	122.8250	AM	Bruntingthorpe	Tower
122.8500	122.8500	AM	Cranfield	Approach
		AM	Londonderry	Approach
122.8750	122.8750	AM	Lasham	Dan Air
		AM	North Sea	Phillips Hewett Field
122.9000	122.9000	AM	Battersea Heliport	Tower
		AM	Birr	Tower
		AM	Cheltenham	Radar
		AM	Coonagh	Tower
		AM	Doncaster	Tower
		AM	Dundee	Tower
		AM	Long Marston	Tower
		AM	Old Sarum	Tower
		AM	RAF Innsworth	Tower

Base	Mobile	Mode	Location	User & Notes
122.9250	122.9250	AM	Beccles Heliport	Tower
		AM	North Sea	Phillip Ekofisk Field
		AM	Fenland	Tower
122.9500	122.9500	AM	Aberdeen/Dyce	Bristow Helicopters
		AM	Hilcote Heliport	Tower
		AM	Nationwide	Freemans Aviation
		AM	North Sea	Penzoil Noordwinning Field
		AM	North Sea	Petroland Petroland Field
		AM	North Sea	Placid Placid Field
		AM	North Sea	Nam Nam Field
		AM	North Sea	Nam Noordwinning
		AM	North Sea	Zanddijk
		AM	Southend Heliport	Tower
122.9750	122.9750	AM	Marston Moor	Tower
123.0000	123.0000	AM	Conington	Tower
		AM	Eaglescott	Tower
		AM	Halfpenny Green	ATIS
		AM	Hull	Tower
		AM	Inisheer	Tower
		AM	Inishmore	Tower
		AM	Lasham	Tower
123.0250	123.0250	AM	North Sea	Hamilton Ravenspurnn
123.0500	123.0500	AM	Gigha	Tower
		AM	Old Warden	Tower
		AM	Peterhead Heliport	Tower
		AM	North Sea	Shell/Esso Field Common
		AM	Brent	Air-Ground
		AM	Cormorant	Air-Ground
		AM	Dunlin	Air-Ground
		AM	Nuthampstead	Tower
		AM	Shipdham	Tower
		AM	Stevenage	BAe
		AM	Wigtown	Tower
123.1000	123.1000	AM	Baldonnel	Ground
		AM	Nationwide	Search and Rescue
		AM	Scotland	Air Mountain Rescue
		AM	Unst	Approach
123.1500	123.1500	AM	Humberside	Approach
		AM	Islay	AFIS
		AM	Scilly Isles	Tower
		AM	Shoreham	Approach
		AM	Sumburgh	Approach
123.2000	123.2000	AM	Barrow	Tower
		AM	Cranfield	Tower
		AM	Eniskillen	Tower
		AM	Plymouth	Approach
		AM	St Angelo	Tower
123.2250	123.2250	AM	Fadmoor	Tower
		AM	North Sea	Arco Thames Field

Base	Mobile	Mode	Location	User & Notes
		AM	Redhill	Tower
		AM	Wroughton	Tower
123.2500	123.2500	AM	Bagby	Tower
		AM	Bantry	Tower
		AM	Bembridge	Tower
		AM	Bitteswell	Tower
		AM	Bridlington	Tower
		AM	Fowlmere	Tower
		AM	Grindale	Tower
123.2750	123.2750	AM	Netherthorpe	Tower
123.3000	123.3000	AM	Barrow	Tower
		AM	Dublin	Dublin Military ATC
		AM	Headfort	Tower
		AM	Nationwide	Military Airfield Radar
		AM	RAF Benson	Radar
		AM	RAF Brawdy	Radar
		AM	RAF Brize Norton	Radar
		AM	RAF Chivenor	Radar
		AM	RAF Coltishall	Radar
		AM	RAF Coningsgy	Radar
		AM	RAF Cottesmore	Radar
		AM	RAF Cranwell	Radar
		AM	RNAS Culdrose	Radar
		AM	RAF Finningley	Radar
		AM	RAF Honington	Radar
		AM	RAF Kinloss	Radar
		AM	RAF Lakenheath	Radar
		AM	RAF Leuchars	Radar
		AM	RAF Linton-on-Ouse	Radar
		AM	RAF Lossiemouth	Radar
		AM	RAF Lyneham	Radar
		AM	RAF Machrihanish	Radar
		AM	RAF Manston	Radar
		AM	RAF Marham	Radar
		AM	RAF Odiham	Radar
		AM	RAF St Athan	Radar
		AM	RAF St Mawgan	Radar
		AM	RAF Scampton	Radar
		AM	RAF Shawbury	Radar
		AM	RAF Topcliffe	Radar
		AM	RAF Valley	Radar
		AM	RAF Waddington	Radar
		AM	RAF Wattisham	Radar
		AM	RAF Wittering	Radar
		AM	RAF Wyton	Radar
		AM	RNAS Yeovilton	Radar
		AM	Spanish Point	Tower
		AM	Trim	Tower
123.3500	123.3500	AM	Chester	Tower

Base	Mobile	Mode	Location	User & Notes
		AM	Hatfield	Tower
		AM	Hawarden	Approach
		AM	UK Waters	Helicopter Air-Ground
123.3750	123.3750	AM	Morecambe Bay	British Gas Morecambe
123.4000	123.4000	AM	RAF Lyneham	Radar
		AM	RAF St Mawgan	Tower
123.4500	123.4500	AM	Errol	DZ Control
		AM	Fairoaks	Tower
		AM	Nationwide	Air-Air
		AM	North Sea	Amoco Indefatigable Field
		AM	North Sea	Dab Duc Dan Field
		AM	North Sea	Dab Duc Gorm Field
		AM	North Sea	Dab Duc Skjold Field
		AM	North Sea	Marathon East Kinsale
		AM	North Sea	Marathon West Kinsale
		AM	North Sea	Mobil Camelot Field
		AM	North Sea	Shell/Esso Clipper Field
123.4750	123.4750	AM	Dunkeswell	Tower
123.5000	123.5000	AM	Baldonnel	Tower
		AM	Banff	Tower
		AM	Berwick-On-Tweed	Tower
		AM	Coal Aston	Tower
		AM	Denham	Tower
		AM	Dunkeswell	Tower
		AM	Duxford	Tower
		AM	Eggesford	Tower
		AM	Felthorpe/Norwich	Tower
		AM	Fenland	Tower
		AM	Isle Of Wight	Tower
		AM	Netherthorpe	Tower
		AM	Newtownards	Tower
		AM	Sandown	Tower
		AM	Shobdon	Tower
		AM	Strathallan	Tower
		AM	Stornoway	Tower
		AM	Swanton Morley	Tower
		AM	Woodvale	Tower
123.5250	123.5250	AM	North Weald	Tower
123.5500	123.5500	AM	North Sea	Phillips Maureen Field
		AM	North Sea	Sun Balmoral Field
123.5750	123.5750	AM	North Sea	Viking Conoco Common
		AM	Old Sarum	Tower
123.6000	123.6000	AM	Belmullet	Tower
		AM	Cambridge	Approach
		AM	Carlisle	Approach/Tower
		AM	Scatsa	Tower
123.6250	123.6250	AM	Londonderry	Approach
		AM	North Sea	Amoco Indefatigable Field
		AM	North Sea	Amoco Leman Field

Base	Mobile	Mode	Location	User & Notes
		AM	North Sea	Shell/Esso Indefatigable
		AM	North Sea	Shell/Esso Leman Field
		AM	North Sea	Shell/Esso Sean Field
123.6500	123.6500	AM	Gamston	Tower
		AM	Heathrow	British Airways
		AM	Nationwide	Brymon Airways
		AM	North Sea	Britoil Beatrice Field
		AM	North Sea	Marathon Beryl Field
123.7000	123.7000	AM	Amsterdam	ACC
123.7250	123.7250	AM	Epson	Tower
123.7500	123.7500	AM	Leeds/Bradford	Approach
123.8000	123.8000	AM	Stansted	Radar
123.8500	123.8500	AM	Amsterdam	ACC
123.9000	123.9000	AM	London	TMA
123.9500	123.9500	AM	Shanwick	Oceanic ACC
124.0000	124.0000	AM	East Midlands	Tower
		AM	Paris	ACC/UACC
124.0500	124.0500	AM	Paris	ACC/UACC
		AM	Prestwick	Scottish ATC
124.1500	124.1500	AM	Nationwide	Army Helicopter Common
		AM	RAF Marham	Approach
		AM	RNAS Portland	Approach
		AM	RAF Shawbury	Approach
124.2000	124.2000	AM	Manchester	Air Traffic Control
124.2250	124.2250	AM	Gatwick	Tower
124.2500	124.2500	AM	Norwich	Radar
124.2750	124.2750	AM	London	ATCC
124.3000	124.3000	AM	Amsterdam	ACC
124.3250	124.3250	AM	Dunsfold	Tower
124.3500	124.3500	AM	Bristol Lulsgate	Radar
		AM	Chester	Tower
124.4000	124.4000	AM	MoD Bedford	Approach
		AM	RAF Brawdy	Approach
124.4500	124.4500	AM	Warton	Approach
124.4750	124.4750	AM	Heathrow	Stand-by Ground
124.5000	124.5000	AM	Guernsey	Approach
		AM	Prestwick	Scottish Air Traffic Control
124.5500	124.5500	AM	RAF Cranfield	Approach
124.6000	124.6000	AM	West Drayton	Air Traffic Control/FIS
124.6500	124.6500	AM	Dublin	Area Control Centre
124.6750	124.6750	AM	Humberside	Humberside Radar
124.7000	124.7000	AM	Shannon	ACC
		AM	Stavanger	Radar
124.7500	124.7500	AM	West Drayton	Air Traffic Control/FIS
124.8000	124.8000	AM	Coventry	Tower
		AM	Rennes	Approach
124.8500	124.8500	AM	Paris	ACC/UACC
124.8750	124.8750	AM	Amsterdam	ACC
124.9500	124.9500	AM	Chester Garrison	Army Helicopter

Base	Mobile	Mode	Location	User & Notes
		AM	Filton	Tower
		AM	Hawarden	Tower
124.9750	124.9750	AM	RAF Northolt	Radar
125.0000	125.0000	AM	Brussels	ACC
		AM	Cardiff	Tower
		AM	Hamble	Approach
		AM	RAF Topcliffe	Approach
125.0500	125.0500	AM	Southend	Radar
125.1000	125.1000	AM	Manchester	Air Traffic Control
125.2000	125.2000	AM	Jersey	Air Traffic Control
125.2500	125.2500	AM	RAE Farnborough	Approach
		AM	RAF Odiham	Approach
125.2750	125.2750	AM	Anglia	Anglia Radar
125.3000	125.3000	AM	Isle Of Man	Radar
125.3250	125.3250	AM	Oxford	Approach
125.3500	125.3500	AM	Alderney	Tower
		AM	RAF Binbrook	Tower
125.4000	125.4000	AM	Chalgrove	Tower
		AM	Shoreham	Tower
		AM	Yeovil	Tower
125.5500	125.5500	AM	Andrewsfield	Approach
		AM	Audley End	Approach
		AM	Bournemouth/Hurn	Tower
		AM	Brest	IFR Control
125.5500	125.5500	AM	Stansted	Approach
125.6000	125.6000	AM	Bournemouth/Hurn	Tower
125.6500	125.6500	AM	Cheltenham	Tower
		AM	RAF Innsworth	Helicopter Air-Ground
		AM	RAF St Mawgan	Radar
125.7000	125.7000	AM	Paris	VFR Control
125.7250	125.7250	AM	Prestwick	Scottish VOLMET
125.7500	125.7500	AM	Amsterdam	ACC
125.8000	125.8000	AM	West Drayton	Air Traffic Control/TMA
125.8500	125.8500	AM	Cardiff	Approach
		AM	RAF St Athan	Radar
		AM	Sumburgh	ATIS
125.8750	125.8750	AM	Gatwick	Approach
125.9000	125.9000	AM	RAF Coltishall	Approach
		AM	RAF Machrihanish	Approach
125.9500	125.9500	AM	West Drayton	London TMA
126.1000	126.1000	AM	RAF Buchan	Highland Radar
126.2500	126.2500	AM	Prestwick	Scottish Air Traffic Control
126.3000	126.3000	AM	West Drayton	London TMA
126.3500	126.3500	AM	Newcastle	Approach
		AM	RAF Manston	Approach
126.4000	126.4000	AM	Bordeaux	VOLMET
126.4500	126.4500	AM	West Drayton	London TMA
126.5000	126.5000	AM	RAF Brize Norton	Tower/Ground
		AM	RAF Church Fenton	Approach

Base	Mobile	Mode	Location	User & Notes
		AM	RAF Elvington	Approach
		AM	RAF Leuchars	Approach
		AM	RAF St Mawgan	Approach
126.5500	126.5500	AM	Booker/Wycombe	Air Park Tower/AFIS
126.6000	126.6000	AM	West Drayton	London VOLMET North
126.6500	126.6500	AM	Manchester	Air Traffic Control
126.7000	126.7000	AM	Lydd	Tower
		AM	Middle Wallop	Army Air-Ground
		AM	RAF Boscombe Down	Approach
		AM	RAF Greenham Common	Radar
126.7500	126.7500	AM	Brussels	ACC
126.8250	126.8250	AM	West Drayton	London TMA
126.8500	126.8500	AM	Prestwick	Scottish ATCC
126.9000	126.9000	AM	Brussels	ACC
129.6250	129.6250	AM	Woodford	Tower
126.9500	126.9500	AM	Stansted	Approach
127.0000	127.0000	AM	Dublin	VOLMET
127.0500	127.0500	AM	Nationwide	CAA Test Flights
127.1000	127.1000	AM	West Drayton	Air Traffic Control
127.1250	127.1250	AM	Prestwick	ATIS
127.1750	127.1750	AM	Stansted	ATIS
127.2750	127.2750	AM	Prestwick	Scottish ATCC
127.3000	127.3000	AM	Cherbourg	Approach/Tower
		AM	Luton	Approach
127.3500	127.3500	AM	RAF Digby	Approach
		AM	RAF Scampton	Approach
		AM	RAF Waddington	Approach
		AM	RN Yeovilton	Approach
127.4500	127.4500	AM	West Drayton	London Mil
127.5000	127.5000	AM	Shannon	ACC
127.5500	127.5500	AM	Heathrow	Approach
127.6500	127.6500	AM	Shanwick	Oceanic ACC
127.7000	127.7000	AM	West Drayton	Air Traffic Control
127.7250	127.7250	AM	Southend	Tower
127.7500	127.7500	AM	Norwich	Air UK Company Channel
		AM	RAF Leeming	Approach
127.8500	127.8500	AM	Reims	ACC/UACC
127.9000	127.9000	AM	Shanwick	Oceanic ACC
127.9500	127.9500	AM	West Drayton	Air Traffic Control
127.9750	127.9750	AM	Filton	Approach
128.0000	128.0000	AM	Dublin	Approach
128.0250	128.0250	AM	London Docklands	Approach
128.0500	128.0500	AM	West Drayton	Air Traffic Control
128.0750	128.0750	AM	Duxford	ATIS
128.1000	128.1000	AM	Paris	ACC/UACC
		AM	St Kilda	Tower
128.1250	128.1250	AM	West Drayton	ATCC
128.1500	128.1500	AM	Exeter	Approach
128.1750	128.1750	AM	Manchester	ATIS

Base	Mobile	Mode	Location	User & Notes
128.2000	128.2000	AM	Brussels	ACC
128.2500	128.2500	AM	West Drayton	London Mil
128.3000	128.3000	AM	Aberdeen/Dyce	Radar Secondary
		AM	Netheravon	Army Tower
128.3500	128.3500	AM	Holland	Dutch Military
		AM	Newcastle	Army Tower
128.4000	128.4000	AM	West Drayton	Air Traffic Control/TMA
128.4250	128.4250	AM	Watton	Border Radar
128.4500	128.4500	AM	Brussels	ACC
128.4750	128.4750	AM	Gatwick	ATIS
128.5000	128.5000	AM	Caen	Tower
		AM	Prestwick	Scottish Air Traffic Control
128.5500	128.5500	AM	RAF Upper Heyford	Approach
128.6000	128.6000	AM	West Drayton	London VOLMET South
128.6500	128.6500	AM	Alderney	Approach
		AM	Guernsey	Approach
128.6750	128.6750	AM	Pennine	Pennine Radar
128.7000	128.7000	AM	West Drayton	London Mil
128.7500	128.7500	AM	Luton	Approach
128.7750	128.7750	AM	RAF Manston	Tower
128.8000	128.8000	AM	Brussels	ACC
128.8500	128.8500	AM	Nationwide	Eastern Airlines Packet Ch
		AM	Nationwide	Jetstream Company Channel
		AM	Southampton	Approach
		AM	Tees-side	Radar
128.9000	128.9000	AM	West Drayton	Air Traffic Control/TMA
128.9250	128.9250	AM	Anglia	Anglia Radar
128.9500	128.9500	AM	Southend	Approach
128.9750	128.9750	AM	Edinburgh	Air Traffic Control
129.0000	129.0000	AM	Aldernay	Air-Ground
		AM	Brest	ACC/UACC
		AM	Paris	UIR Control
129.0250	129.0250	AM	Nationwide	Air France Company Chan
129.0500	129.0500	AM	RAF Honington	Approach
		AM	RAF Lakenheath	Approach/Departure Control
		AM	RAF Mildenhall	Departure Control
129.1000	129.1000	AM	West Drayton	ATCC
129.1250	129.1250	AM	RAF Linton-On-Ouse	Approach
		AM	RAF Marham	Talkdown
		AM	RAF Northolt	Radar
129.1500	129.1500	AM	Linton	Approach
129.2000	129.2000	AM	Nationwide	American Airlines Packet
		AM	West Drayton	ATCC
129.2250	129.2250	AM	Weston-super-Mare	Approach
		AM	Wroughton	PFA Circuit
129.3000	129.3000	AM	Amsterdam	ACC
129.3500	129.3500	AM	Paris	ACC/UACC
129.4000	129.4000	AM	Biggin Hill	Approach/AFIS
129.4500	129.4500	AM	Kent	Kent Radar

Base	Mobile	Mode	Location	User & Notes
129.5000	129.5000	AM	Brest	UACC
		AM	Nationwide	Delta Airlines Packet Freq
129.5500	129.5500	AM	Luton	Approach
129.6000	129.6000	AM	Nationwide	Delta Airlines Packet Freq
		AM	West Drayton	National Air Traffic Control
129.6250	129.6250	AM	Nationwide	TWA Packet Frequency
129.6500	129.6500	AM	Brussels	ACC
		AM	North Sea	Mobil/Statoil Statfjord Field
129.7000	129.7000	AM	Alderney	Trinity Lightship Heliport
		AM	Baldonnel	Radar
		AM	Blackbush	A.T.S.
		AM	Bishops Rock	Trinity Lightship Heliport
		AM	Casquets	Trinity Lightship Heliport
		AM	Glasgow	Northwest Company Chan
		AM	Flatholm	Trinity Lightship Heliport
		AM	Fort William	Heliport
		AM	Hanois	Trinity Lightship Heliport
		AM	Inner Dowsing	Trinity Lightship Heliport
		AM	Longships	Trinity Lightship Heliport
		AM	Lundy South	Trinity Lightship Heliport
		AM	Nationwide	Britannia Ops
		AM	Nationwide	KLM Ops
		AM	Nationwide	Ward Air Ops
		AM	North Sea	Amoco Montrose Field
		AM	North Sea	Amoco Arbroath Field
		AM	Round Island	Trinity Lightship Heliport
		AM	Royal Sovereign	Trinity Lightship Heliport
		AM	Skerries	Trinity Lightship Heliport
		AM	Skokholm	Trinity Lightship Heliport
		AM	Smalls	Trinity Lightship Heliport
		AM	South Bishop	Trinity Lightship Heliport
		AM	St Anns Head	Trinity Lightship Heliport
129.7250	129.7250	AM	Conington	Tower
		AM	Deanland	Tower
129.7500	129.7500	AM	Filton	Rolls Royce Ops
		AM	Gleneagles	Gleneagles Helicopters
		AM	Jersey	Air UK Ops
		AM	Norwich	Air Europe Ops
		AM	Nationwide	Air Express Ops
		AM	Nationwide	BMA Ops
		AM	Nationwide	Bourn Air Ops
		AM	Nationwide	Brymon Airways
		AM	Nationwide	British Isl. Air Bridge Ops
		AM	North Sea	Elf Aquataine Norge Frigg
		AM	North Sea	Kewanee Nordsee Field
		AM	North Sea	Total/Elf Frigg Field
		AM	Norwich	Air UK Ops
		AM	Stansted	Servisair Ops
129.8000	129.8000	AM	Bourn	Tower

Base	Mobile	Mode	Location	User & Notes
		AM	Breighton	Tower
		AM	Carrickfin	Tower
		AM	Clonbulloge	Tower
		AM	Donegal	Tower
		AM	Popham	Tower
		AM	Truro	Tower
129.8250	129.8250	AM	Insch	Tower
		AM	Northrepps	Tower
129.8500	129.8500	AM	Chester Garrison	Army Helicopter Tower
		AM	Hawarden	Radar
		AM	Waterford	Tower
129.8750	129.8750	AM	Ashford	Tower
		AM	Enstone	Tower
		AM	Hethersett	Hethersett Radio
		AM	North Sea	Amethyst Field
		AM	North Sea	BP Cleeton Field
		AM	North Sea	BP Ravenspurn Field
		AM	North Sea	BP West Sole Field
		AM	North Sea	British Gas Rough Field
129.9000	129.9000	AM	Brunton	Tower
		AM	Coonagh	Tower
		AM	High Easter	AFIS
		AM	Langar	Tower
		AM	Lesham	Tower
		AM	Nationwide	Air Ambulance
		AM	Nationwide	Hang Gliding
		AM	Nationwide	Hot Air Ballooning
		AM	Nationwide	RAF Formation Air-Air
		AM	North Sea	Phillips Ekofisk Field
		AM	North Sea	Phillips Emden Field
		AM	Old Sarum	Tower
		AM	Sproatley/Hull	Tower
		AM	Strathallan	Air-to-Ground
129.9250	129.9250	AM	RAF Upper Heyford	Tower
129.9500	129.9500	AM	North Sea	Shell/Esso Field Common
		AM	Cormorant	Air-Ground
		AM	Dunlin	Air-Ground
		AM	North Sea	Viking Oil Field
		AM	West Shetland	Helicopter Information
129.9750	123.0000	AM	North Sea	Helicopter Common
129.9750	129.9750	AM	Nationwide	Gliding
		AM	North Weald	Tower
130.0000	130.0000	AM	RAF Alconbury	Tower
		AM	MoD Bedford	Approach
		AM	RAE Boscombe Down	Tower
130.0500	130.0500	AM	Aberdeen/Dyce	Ground Staff
		AM	Castleforbes	Tower
		AM	RAE Farnborough	Radar
		AM	RAF West Freugh	Approach

Base	Mobile	Mode	Location	User & Notes
		AM	Sumburgh	Radar
		AM	Woodford	Approach
130.0750	130.0750	AM	Gatwick	Servisair Ops
		AM	Heathrow	Air Malta Ops
		AM	RAF Brize Norton	Brize Ops
130.1000	130.1000	AM	Eaglescott	Tower
		AM	Long Marsden	Tower
		AM	Nationwide	Gliders
		AM	Netheravon	Tower
		AM	Pocklington	Tower
		AM	Rufforth	Tower
		AM	Strubby	Tower
		AM	Tibbenham/Norwich	Tower
130.1250	130.1250	AM	Nationwide	Glider Training
130.1500	130.1500	AM	Bitteswell	Approach
		AM	Deptford Down	Approach
		AM	Netheravon	Salisbury Plain Tower
		AM	Unst	Tower
		AM	RAF Upavon	Tower
130.1750	130.1750	AM	Exeter	Handling
		AM	North Weald	Aces High
130.2000	130.2000	AM	North Sea	Chevron Ninian Field
		AM	North Sea	Alwyn Field
		AM	RAF Chivenor	Approach
		AM	RAF Cottesmore	Approach
		AM	RAF Wittering	Approach
130.2500	130.2500	AM	Nationwide	American Airlines Packet
		AM	RAF Abingdon	Tower
130.3000	130.3000	AM	Oxford/Kidlington	Approach
		AM	Sturgate	Tower
130.3500	130.3500	AM	RAF Northolt	Talkdown
		AM	Samlesbury/Warton	Tower
		AM	Unst	Tower
130.3750	130.3750	AM	Blackbush	Air Hanson
		AM	Bognor Regis	Company Channel
130.4000	130.4000	AM	Edinburgh	Approach
		AM	Nationwide	Gliders Channel 1
		AM	Rufforth	Tower
		AM	Strubby	Tower
		AM	Tibbenham/Norwich	Tower
130.4250	130.4250	AM	Badminton	Tower
		AM	Halton	Tower
		AM	Sandtoft	Tower
		AM	South Marston	Ground
130.4500	130.4500	AM	Glenrothes	Tower
		AM	Skegness	Tower
		AM	Swanton Morley	Tower
		AM	Thruxton	Tower
		AM	Wellesbourne Mountford	Tower

Base	Mobile	Mode	Location	User & Notes
130.4750	130.4750	AM	Gamston	Tower
		AM	Lippits Hill	Tower
130.5000	130.5000	AM	Nationwide	Aquilla Air-to-Air
		AM	Castleforbes	Tower
130.5500	130.5500	AM	Andrewsfield	Tower
		AM	Brough	Tower/Approach
		AM	Holme-On-Spalding	Approach
		AM	North Sea	Amoco Vauxhall Field
		AM	North Sea	Phillips Albuskjell Field
		AM	North Sea	Phillips Cod Field
		AM	North Sea	Phillips Edda Field
		AM	North Sea	Phillips Ekofisk Field
		AM	North Sea	Phillips Eldfisk Field
		AM	North Sea	Phillips Tor Field
130.5750	130.5750	AM	Stansted	Universal Air Handling
130.6000	130.6000	AM	Aberdeen/Dyce	Servisair
		AM	Belfast/Aldergrove	Servisair
		AM	Birmingham	Servisair
		AM	Blackpool	Servisair
		AM	Bournemouth/Hurn	Channel Express
		AM	Bristol Lulsgate	Servisair
		AM	Cardiff	Servisair
		AM	Channel Isles	Air UK
		AM	Channel Isles	Channel Express
		AM	East Midlands	Ground
		AM	Edinburgh	Ground
		AM	Gatwick	British Caledonian
		AM	Heathrow	Fields Aviation Ops
		AM	Nationwide	Brymon Airways Ops
		AM	Nationwide	Delta Airlines Ops
		AM	Nationwide	Servisair
		AM	RAF Manston	KIA Ops
130.6250	130.6250	AM	Aberdeen	Granite Ops
		AM	Bristol Lulsgate	Clifton Ops
		AM	East Midlands	Donington Aviation Ops
		AM	Horsham	Tower
		AM	Southend	British Air Ferries Ops
130.6500	130.6500	AM	Barra	Tower
		AM	Bournemouth/Hurn	Services
		AM	Cowick Hall	Tower
		AM	Foulsham	Tower
		AM	Gatwick	American Airlines
		AM	Gatwick	China Airlines
		AM	Gatwick	Dan Air Ops
		AM	Gatwick	Euro Air Ops
		AM	Gatwick	Handling
		AM	Gatwick	Korean Air
		AM	Gatwick	Northwest
		AM	RN Kyle Of Lochalsh	Heliport

Base	Mobile	Mode	Location	User & Notes
		AM	Skye	Tower
130.7000	130.7000	AM	Bedford	Approach
		AM	Cleeton	Tower
		AM	Connaught	Tower
		AM	Lands End	Tower
		AM	Wroughton	Tower
130.7250	130.7250	AM	Denham	Tower
		AM	North Sea	Total/Elf Frigg/Fergus Field
		AM	RAF West Freugh	Approach
130.7500	130.7500	AM	Belfast Harbour	Tower
		AM	Cambridge	Radar
		AM	RAE Boscombe Down	Talkdown
		AM	Sydenham	Tower
		AM	Woodford	Radar
130.8000	130.8000	AM	Hatfield	Tower
		AM	Hucknall	Tower
		AM	North Sea	Amoco/Conoco Hutton Field
		AM	South West UK	Fisheries Protection
		AM	Warton	Tower
		AM	Yeovil Judwin	Approach
130.8500	130.8500	AM	Belfast Harbour	Approach
		AM	Little Gransden	Tower
		AM	Sandy	Tower
		AM	Sydenham	Approach
130.8750	130.8750	AM	West Malding	Tower
130.9250	130.9250	AM	RAF West Drayton	London TMA
130.9500	130.9500	AM	Shannon	ATIS
131.0000	131.0000	AM	Southampton	Radar
131.0500	131.0500	AM	West Drayton	Air Traffic Control
131.0750	131.0750	AM	Gatwick	Servisair
131.1000	131.1000	AM	Nationwide	British Airways Packet
		AM	Brussels	ACC
131.1500	131.1500	AM	Shannon	ACC
131.1750	131.1750	AM	Brest	UACC
131.2000	131.2000	AM	West Drayton	Air Traffic Control
131.2250	131.2250	AM	Watton	Eastern Radar
131.2500	131.2500	AM	Paris	ACC/UACC
131.3000	131.3000	AM	Lydd	Approach
		AM	Prestwick	Scottish ATC
131.3250	131.3250	AM	Birmingham	Approach
131.3750	131.3750	AM	Glasgow	Air Canada
131.4000	131.4000	AM	Heathrow	Air India Ops
		AM	Heathrow	Bangladesh Biman
		AM	Heathrow	CSA
		AM	Heathrow	Kenya Airways
		AM	Heathrow	Trans Mediteranian
		AM	Heathrow	Zambian Airlines
131.4250	131.4250	AM	Heathrow	Saudia
131.4500	131.4500	AM	Heathrow	Air Canada Ops

Base	Mobile	Mode	Location	User & Notes
		AM	Heathrow	Air Jamaica Ops
		AM	Heathrow	Air Malta
		AM	Heathrow	Alitalia Ops
		AM	Heathrow	KLM Ops
		AM	Heathrow	Pakistan International Ops
		AM	Heathrow	Royal Jordanian
		AM	Heathrow	Thai Airways Ops
		AM	Prestwick	Air Canada Ops
131.4750	131.4750	AM	Gatwick	British Airways Mainten.
		AM	Gatwick	Virgin Atlantic
		AM	Heathrow	Sebena
		AM	Heathrow	Speedbird Control North
		AM	Nationwide	Canadian Armed Forces
131.5000	131.5000	AM	Cork	Aer Lingus Ops
		AM	Dublin	Aer Lingus
		AM	Heathrow	Air France Ops
		AM	Heathrow	British Airways Ops
		AM	Heathrow	Kuwait Airways Ops
131.5250	131.5250	AM	Heathrow	Ryanair
		AM	Heathrow	Wardair
		AM	Luton	London European Airways
		AM	Luton	Monarch Airlines Ops
131.5500	131.5500	AM	Heathrow	British Airways Parking
131.5750	131.5750	AM	Belfast/Aldergrove	British Midlands
		AM	Birmingham	TEA Operations
		AM	East Midlands	British Midlands Comp Ch
		AM	Heathrow	El Al
		AM	Heathrow	Iran Air Ops
131.6000	131.6000	AM	Heathrow	Air India Ops
		AM	Heathrow	TWA Ops
131.6250	131.6250	AM	Gatwick	British Caledonian Ops
		AM	Gatwick	Canadian Pacific Ops
		AM	Heathrow	British Airways
		AM	Heathrow	Royal Jordanian Ops
		AM	Heathrow	Sabena
		AM	Portishead	Portishead Radio Telephone
131.6500	131.6500	AM	Heathrow	Air Malta Ops
		AM	Heathrow	Dan Air Ops
		AM	Heathrow	KLM Ops Terminal 4
		AM	Heathrow	Japan Airlines Ops
		AM	Heathrow	Lufthansa Ops
		AM	Schiphol	KLM Ops
131.6750	131.6750	AM	Luton	Britannia Airways Comp Ch
131.7000	131.7000	AM	Heathrow	SAS Ops
		AM	Heathrow	Swissair Ops
		AM	Heathrow	KLM
		AM	Heathrow	Sabena
131.7250	131.7250	AM	Heathrow	Lufthansa Company Chan
131.7500	131.7500	AM	Heathrow	Aer Lingus Ops

Base	Mobile	Mode	Location	User & Notes
		AM	Heathrow	Lufthansa Ops
		AM	Heathrow	Air UK
		AM	Heathrow	Kenya Airways
		AM	Heathrow	TAP Air Portugal
131.7750	131.7750	AM	East Midlands	Orion Airways Ops
		AM	Heathrow	Aeroflot
		AM	Heathrow	British Airways Ops
		AM	Heathrow	CSA
		AM	Heathrow	Icelandair
		AM	Heathrow	JAT
		AM	Heathrow	LOT
		AM	Heathrow	Malev
		AM	Heathrow	Olympic
		AM	Heathrow	Sabena
		AM	Luton	Britannia Airways Mainten.
131.8000	131.8000	AM	Heathrow	British Airways
		AM	Heathrow	Olympic
		AM	International	Air-Air Channel
		AM	Nationwide	Fisheries Protection
131.8250	131.8250	AM	Heathrow	Olympic
131.8500	131.8500	AM	Aberdeen/Dyce	British Airways
		AM	Belfast/Aldergrove	British Airways
		AM	Benbecula	British Airways
		AM	Birmingham	British Airways
		AM	Cork	Aer Lingus Company Chan.
		AM	Heathrow	Malaysian Airlines
		AM	Heathrow	Zambian Airlines
		AM	Nationwide	British Airways Ops
131.8750	131.8750	AM	Belfast/Aldergrove	Paramount Executive
		AM	Heathrow	Quantas Ops
131.9000	131.9000	AM	Heathrow	BA Speedbird Ops
		AM	Heathrow	Qantas
		AM	Heathrow	South African Airlines
		AM	Prestwick	Eastern Airlines Ops
131.9250	131.9250	AM	Heathrow	Lufthansa
131.9500	131.9500	AM	Brussels	Sabena Ops
		AM	Heathrow	El Al Ops
		AM	Heathrow	Iberia Airlines Ops
		AM	Heathrow	Olympic Airways Ops
		AM	Heathrow	Flying Tigers Ops
		AM	Nationwide	Air France Company Chan
131.9750	131.9750	AM	Glasgow	British Airways Ops
		AM	Heathrow	El Al
		AM	Heathrow	Finnair
		AM	Heathrow	Flying Tigers
		AM	Heathrow	Iberia
		AM	Heathrow	Nigerian Airlines Ops
132.0000	132.0000	AM	English Channel	French-London Handover
		AM	Paris	ACC/UACC

Base	Mobile	Mode	Location	User & Notes
132.0500	132.0500	AM	Brest	UACC
		AM	Heathrow	Departure/TMA
132.0750	132.0750	AM	Edinburgh	ATIS
132.1000	132.1000	AM	Paris	ACC/UACC
132.1250	132.1250	AM	Brest	UACC
132.1500	132.1500	AM	Shannon	Shannon ACC
132.2000	132.2000	AM	Maastricht	UAC
132.3000	132.3000	AM	West Drayton	ATCC
132.3500	132.3500	AM	Filton	Filton Director
		AM	Holland	Dutch Military
132.3750	132.3750	AM	Paris	ACC/UACC
132.4000	132.4000	AM	Bristol Lulsgate	Approach
132.4500	132.4500	AM	West Drayton	ATCC
132.5000	132.5000	AM	Reims	ACC/UACC
132.5250	132.5250	AM	Holland	Dutch Military
132.6000	132.6000	AM	West Drayton	ATCC
132.6250	132.6250	AM	Reims	ACC/UACC
132.6500	132.6500	AM	Kent	Kent Air Ambulance
		AM	London	Medivac
		AM	Oxford	Churchill Hospital
132.7000	132.7000	AM	London Docklands	Thames Radar
132.8000	132.8000	AM	West Drayton	ATCC
132.8250	132.8250	AM	Paris	ACC/UACC
132.8500	132.8500	AM	Maastricht	UAC
132.9000	132.9000	AM	Pennines	Pennine Radar
		AM	Wroughton	PFA Arrivals
132.9500	132.9500	AM	West Drayton	ATCC
133.0000	133.0000	AM	Brest	ACC/UACC
133.0500	133.0500	AM	Manchester	Manchester ATC
133.0750	133.0750	AM	Heathrow	ATIS
133.1000	133.1000	AM	Amsterdam	ACC
133.2500	133.2500	AM	Maastricht	UAC
133.3000	133.3000	AM	West Drayton	London Mil
133.3250	133.3250	AM	Watton	Eastern Radar
133.3500	133.3500	AM	Maastricht	UAC
133.4000	133.4000	AM	Manchester	Manchester ATC
133.4500	133.4500	AM	West Drayton	London ATCC
133.4750	133.4750	AM	Brest	UACC
133.5000	133.5000	AM	Paris	ACC/UACC
133.5350	133.5350	AM	West Drayton	London ATCC
133.5500	133.5500	AM	Plymouth	Approach
133.6000	133.6000	AM	West Drayton	ATCC
133.6500	133.6500	AM	Wroughton	Tower
133.7000	133.7000	AM	West Drayton	Air Traffic Control
133.7500	133.7500	AM	RAF Brize Norton	Approach
133.8000	133.8000	AM	Shanwick	Shanwick Track Broadcast
133.8250	133.8250	AM	Reims	ACC/UACC
133.8500	133.8500	AM	Bristol Lulsgate	Tower
		AM	Maastricht	UAC

Base	Mobile	Mode	Location	User & Notes
133.8750	133.8750	AM	Borders	Border Radar
133.9000	133.9000	AM	West Drayton	London Mil
133.9250	133.9250	AM	Paris	ACC/UACC
133.9500	133.9500	AM	Maastricht	UAC
133.9750	133.9750	AM	Heathrow	ATIS
134.0500	134.0500	AM	RAF Alconbury	Approach
		AM	RAF Wyton	Approach
		AM	RNAS Culdrose	Approach
		AM	RNAS Predannack	Approach
134.1000	134.1000	AM	RAF Buchan	Highland Radar
134.1500	134.1500	AM	RAF Northolt	Approach/Tower
134.1750	134.1750	AM	West Drayton	ATCC
134.2000	134.2000	AM	Brest	ACC
134.2250	134.2250	AM	Gatwick	Approach
134.2500	134.2500	AM	Watton	Eastern Radar
		AM	West Drayton	ATCC
134.3000	134.3000	AM	RAF Abingdon	Centralised Approach Cont.
		AM	RAF Benson	Centralised Approach Cont.
		AM	RAF Brize Norton	Brize Radar
		AM	RAF Buchan	Highland Radar
		AM	RAF Fairford	Centralised Approach Cont.
		AM	RAF Greenham Common	Centralised Approach Cont.
		AM	RAF Kemble	Radar
		AM	RAF Lyneham	Centralised Approach Cont.
		AM	Prestwick	Scottish Mil
		AM	Watton	Eastern Radar
134.3500	134.3500	AM	RAE Farnborough	Approach
		AM	RAF Valley	Approach
134.3750	134.3750	AM	Maastricht	UAC
134.4000	134.4000	AM	Reims	ACC/UACC
134.4250	134.4250	AM	West Drayton	ATCC
134.4500	134.4500	AM	Heathrow	London Zone
134.5000	134.5000	AM	Filton	Filton Ops
134.6000	134.6000	AM	Beccles	Tower
134.7000	134.7000	AM	RAF Brize Norton	Brize Radar
		AM	West Drayton	Control Information
134.7750	134.7750	AM	Prestwick	Scottish ATCC
134.8000	134.8000	AM	Belfast Harbour	Radar
		AM	Biggin Hill	Tower/Ground
134.8250	134.8250	AM	Brest	UACC
134.8500	134.8500	AM	Watton	Eastern Radar
134.8750	134.8750	AM	Brest	UACC
134.9000	134.9000	AM	West Drayton	Air Traffic Control
134.9750	134.9750	AM	Nationwide	CAA Flight Tests
135.0000	135.0000	AM	Nationwide	CAA Flight Tests
135.0500	135.0500	AM	West Drayton	London ATCC
135.0750	135.0750	AM	Watton	Eastern Radar
135.1500	135.1500	AM	West Drayton	London Mil
		AM	Maastricht	UAC

Base	Mobile	Mode	Location	User & Notes
135.2000	135.2000	AM	RAF Wattisham	Approach MATZ
135.2500	135.2500	AM	West Drayton	London ATCC
135.2750	135.2750	AM	Watton	Eastern Radar
135.3000	135.3000	AM	Paris	ACC/UACC
135.3750	135.3750	AM	West Drayton	London VOLMET Main
135.4250	135.4250	AM	West Drayton	London ATCC
135.4500	135.4500	AM	Maastricht	UAC
135.4750	135.4750	AM	Nationwide	CAA Flight Tests
135.5000	135.5000	AM	Reims	ACC/UACC
135.5250	135.5250	AM	West Drayton	ATCC
		AM	Shanwick	Oceanic ACC
135.5750	135.5750	AM	West Drayton	London TMA
		NFM	Space	US ATS-3 Satellite
135.6000	135.6000	AM	Shannon	ACC
		NFM	Space	US ATS-3 Satellite FM
135.6250	135.6250	AM	Watton	Eastern Radar
		NFM	Space	US ATS-3 Satellite FM
135.6500	135.6500	AM	Brest	UACC
135.6750	135.6750	AM	Prestwick	Scottish ATCC
135.7000	135.7000	AM	Fleetlands	Tower
		AM	RN Lee-On-Solent	Tower
135.7500	135.7500	AM	Nationwide	CAA Flight Tests
135.8000	135.8000	AM	Paris	ACC/UACC
135.8500	135.8500	AM	Prestwick	Scottish Air Traffic Control
135.9000	135.9000	AM	Paris	ACC/UACC
135.9250	135.9250	AM	Watton	Eastern Radar
135.9500	135.9500	AM	Blackpool	Approach
135.9750	135.9750	AM	Maastricht	UACC
136.8000	136.8000	AM	Manchester	Airtours Ops

136.0000 - 138.0000 MHz   Satellite Down Link Band

Base	Mobile	Mode	Location	User & Notes	
136.0500		NFM	Nongeostationary	Canada	Isis 1
136.0800		NFM	Nongeostationary	Canada	Isis 2
136.1000		NFM	Nongeostationary	NASA	Explorer 15
136.1100		NFM	Nongeostationary	NASA	Explorer 35
136.1110		NFM	Nongeostationary	NASA	Explorer 18
136.1120		NFM	Nongeostationary	France/US	Ayame 2
		NFM	Nongeostationary	Japan	MOS-1
136.1250		NFM	Nongeostationary	NASA	Explorer 28
136.1410		NFM	Nongeostationary	NASA	Explorer 34
136.1420		NFM	Nongeostationary	NASA	Explorer 21
136.1450		NFM	Nongeostationary	NASA	Explorer Series
136.1590		NFM	Nongeostationary	Japan	Ohsumi 1
136.1600		NFM	Nongeostationary	ESRO	Aurorae
136.1700		NFM	Nongeostationary	NASA	Explorer 42
		NFM	Nongeostationary	US	Echo 2
136.1710		NFM	Nongeostationary	NASA	Explorer 22
136.2000		NFM	Nongeostationary	US	Cameo 1
		NFM	Nongeostationary	US	Injun SR3

Base	Mobile	Mode	Location	User & Notes	
		NFM	Nongeostationary	US	Nimbus 2
136.2200		NFM	Nongeostationary	US	OAO 1
136.2300		NFM	Nongeostationary	US	ESSA 1
		NFM	Nongeostationary	US	SERT 2
136.2310		NFM	Nongeostationary	US	Tiros 9
136.2330		NFM	Nongeostationary	US	Tiros 8
136.2340		NFM	Nongeostationary	US	Tiros 7
136.2500		NFM	Nongeostationary	France	Castor
136.2600		NFM	Nongeostationary	US	ERS 20
		NFM	Nongeostationary	NASA	OV5-3
136.2730		NFM	Nongeostationary	NASA	Explorer Series
136.2750		NFM	Nongeostationary	NASA	Explorer 26
136.2900		NFM	Nongeostationary	NASA	Explorer 40
		NFM	Nongeostationary	NASA	Hawkeye
136.2930		NFM	Nongeostationary	NASA	Explorer 25
136.3000		NFM	Nongeostationary	NASA	SMS 1
136.3190		NFM	Nongeostationary	USAF	GGSE 1
136.3200		NFM	Nongeostationary	USAF	Ferret
136.3200		NFM	Nongeostationary	USAF	Ferret
		NFM	Nongeostationary	NASA	GEOS 3
136.3482		NFM	Nongeostationary	Australia	WRESAT 1
136.3500		NFM	Nongeostationary	France	EOLE 1
		NFM	Nongeostationary	France	FR 1
		NFM	Nongeostationary	USAF	SR 11B
136.3800		NFM	Nongeostationary	US	ERS 27
136.4100		NFM	Nongeostationary	ITSO	Intelsat
		NFM	Nongeostationary	Canada	Isis 1
136.4150		NFM	Nongeostationary	USAF	ERS 6
136.4300		NFM	Nongeostationary	India	Bhaskara
136.4400		NFM	Nongeostationary	USAF	ERS 15
136.4680		NFM	Nongeostationary	NASA	SYNCOM 2
136.5000		NFM	Nongeostationary	NASA	ATS Series
		NFM	Nongeostationary	NASA	Injun
		NFM	Nongeostationary	NASA	SR 3
		NFM	Nongeostationary	US	NOAA 10
136.5100		NFM	Nongeostationary	NASA	OVS 9
136.5210		NFM	Nongeostationary	US	SOLRAD 11B
136.5300		NFM	Nongeostationary	US	OV 5-9
		NFM	Nongeostationary	US	SOLRAD 11B
		NFM	Nongeostationary	US	Vela Hotel 8
136.5600		NFM	Nongeostationary	Germany	GRS-A
136.5630		NFM	Nongeostationary	US	RADSAT 43
136.5900		NFM	Nongeostationary	Canada	Alouette 1
		NFM	Nongeostationary	Canada	Isis 1 & 2
136.6100		NFM	Nongeostationary	ESA	Arian LO3
		NFM	Nongeostationary	ESA	CAT 1
136.6200		NFM	Nongeostationary	USAF	OV 5
		NFM	Nongeostationary	Italy	Sirio 1
136.6300		NFM	Nongeostationary	France	Signe 3

Base	Mobile	Mode	Location	User & Notes	
136.6500		NFM	Nongeostationary	USAF	OV 5-5
		NFM	Nongeostationary	US	TRAAC
		NFM	Nongeostationary	US	Transit 5B5
136.6510		NFM	Nongeostationary	USAF	SN-43
136.6780		NFM	Nongeostationary	US	SMS
136.6940		NFM	Nongeostationary	Japan	Shinsei
136.6950		NFM	Nongeostationary	Japan	Jiki'ken
136.7100		NFM	Nongeostationary	US	OSO-4
136.7120		NFM	Nongeostationary	US	OGO-2
136.7130		NFM	Nongeostationary	NASA	OSO-2
		NFM	Nongeostationary	Japan	Tansei
136.7250		NFM	Nongeostationary	Japan	CORSA B
136.7400		NFM	Nongeostationary	France	ERS-A
136.7680		NFM	Nongeostationary	ESA	ERS-17 ORS3
136.7700		NFM	Nongeostationary	US	NOAA 6
		NFM	Nongeostationary	US	NOAA 8
		NFM	Nongeostationary	US	NOAA 9
136.7710		NFM	Nongeostationary	USAF	ERS-13 TRS6
136.8000		AM	Nationwide	Kestrel Ops	
136.8010		NFM	Nongeostationary	USAF	SOLRAD 7B
136.8040		NFM	Nongeostationary	US	EGRS SECOR
136.8097		NFM	Nongeostationary	Japan	UME 1 & 2
136.8100		NFM	Nongeostationary	Japan	ETS-1 KIKU
136.8300		NFM	Nongeostationary	USAF	EGRS 8
		NFM	Nongeostationary	USAF	ERS 28
136.8400		NFM	Nongeostationary	USAF	EGRS 9
		NFM	Nongeostationary	USAF	TOPO 1
136.8600		NFM	Nongeostationary	USAF	Cannonball 2
		NFM	Nongeostationary	US	ERS 21
		NFM	Nongeostationary	NASA	IUE TETR 2
		NFM	Nongeostationary	USA	Landsat 2
		NFM	Nongeostationary	USAF	OV5-4
		NFM	Nongeostationary	NASA	RMS
136.8700		NFM	Nongeostationary	US	Injun 3
136.8870		NFM	Nongeostationary	USAF	SOLRAD 7A
136.8900		NFM	Nongeostationary	USAF	ERS 9 TRS4
		NFM	Nongeostationary	NASA	Explorer 47
		NFM	Nongeostationary	USAF	SOLRAD 6
136.8910		NFM	Nongeostationary	USAF	ERS 9
136.8920		NFM	Nongeostationary	USAF	ERS 5
136.9190		NFM	Nongeostationary	US	Tiros 9
136.9200		NFM	Nongeostationary	USAF	SERT 28
		NFM	Nongeostationary	USAF	OSO 8
136.9500		NFM	Nongeostationary	ESA	COS B1
137.0400		NFM	Nongeostationary	USAF	Ferret
137.0800		NFM	Nongeostationary	ESA	Meteorsat 1/2
137.1100		NFM	Nongeostationary	US	ATS 6
137.1400		NFM	Nongeostationary	ERS	ECS 2
137.1500		NFM	Nongeostationary	USSR	Meteor

Base	Mobile	Mode	Location	User & Notes	
137.1700		NFM	Nongeostationary	ERS	MARECS A
		NFM	Nongeostationary	France	MAROTS
137.1900		NFM	Nongeostationary	US	GEOS 3
137.2000		NFM	Nongeostationary	USSR	Meteor
137.2300		NFM	Nongeostationary	India/USSR	Bhaskara 2
		NFM	Nongeostationary	US	NOAA 61
137.2600		NFM	Nongeostationary	US	OAO-A2
137.3000		NFM	Nongeostationary	US	Meteor 2-17
		NFM	Nongeostationary	USSR	Meteor 2-18
		NFM	Nongeostationary	USSR	Meteor 2-5
		NFM	Nongeostationary	US	Meteor 3-2
		NFM	Nongeostationary	US	Timation 2
137.3800		NFM	Nongeostationary	USAF	OVS 3
137.4000		NFM	Nongeostationary	USSR	Meteor 2-16/17
		NFM	Nongeostationary	USAF	SMS-2
137.4100		NFM	Nongeostationary	USAF	Explorer 30
		NFM	Nongeostationary	USSR	Meteor 3-1
137.4200		NFM	Nongeostationary	India	Rohini
137.4400		NFM	Nongeostationary	India	Aryabhata
		NFM	Nongeostationary	India	Bhaskari 3
137.5000		NFM	Nongeostationary	USSR	Meteor 3-1
		NFM	Nongeostationary	US	NOAA 6
		NFM	Nongeostationary	US	NOAA 10
137.5600		NFM	Nongeostationary	UK	UK 6
137.5700		NFM	Nongeostationary	NASA	Explorer Series
137.6200		NFM	Nongeostationary	NASA	NOAA 11
		NFM	Nongeostationary	NASA	NOAA 9
137.6760		NFM	Nongeostationary	US	P76-5
137.7700		NFM	Nongeostationary	US	NOAA 9
137.8000		NFM	Nongeostationary	USAF	SOLRAD 11
137.8500		NFM	Nongeostationary	USSR	Intercosmos 18
		NFM	Nongeostationary	USSR	Meteor 2-15
		NFM	Nongeostationary	USSR	Meteor 2-16
		NFM	Nongeostationary	USSR	Meteor 2-19
		NFM	Nongeostationary	USSR	Meteor 3-3
137.8600		NFM	Nongeostationary	US	Landsat 2
137.8900		NFM	Nongeostationary	US	ANS-1
		NFM	Nongeostationary	NASA	RMS
137.9500		NFM	Nongeostationary	NASA	Explorer 45
		NFM	Nongeostationary	Canada	Isis
137.9800		NFM	Nongeostationary	NASA	Explorer 50
138.0000		NFM	Nongeostationary	USAF	Hilat 1

138.00625 - 140.49375 MHz   PMR VHF High Band 12.5 kHz Public Utiliy Repeaters & Police Heli-teli

Base	Mobile	Mode	Location	User & Notes
138.01250	105.01250	NFM	Reading	Bus Company
		NFM	Windermere	Bus Company
138.03125	105.00000	NFM	Norfolk	BR Train Link
138.04387	105.00000	NFM	Kent	BR Train Link

Base	Mobile	Mode	Location	User & Notes
		NFM	Beccles, Suffolk	BR Train Link
		NFM	Oulton Broad South	BR Train Link
138.06875	105.00000	NFM	Suffolk	BR Train Link
138.07500		NFM	Nationwide	Mercury Paging
138.09375	138.09375	NFM	Nationwide	Police Heli-Teli Chan 1
138.10000	138.10000	NFM	London	Met Police Helicopter
138.10500	138.10500	NFM	Great Lippits Hill	Police Heli-Teli Chan 40
138.10625	138.10625	NFM	Nationwide	Police Heli-Teli Chan 2
138.16125		NFM	Windermere	Data Link
138.17500		NFM	Nationwide	Mercury Personal Pagers
138.24375	105.24375	NFM	Suffolk	BR Train Link
		NFM	Brampton	BR Train Link
		NFM	Darsham	BR Train Link
		NFM	Halesworth	BR Train Link
		NFM	Saxmundham	BR Train Link
		NFM	Woodbridge	BR Train Link
138.25625		NFM	Perth	Data Link
138.29375	138.29375	NFM	Nationwide	Police Heli-Teli Chan 3
138.30625	138.30625	NFM	Nationwide	Police Heli-Teli Chan 4
138.32500	105.00000	NFM	Nationwide	British Rail RETD
		NFM	Woodbridge	East Suffolk Line RETB
138.33125		NFM	Tayside	Data Link
138.34375	105.34375	NFM	Anglia	BR Data Link
138.35625	105.35625	NFM	Nationwide	BR Radiophone
138.36875	105.36875	NFM	Anglia	Data Link
138.50000	138.50000	AM	RAF Alconbury	USAF Air-Ground
138.50000	105.50000	NFM	Nationwide	BR Radiophone
138.75625	105.75625	AM	Suffolk	Gas Board
		AM	Wales	Welsh Gas Channel 1
138.83125	105.83125	AM	Brecknock East	Welsh Gas Channel 8
		AM	Bury St Edmunds	Gas Board
		AM	Luton	Gas Board Channel 2
138.84375	105.84375	AM	Peterborough	Gas Board Channel 1
138.85625	105.85625	AM	Anglia	Eastern Gas
		AM	Anglia	Norfolk Gas
		AM	Wales	Welsh Gas Channel 7
138.86875	105.86875	AM	Norwich	Gas Board Channel 2
138.95000	138.95000	AM	RAF Alconbury	USAF Air-Ground
138.95625	105.95625	AM	Nationwide	PLC Plant Hire
138.96875	105.96875	AM	North Humberside	Gas Board
138.98000		AM	Aberdeen	Gas Board
138.98125	105.98125	AM	Peterborough	Gas Board Channel 2
		AM	Wales	Welsh Gas Channel 2
138.99375	105.99375	AM	Brecknock East	Welsh Gas Channel 9
		AM	Luton	Gas Board Channel 1
		AM	South Humberside	Gas Board
139.00625	106.00625	AM	Ipswich	Gas Board
		AM	East Midlands	Gas Board
		AM	Norwich	Gas Board Channel 1

Base	Mobile	Mode	Location	User & Notes
		AM	North Humberside	Gas Board
139.01875	106.01875	AM	Oxfordshire	Gas Board
139.03125	106.03125	AM	Bradford	North East Gas
		AM	Luton	Gas Board
		AM	Peterborough	Gas Board Channel 3
139.04375	106.04375	AM	Bradford	North East Gas Yellow Ch
139.05600		NFM	Nongeostationary	US ATS 6
139.05626	106.05626	AM	Bradford	North East Gas
139.06875	106.06875	NFM	Nationwide	Press Construction Ltd
		AM	Suffolk	Gas Board
139.08125	106.08125	AM	Anglia	Essex Gas
		AM	Radnor	Welsh Gas Channel 3
		AM	Reading	Gas Board
139.10625	106.10625	AM	Balshall	Gas Board
		AM	Grimsby	Gas Board
		AM	Ipswich	Gas Board
		AM	Kingsheath	Gas Board
139.11875	106.11875	AM	Oxford	Gas Board
139.12500	139.12500	AM	Nationwide	USAF 496 TFW Air-Air
		AM	Kent	Gas Board
139.13125	106.13125	AM	Bath	Gas Board
		AM	Brecon	South Wales Electric Ch 91
139.13750	106.13750	AM	London	Gas Board
		AM	Sussex	Gas Board
139.15000	106.15000	AM	Leicester	Gas Board
139.15500	106.15500	AM	Crewe	Gas Board
139.16250	106.16250	AM	Surrey	Gas Board
139.19500	106.19500	AM	South Yorkshire	Gas Board
139.25625	106.25625	AM	Clacton	Gas Board
139.26875	106.26875	AM	Midlands	Electric Company
139.27500	106.27500	AM	London	Gas Board
139.28750	106.28750	AM	London	Gas Board
139.31875	106.31875	AM	Wales	S Wales Electric Chan 106
139.36250	106.36250	AM	Surrey	Gas Board
139.36875	106.36875	AM	Woodbridge	Gas Board
139.39375	106.39375	AM	Anglia	Gas Board
139.50000	106.50000	AM	Guernsey	BBC Radio Guernsey OB
139.54375	106.54375	AM	Windsor	Electric Company
139.55000	106.55000	AM	Jersey	BBC Radio Jersey OB
139.55500	106.55500	AM	Nationwide	CEGB Line Faults
139.56250		NFM	Arnside	Voice Link
139.56250	106.56250	AM	Hampshire	Electricity Board
139.56250	139.56250	AM	Nationwide	Military Test Flights
139.56875		AM	Perth	Hydro Electric
139.57500	106.57500	AM	Jersey	BBC Radio Jersey OB
139.58125	106.58215	AM	Thames Valley	Electric Company
139.60000		NFM	Nationwide	Illegal Bugging Devices
139.62500		AM	W Perthshire	Hydro Electric Board
139.63125	106.63125	AM	Thames Valley	Electric Company

Base	Mobile	Mode	Location	User & Notes
139.63750		AM	Aberdeen	Hydro Electric
139.63750	106.63750	AM	Kent	Electricity Board
139.64375	106.64375	AM	Cambridge	Electric Company
139.65625	106.65625	AM	Norwich	Electric Company
139.66875	106.66875	AM	Clacton	Eastern Electricity
139.67500		AM	W Perthshire	Hydro Electric Board
139.67500	106.67500	AM	Surrey	Electricity Board
139.68000		AM	Aberdeen	Hydro Electric
139.68125	106.68125	AM	Norfolk	Electric Company
139.68200		AM	Perth	Hydro Electric
139.68500		AM	Aberdeen	Hydro Electric
139.68750	106.68750	AM	Sussex	Electricity Board
139.69375	106.69375	AM	Norwich	Electric Company
139.70000	106.70000	AM	Kent	Electricity Board
139.70625	106.70625	AM	Chelmsford	Electric Company
139.70625		AM	Dundee	Hydro Electric
139.71250		AM	Aberdeen	Hydro Electric
139.71625		AM	Perth	Hydro Electric
139.71875	106.71875	AM	Harlow	Electric Company
139.72500	106.72500	AM	Hampshire	Electricity Board
		AM	London	Electricity Board
139.73000	106.73000	AM	South Yorkshire	Electricity Board
139.73125	106.73126	AM	Kings Lynn	Electric Company
139.73750	106.73750	AM	Surrey	Electricity Board
139.74375	106.74375	AM	Yorkshire	Electric Company
139.75000	106.75000	NFM	Jersey	BBC Radio Jersey O/B
		AM	Hertfordshire	Electricity Board
139.76250		AM	W Perthshire	Hydro Electric Board
139.76250	106.76250	AM	Shepway	Electric Company
139.76875	106.76875	AM	North Humberside	Electric Company
139.78125	106.78125	AM	Reading	Electric Company
139.78500	106.78500	AM	Leicester	Electric Company
139.78750		AM	Aberdeen	Hydro Electric
139.78750	106.78750	AM	Loughborough	Electric Company
139.79375	106.79375	AM	Aldershot	Electric Company
		AM	Chilterns	Electric Company
		AM	Montgomery	MANWEB
		AM	Salisbury	Electric Company
139.80000		NFM	Nationwide	Illegal Bugging Devices
139.81250	106.81250	AM	Tunbridge Wells	Electric Company
139.81875	106.81875	AM	Chilterns	Electric Company
139.83125	106.83125	AM	North Humberside	Electric Company
139.83750	106.83750	AM	East Sussex	Electric Company
139.84375	106.84375	AM	Bury St Edmunds	Eastern Electricity
139.85000	106.85000	AM	Leicester	Electric Company
		AM	Kent	Electricity Board
139.86875	106.86875	AM	Kings Lynn	Electric Company
		AM	Perth	Hydro Electric
139.87250	139.87250	AM	Nationwide	USAF Air-Air
139.89375	106.89375	AM	Midlands	Electric Company

Base	Mobile	Mode	Location	User & Notes
139.93125	106.93125	AM	South Hertfordshire	Electric Company
139.94375	106.94375	AM	Midlands	Electric Company
139.98125		AM	Perthshire	Hydro Electric Data Link
140.00000	140.00000	AM	Nationwide	USAF Air-Air
140.00000		NFM	Nationwide	Illegal Bugging Devices
140.04400		AM	Perthshire	Data Link
140.05000	107.50500	AM	Nationwide	Mine Rescue Channel
140.05600		NFM	Nongeostationary	US ATS 6
140.05625		AM	Nationwide	NCB Mine Rescue
140.14375		AM	Nationwide	NCB Ambulance Channel
140.16875	140.10625	AM	Perthshire	Hydro Electric
140.18125	107.18125	AM	Nationwide	CEGB Line Faults
140.18250	107.18125	AM	Nationwide	CEGB Line Faults
140.18750	107.18750	AM	Nationwide	CEGB Line Faults
140.19375	107.19375	AM	Nationwide	CEGB Line Faults
140.20000	107.20000	AM	Nationwide	CEGB Line Faults
140.20625	107.20625	AM	Nationwide	CEGB Line Faults
140.21875	107.21875	AM	Clacton	CEGB Repairs Depot
140.23125	107.23125	AM	Ipswich	CEGB Repairs Depot
140.24375	107.24375	AM	Bury St Edmunds	CEGB Repairs Depot
140.25625	107.25625	AM	Ipswich	Gas Board Repairs
140.29500		AM	Gt Manchester	Gt Manchester Buses
140.35000		AM	Midlands	Bus Company
140.35625	140.10625	AM	Perthshire	Hydro Electric
140.38000		AM	Gt Manchester	Gt Manchester Buses
140.40000	148.80000	AM	Croydon	Gas Board
140.40625	107.40625	AM	Colchester	CEGB Repairs Depot
		AM	South Yorkshire	Greenland Bus Company
140.45000		AM	Midlands	Bus Company

140.45625 - 140.99500 MHz   ITN Live Talkback, DTI 28 Day Hire, Bus Companies & CEGB Engineering

Base	Mobile	Mode	Location	User & Notes
140.45625	107.45625	AM	Ipswich	CEGB Engineering
		NFM	Wolverhampton	Bus Company
140.46875	107.46875	AM	Ipswich	CEGB Repairs Depot
140.46875	107.46875	NFM	Gt Manchester	Gt Manchester Buses
140.48750	107.48750	NFM	London	London Transport
140.50000	107.50000	NFM	London	London Transport
140.53125	107.53125	NFM	Gt Manchester	Gt Manchester Buses
140.54375	107.54375	NFM	Gt Manchester	Gt Manchester Buses
140.63125	107.63125	NFM	Newcastle	City Transport Channel 1
140.70625	107.70625	NFM	South Yorkshire	Bus Company
140.71875	107.71875	NFM	Newcastle	City Transport Channel 2
140.73750	107.73750	NFM	London	London Transport
140.76250	107.76250	NFM	London	London Transport
140.85625	107.85625	NFM	South Yorkshire	Bus Company
140.94375	107.94375	NFM	Brighton & Hove	Bus Company
		NFM	Nationwide	LWT Engineering Talkback
		NFM	Nationwide	NCT Bus Channel
140.96875	107.96875	NFM	Nationwide	DTI Short Term 28 day Hire

Base	Mobile	Mode	Location	User & Notes
140.99375	107.99375	NFM	London	ITN 6 O'Clock News
140.99500	107.99500	NFM	London	ITN London Weekend TV

141.0000 - 141.2000 MHz   ILR, BBC and Local Radio Talkback

Base	Mobile	Mode	Location	User & Notes
141.01250		NFM	Nationwide	ILR Talkback Channel 1
141.01875		NFM	Inverness	Moray Firth Radio
141.02500		NFM	Nationwide	ILR Common Talkback Ch 5
141.03125		NFM	Stoke-on-Trent	Signal Radio
141.03750		NFM	Nationwide	ILR Primary O/B Channel 2
		NFM	Nationwide	ILR Engineering Ch 4
141.04375		NFM	Wolverhampton	Beacon Radio
141.0500	141.1875	NFM	Andover	ILR
		NFM	Basingstoke	ILR
		NFM	Bristol	ILR
		NFM	Chelmsford	ILR
		NFM	Cornwall	ILR
		NFM	Hereward	ILR
		NFM	Inverness	ILR
		NFM	Liverpool	ILR
		NFM	Newcastle	ILR
		NFM	Oxford	ILR
		NFM	Peterborough	ILR
		NFM	Reigate	ILR
		NFM	Sheffield	ILR
		NFM	Wolverhampton	ILR
141.0560		NFM	Nongeostationary	US ATS 6
141.05625		NFM	Preston	Red Rose Radio
141.0625	141.1125	NFM	Ayr	ILR
		NFM	Exeter	ILR
		NFM	Gloucester	ILR
		NFM	Great Yarmouth	ILR
		NFM	Hereford	ILR
		NFM	Huddersfield	ILR
		NFM	Leicester	ILR
		NFM	Maidstone	ILR
		NFM	Nationwide	2CR Eye in the Sky
		NFM	NW Wales	ILR
		NFM	Reading	ILR
		NFM	Shrewsbury	ILR
141.06875		NFM	Liverpool	Radio City
141.0750	141.1375	NFM	Aberdeen	ILR
		NFM	Barnsley	ILR
		NFM	Berwick	ILR
		NFM	Cardiff	ILR
		NFM	Coventry	ILR
		NFM	Glasgow	ILR
		NFM	London	ILR
		NFM	Portsmouth	ILR
		NFM	Stoke	ILR
		NFM	Swindon	ILR

Base	Mobile	Mode	Location	User & Notes
141.08125		NFM	Manchester	Piccadilly Radio
141.0875	141.1500	NFM	Bedford	ILR
		NFM	Bournemouth	ILR
		NFM	Eastbourne	ILR
		NFM	Edinburgh	ILR
		NFM	Guildford	ILR
		NFM	Hereford	ILR
		NFM	Humberside	ILR
		NFM	Ipswich	ILR
		NFM	Manchester	ILR
		NFM	Nottingham	ILR
		NFM	Plymouth	ILR
		NFM	Swansea	ILR
		NFM	Trent	ILR
		NFM	Whitehaven	ILR
141.1000	141.2000	NFM	Bradford	ILR
		NFM	Brighton	ILR
		NFM	Bury St Edmunds	ILR
		NFM	Derby	ILR
		NFM	Dorchester	ILR
		NFM	Dumfries	ILR
		NFM	Gwent	ILR
		NFM	London	Capitol Radio Link
		NFM	London	ILR
		NFM	Londonderry	ILR
		NFM	Milton Keynes	ILR
		NFM	Newport	ILR
		NFM	Northampton	ILR
		NFM	Stranraer	ILR
		NFM	Weymouth	ILR
		NFM	Wrexham	ILR
141.1250	141.1750	NFM	Aylesbury	ILR
		NFM	Belfast	ILR
		NFM	Birmingham	ILR
		NFM	Blackpool	ILR
		NFM	Cambridge	ILR
		NFM	Canterbury	ILR
		NFM	Dover	ILR
		NFM	Dundee	ILR
		NFM	Leeds	ILR
		NFM	Middlesborough	ILR
		NFM	Newmarket	ILR
		NFM	Perth	ILR
		NFM	Preston	ILR
		NFM	Southampton	ILR
		NFM	Taunton	BBC
		NFM	Yeovil	ILR
141.13750		NFM	Nationwide	IBA Local Radio Engineers
141.15000		NFM	Nationwide	IBA Local Radio Engineers
		NFM	Trent	ILR Talkback

Base	Mobile	Mode	Location	User & Notes
141.16875		NFM	Coventry	Mercia Sound
141.19375		NFM	Birmingham	BRMB/Xtra AM
141.19500		NFM	London	Radio Piccadilly Studio Link
141.20625		NFM	Birmingham	BBC Radio West Midlands
		NFM	Northamptonshire	BBC Radio Northants
141.21875		NFM	Hereford/Worcester	BBC Radio Worcester
141.23125		NFM	Surrey	BBC Radio Surrey
141.24375	224.16875	NFM	Coventry	BBC CWR
141.25000		WFM	London	ITN Music Link
141.25500		NFM	Stoke	Radio Stoke Engineering
141.25625	224.10625	NFM	Stoke-on-Trent	BBC Radio Stoke
141.29375	224.15625	NFM	Shrewsbury	BBC Radio Shrewsbury
141.30000	446.93750	NFM	Leicester	BBC Radio Leicester O/B
141.30625	224.13125	NFM	Leicester	BBC Radio Leicester
141.32000		NFM	Nationwide	BBC O/B Link
141.35000		NFM	Nationwide	BBC Radio 2 Engineering
		WFM	London	ITN Music Links
141.37500		NFM	London	BBC1 TV Studio Sound Link
141.44500		NFM	London	BBC2 TV Studio Sound Link
141.45000		WFM	London	ITN Music Link
141.46250		NFM	London	BBC1 Link
141.55000		NFM	Nationwide	BBC Radio 2 O/B
141.61875		NFM	Trent	BBC Radio Trent O/B
141.64375		NFM	Nationwide	BBC Outside Broadcasts
141.66875		NFM	Nationwide	BBC Engineering Talkback
141.67000		NFM	Nationwide	TV News ENG
141.68125		NFM	Lincolnshire	BBC TV O/B
141.69375		NFM	Nationwide	GLR Talkback
141.72500		NFM	Sheffield	BBC Radio Sheffield O/B
141.74375		NFM	Nationwide	BBC Radio Cue
141.75500		NFM	Stoke	Radio Stoke Engineering
141.76875		NFM	Nationwide	BBC Radio Cue
141.77500		NFM	Sheffield	BBC Radio Sheffield O/B
141.78750		NFM	Manchester	Radio Manchester O/B
141.79500		NFM	Manchester	Radio Manchester O/B
141.80000		NFM	Lincolnshire	BBC TV O/B
141.81875		NFM	Nationwide	BBC News ENG
141.83875		NFM	Nationwide	BBC TV O/B's Data
141.85750		NFM	Nationwide	BBC Transmitter Group
141.87500	141.87500	AM	Nationwide	BBC TV O/B's Data
141.88750	141.88750	AM	Nationwide	BBC TV Air-Ground
141.89250	141.89250	AM	Nationwide	BBC TFS Air-Ground

141.9000 - 142.0000 MHz   Government Agencies NFM

142.0000 - 142.9750 MHz   MoD, USAF & Soviet Space Communications

Base	Mobile	Mode	Location	User & Notes
142.2950	142.2950	AM	RAF Coltishall	Approach
		AM	RAF Wittering	Approach
142.4000		NFM	Space	Soviet Mir Space Station
142.4125		NFM	Nationwide	Mould

Base	Mobile	Mode	Location	User & Notes
142.4170		NFM	Space	Soviet Mir Space Station
142.4200		NFM	Space	Soviet Mir Space Station
142.4200		NFM	Space	Soviet Mir Space Station
142.5000		NFM	London	Rimington Minicabs
142.6000		NFM	Space	Soviet Mir Space Station
142.6750		NFM	Nationwide	MoD Paging
142.7280	142.7250	AM	Nationwide	USAF Air to Air
142.8250	142.8250	AM	Nationwide	USAF Air to Air
142.8500	142.8500	AM	RAF Mildenhall	Command Post
142.8975		NFM	Nationwide	Mould

143.0000 - 144.0000 MHz Metropolitan And SW Scottish Police Mobiles
12.5 kHz (Repeater/Base Duplex + 9MHz)
Soviet Space Communications

Base	Mobile	Mode	Location	User & Notes
143.0125	143.0125	NFM	Strathclyde Police	Channel 32
143.0750	143.0750	NFM	Strathclyde Police	Channel 31
143.1125	143.1125	NFM	Strathclyde Police	Channel 33
143.1440		NFM	Nongeostationary	Soviet Voice Channel
143.1500	143.1500	NFM	Strathclyde Police	Channel 37
143.2125	143.2125	NFM	Strathclyde Police	Channel 39
143.3000		AM	RAF Sculthorpe	USAF
143.3000	143.3000	NFM	Strathclyde Police	Channel 34
143.3500	143.3500	NFM	Strathclyde Police	Channel 35
143.4250	143.4250	NFM	Strathclyde Police	Channel 36
143.4500		AM	RAF West Drayton	London Radar
143.5500	143.5500	NFM	Strathclyde Police	Channel 41
143.5625	143.5625	NFM	Strathclyde Police	Channel 43
143.6000		AM	Nationwide	USAF Air-Air
143.6125	143.6125	NFM	Strathclyde Police	Channel 38
143.6250		NFM	Nongeostationary	Soviet Voice Channel (Myr)
143.6250	143.6250	NFM	Strathclyde Police	Channel 42
143.6375	143.6375	NFM	Strathclyde Police	Channel 40
143.8250		NFM	Nongeostationary	Soviet Military Coded Ch

144.0000 - 146.0000 MHz 2m Amateur Radio

Base	Mobile	Mode	Location	User & Notes
144.0500	144.0500	CW	Nationwide	CW Calling Frequency
144.2600	144.2600	NFM	Nationwide	Raynet
144.3000	144.3000	SSB	Nationwide	SSB Calling Frequency
144.5000	144.5000	CW	Nongeostationary	OSCAR 5 Telemetry Bcn
		NFM	Nationwide	SSTV Calling
144.6000	144.6000	NFM	Nationwide	RTTY Calling
144.6750	144.6750	NFM	Nationwide	Data Calling
144.7000	144.7000	NFM	Nationwide	FAX Calling
144.7500	144.7500	NFM	Nationwide	ATV Calling
144.7750	144.7750	NFM	Nationwide	Raynet
144.8000	144.8000	NFM	Nationwide	Raynet
144.8250	144.8250	NFM	Nationwide	Raynet
144.9125		CW	St Austell	GB3MCB Beacon
144.9175		CW	Portlaw	EI2WRB Beacon
144.9250		CW	Wrotham	GB3VHF Beacon

Base	Mobile	Mode	Location	User & Notes
144.9625		CW	Lerwick	GB3LER Beacon
144.9750		CW	Dundee	GB3ANG Beacon
144.9830	144.9830	CW	Nongeostationary	OSCAR 1 & 2 Beacon
145.2000	145.2000	NFM	Nationwide	Raynet S08
145.2250	145.2250	NFM	Nationwide	Raynet S09
145.2500	145.2500	NFM	Nationwide	Raynet S10
145.2750	145.2750	NFM	Nationwide	Raynet S11
145.3000	145.3000	NFM	Nationwide	Raynet S12
145.3250	144.7250	NFM	Caen Repeater	FZ2VHB
145.3250	145.3250	NFM	Nationwide	Raynet S13
145.3500	145.3500	NFM	Nationwide	Raynet S14
145.3750	145.3750	NFM	Nationwide	Raynet S15
145.4000	145.4000	NFM	Nationwide	Raynet S16
145.4250	145.4250	NFM	Nationwide	Raynet S17
145.4500	145.4500	NFM	Nationwide	Raynet S18
145.4750	145.4750	NFM	Nationwide	Raynet S19
145.5000	145.5000	NFM	Nationwide	Raynet S20
145.5250	145.5250	NFM	Nationwide	Raynet S21
145.5500	145.5500	NFM	Nationwide	Raynet S22
145.5750	145.5750	NFM	Nationwide	Raynet S23
145.6000	145.0000	NFM		Raynet Channel R0
		NFM	FZ3VHF	St. Brieuc
		NFM	GB3AS	Calbeck, Cumbria
		NFM	GB3CF	Charnwood Forest, Leics
		NFM	GB3EL	London
		NFM	GB3FF	Burntisland, Fife
		NFM	GB3LY	Limavady, Co Londonderry
		NFM	GB3MB	Bury, Manchester
		NFM	GB3SR	Brighton
		NFM	GB3SS	Knockmore, Elgin
		NFM	GN3WR	Mendip Hills, Somerset
145.6250	145.0250	NFM		Raynet Channel R1
		NFM	GB3GD	Shaefell, Isle of Man
		NFM	GB3HG	Bilsdale, Yorkshire
		NFM	GB3KS	Dover, Kent
		NFM	GB3MH	Malvern Hills, Worcs
		NFM	GB3NB	Wymondham, Norfolk
		NFM	GB3NG	Peterhead
		NFM	GB3PA	Gleniffer Braes, Paisley
		NFM	GB3SC	Bournemouth, Dorset
		NFM	GB3SI	St Ives, Cornwall
		NFM	GB3WL	Hilingdon, W London
145.6500	145.0500	NFM		Raynet Channel R2
		NFM	GB3AY	Patna, Ayrshire
		NFM	GB3BX	North Birmingham
		NFM	GB3GJ	Jersey
		NFM	GB3HS	Little Weighton, Humber
		NFM	GB3MN	Stockport, Cheshire
		NFM	GB3OC	Kirkwall, Orkney
		NFM	GB3SB	Duns, Berwickshire

Base	Mobile	Mode	Location	User & Notes
		NFM	GB3SL	Crystal Palace, London
		NFM	GB3TR	Torquay, Devon
		NFM	GB3WH	Swindon, Wiltshire
145.6750	145.0750	NFM		Raynet Channel R3
		NFM	GB3BM	Birmingham
		NFM	GB3DR	Dorset
		NFM	GB3ES	Hastings, E Sussex
		NFM	GB3LD	Morecambe Bay, Lancs
		NFM	GB3LU	Lerwick, Shetland Isl
		NFM	GB3NA	Barnsley, Yorks
		NFM	GB3PE	Peterborough
		NFM	GB3PO	Martlesham Heath
		NFM	GB3PR	Perth, Tayside
		NFM	GB3RD	Reading, Berks
		NFM	GB3SA	Swansea
145.7000	145.1000	NFM		Raynet Channel R4
		NFM	FZ3THF	Northern France
		NFM	GB3AR	Arfon, Gwynedd
		NFM	GB3BB	Brecon, Powys
		NFM	GB3BT	Berwick-on-Tweed
		NFM	GB3EV	Appleby, Cumbria
		NFM	GB3HH	Buxton, Derby
		NFM	GB3HI	Torosay, Isle of Mull
		NFM	GB3KN	Maidstone, Kent
		NFM	GB3VA	Aylesbury, Bucks
		NFM	GB3WD	Dartmoor, Devon
145.7250	145.1250	NFM		Raynet Channel R5
		NFM	FZ2VHT	Le Havre
		NFM	GB3BI	Mounteagle, Inverness
		NFM	GB3DA	Danbury, Essex
		NFM	GB3LM	Lincoln
		NFM	GB3NC	St Austell, Cornwall
		NFM	GB3NI	Belfast
		NFM	GB3SN	Fourmarks, Hants
		NFM	GB3TP	Keighley, W Yorkshire
		NFM	GB3TW	Burnhope, Tyne & Wear
145.7500	145.1500	NFM		Raynet Channel R6
		NFM	GB3AM	S Birmingham
		NFM	GB3BC	Mynydd Machen, Gwent
		NFM	GB3CS	Blackhill, Central Scotland
		NFM	GB3IB	Aviemore, Highlands
		NFM	GB3MP	Moel-Y-Parc, Clwyd
		NFM	GB3PI	Barkway, Herts
		NFM	GB3TY	Hexham, Northumberland
		NFM	GB3WS	Horsham, W Sussex
145.7750	145.1750	NFM		Raynet Channel R7
		NFM	FZ3VHF	Northern France
		NFM	GB3FR	Old Bolingbroke, Lincs
		NFM	GB3DG	Dumfries & Galloway
		NFM	GB3GN	Durris, Aberdeen

Base	Mobile	Mode	Location		User & Notes	
		NFM	GB3NL		North London	
		NFM	GB3PC		Portsmouth	
		NFM	GB3PW		Newton, Powys	
		NFM	GB3RF		Blackburn, Lancs	
		NFM	GB3SR		Worthing, W Sussex	
		NFM	GB3WK		Leamington Spa	
		NFM	GB3WT		West Tyrone, N Ireland	
		NFM	GB3WW		Crosshands, Dyfed	
145.8100		CW	Nongeostationary		OSCAR 10 Beacon	
145.8175		CW	Nongeostationary		OSCAR 21 Beacon	
145.8250		CW	Nongeostationary		OSCAR 9 & 11 Telemetry	
145.9500		CW	Nongeostationary		OSCAR Beacons	
145.9750		CW	Nongeostationary		OSCAR 7 Telemetry Bcn	
145.9870		NFM	Nongeostationary		OSCAR 10 Engineering	
145.9875		NFM	Nongeostationary		OSCAR 21 Calling Channel	

146.0000 - 148.0000 MHz Government and Police (Repeaters + 8.0 MHz)

Base	Mobile	Mode	Location		User & Notes	
146.0250		NFM	Surrey	HF	Fire	
146.0250		NFM	Sussex	M2KB	Police Link	
146.1875		NFM	Sussex	M2KB	Police Link	
146.3250		NFM	Sussex	M2KB	Police Link	
146.3875		NFM	Sussex		Mobile Ch 4	
146.4125		NFM	Kent		Fire Brigade	
146.4750		NFM	Sussex	M2KB	Police Link	
146.6450		NFM	Hutton		Police Link	
146.7375		NFM	Hungerford	HU		
146.8250		NFM	Sussex	M2KD	Police Link	
146.8500		NFM	Birkenhead (M53)			
146.8626		NFM	London		Rimington Minicabs	
146.8750		NFM	West Hoathly	M2KB	Police Link	
146.9500		NFM	Sussex	M2KB	Police Link	
147.0125		NFM	Kingston	VK		
147.0600		NFM	Sussex	M2KB	Police Link	
147.0625		NFM	Sussex	M2KB	Police Link	
147.1000		NFM	Sussex	M2KB	Police Link	
147.1250		NFM	Sussex	M2KB	Police Link	
147.1500		NFM	Barkingside	JB		
147.2000		NFM	London		Ranger Diplo. Protect. Ch 25	
147.2125		NFM	Bethnal Green	HB	Ch 26	
		NFM	Brick Lane	HR		
		NFM	Croydon	ZD		
		NFM	Kenley	ZK		
		NFM	Notting Hill	BZ		
147.2250		NFM	Banstead	ZB	Ch 27	
		NFM	Brockley	PK		
		NFM	Deptford	PP		
		NFM	Epsom	ZP		
		NFM	Fulham	FF		
		NFM	Hainault	JT		
		NFM	Leman Street	HD		

Base	Mobile	Mode	Location	User & Notes	
		NFM	Lewisham	PL	
		NFM	Loughton	JO	
		NFM	Stoneleigh	ZS	
		NFM	Sutton	ZT	
		NFM	Wallington	ZW	
		NFM	Woodford	JF	
147.2375		NFM	Daiston	GA	Ch 28
		NFM	East Dulwich	ME	
		NFM	Stoke Newington	GN	
		NFM	Peckham	MM	
		NFM	Shepherd's Bush	FS	
147.2500		NFM	Barnet	SA	Ch 29
		NFM	Boreham Wood	SD	
		NFM	Bushey	SU	
		NFM	Hackney	GH	
		NFM	Norbury	ZD	
		NFM	North Addington	ZA	
		NFM	Potter's Bar	SP	
		NFM	Radlett	SE	
		NFM	Rochester	AR	
		NFM	Shenley	SY	
		NFM	South Norwood	ZS	
		NFM	South Mimms	SM	
		NFM	Whetstone	ST	
147.2625		NFM	Bromley	PR	Ch 30
		NFM	Chingford	JC	
		NFM	Hampstead	EH	
		NFM	Penge	PG	
		NFM	Vine Street	CV	
		NFM	Waltham Abbey	JA	
		NFM	Walthamstow	JW	
		NFM	West End Central	CD	
		NFM	West Hampstead	EW	
147.2750		NFM	Catford	PD	Ch 31
		NFM	Forest Gate	KF	
		NFM	Lee Road	PE	
		NFM	Sydenham	PS	
		NFM	West Ham	KW	
		NFM	Westminster	AP	
147.2875		NFM	Earlsfield	WF	Ch 32
		NFM	Eltham	RD	
		NFM	Feltham	TF	
		NFM	Greenwich	RG	
		NFM	Harrow Road	DR	
		NFM	Hounslow	TD	
		NFM	Tooting	WD	
		NFM	Westcombe Park	RK	
147.3000		NFM	Hampton	TM	Ch 33
		NFM	Harlesden	QH	
		NFM	Kennington	LK	

Base	Mobile	Mode	Location	User & Notes	
		NFM	Kilburn	OK	
		NFM	Staines	TG	
		NFM	St Annes Road	YA	
		NFM	Sunbury	TY	
		NFM	Teddington	TT	
		NFM	Tottenham	YT	
		NFM	Twickenham	TW	
		NFM	Willesden Green	QL	
		NFM	Wood Green	YD	
147.3125		NFM	Belvedere	RB	Ch 34
		NFM	Chobham	VC	
		NFM	East Molesey	VE	
		NFM	Esher	VH	
		NFM	Gerard Road	AP	
		NFM	Kingsbury	QY	
		NFM	Kingston	VK	
		NFM	New Malden	VN	
		NFM	Plumstead	RP	
		NFM	Shooters Hill	RH	
		NFM	Surbiton	VS	
		NFM	Thamesmead	RT	
		NFM	Wembley	QD	
		NFM	Woolwich	RW	
147.3250		NFM	Cheshunt	YC	Ch 35
		NFM	Collier Row	KL	
		NFM	Enfield	YF	
		NFM	Goffs Oak	YG	
		NFM	Harold Hill	KA	
		NFM	Hornchurch	KC	
		NFM	Ponders End	YP	
		NFM	Putney	WP	
		NFM	Rainham	KM	
		NFM	Romford	KD	
		NFM	Upminster	KU	
		NFM	Wandsworth	WW	
147.3375		NFM	London	GT	Ch 36
147.3500		NFM	Harefield	XF	Ch 37
		NFM	Kensington	BD	
		NFM	Mitcham	VM	
		NFM	Northwood	XN	
		NFM	Notting Hill	BH	
		NFM	Ruislip	XB	
		NFM	Uxbridge	XU	
		NFM	Wimbledon	VW	
147.3625		NFM	Barking	KB	Ch 38
		NFM	Cavendish	LC	
		NFM	Clapham	LM	
		NFM	Dagenham	KG	
		NFM	Finchley	SF	
		NFM	Golders Green	SG	

Base	Mobile	Mode	Location	User & Notes	
		NFM	Leyton	JL	
147.3750		NFM	Arbour Square	HA	Ch 39
		NFM	Cannon Row	AD	
		NFM	City Road	GD	
		NFM	East Ham	KE	
		NFM	Hendon	SN	
		NFM	Hyde Park	AH	
		NFM	Mill Hill	SH	
		NFM	North Woolwich	KN	
		NFM	Plaistow	KO	
		NFM	Wellington	AW	
		NFM	West Drayton	XW	
		NFM	West Hendon	SG	
147.3875		NFM	Bexley Heath	RY	Ch 40
		NFM	Brixton	LD	
		NFM	Erith	RE	
		NFM	Hornsey	NR	
		NFM	Marylebone Lane	PM	
		NFM	Sidcup	RS	
		NFM	Welling	RL	
147.4000		NFM	Barnes	TN	Ch 41
		NFM	Gipsy Hill	LG	
		NFM	Holborn	EO	
		NFM	Richmond	TR	
		NFM	Tottenham Court Rd	ET	
147.4125		NFM	Battersea	WA	Ch 42
		NFM	Caledonian Road	NC	
		NFM	Chadwell Heath	JH	
		NFM	Chigwell	JD	
		NFM	Ilford	JI	
		NFM	Islington	NI	
		NFM	Kings Cross	ND	
		NFM	Lavender Hill	WL	
		NFM	New Southgate	YN	
		NFM	Nine Elms	WN	
		NFM	Wanstead	JN	
147.4250		NFM	Bow Street	CB	Ch 43
		NFM	Edmonton	YE	
		NFM	Southgate	YS	
		NFM	Winchmore Hill	YW	
147.4375		NFM	Bow	HW	Ch 44
		NFM	Chelsea	BC	
		NFM	Ealing	XD	
		NFM	Isle of Dogs	HI	
		NFM	Limehouse	HH	
		NFM	Norwood Green	XW	
		NFM	Southall	XS	
147.4500		NFM	Brentford	TB	Ch 45
		NFM	Chiswick	TC	
		NFM	Paddington	DD	

Base	Mobile	Mode	Location	User & Notes	
		NFM	Rotherhithe	MR	
		NFM	Southwark	MD	
		NFM	St Johns Wood	DS	
		NFM	Tower Bridge	MT	
147.4750		NFM	Albany Street	ED	Ch 46
		NFM	Camberwell	MC	
		NFM	Carter Street	MS	
		NFM	Edgware	QE	
		NFM	Hammersmith	FD	
		NFM	Harrow	QA	
		NFM	Kentish Town	EK	
		NFM	Pinner	QP	
		NFM	Wealdstone	QW	
147.4875		NFM	London	GT	Ch 47
147.5000		NFM	London	GT	Ch 48
147.5125		NFM	London		Ch 50
147.5375		NFM	Streatham	LS	Ch 52
147.6500		NFM	Sussex	M2KB	Police Link
147.6625	147.6625	NFM	Lancashire		Car-Car
147.6750		NFM	Leman Street	HD	
147.7750	147.7750	NFM	London Met	MP	Channel 18
147.8000		NFM	Nationwide		Fire Alert
147.8125	147.8125	NFM	Strathclyde Police		Channel 45
147.8500		NFM	Dartford Tunnel		
147.8500	147.8500	NFM	Scottish Fire Brigades		Channel 51
		NFM	Newcastle (MoD Police)		
147.8750	147.8750	NFM	England & Wales Mobiles		Channel 21
147.8750	147.8750	NFM	London Met	MP	Channel 16
147.8750		NFM	Tilbury	FD	
147.9000		NFM	English Fire Brigade Pagers		
147.9000	147.9000	NFM	Scottish Fire Brigades		Channel 49
147.9125	147.9125	NFM	England & Wales Mobiles		Channel 22
147.9125	147.9125	NFM	London Met	MP	Channel 17
147.9250	147.9250	NFM	Scottish Fire Brigades		Channel 50
147.9375	147.9375	NFM	London Met	MP	Channel 19
147.9500		NFM	London		CID
147.9750		NFM	London		CID
148.2500		NFM	RAF Sculthorpe		Approach
148.2970		NFM	Sussex	M2KB	Police Link
148.4000		NFM	RAF Lakenheath		Approach
148.8250		NFM	Fire Brigade Pagers		

149.0000 - 149.9000 MHz   Government & MoD Mould Repeaters

Base	Mobile	Mode	Location	User & Notes	
149.2375		NFM	Nationwide	MoD Establishments	
149.3625		NFM	Nationwide	MoD Establishments	
149.2875		NFM	East Scotland	Mould	
149.3875		NFM	Nationwide	MoD Establishments	
149.4000	149.4000	NFM	Nationwide	Air Training Corps	
149.4125		NFM	Nationwide	MoD Establishments	
149.6500	149.6500	NFM	Nationwide	USAF Air-Air	

Base	Mobile	Mode	Location		User & Notes
149.7375		NFM	US Navy Edzell		Base Security
149.7375		NFM	East Scotland		Mould
149.7625		NFM	Nationwide		MoD Establishments
149.7750		NFM	Nationwide		MoD Regional Police Ch 1
149.7750	149.7750	NFM	Nationwide		USAF Air-Air
149.8250		NFM	London		MoD Police Channel 2
149.8500	149.8500	NFM	Nationwide		Common Base Channel
149.8500		NFM	Nationwide		MoD Police Channel 3
149.8875		NFM	East Scotland		Mould
149.9000	149.0000	NFM	Nationwide		Air Training Corps Ch 2

149.9500 - 150.0500 MHz   Soviet Satellite Beacons & Telemetry

Base	Mobile	Mode	Location		User & Notes
149.9700		NFM	Space		Polar Bear 8688A
149.9800		NFM	Space		Soviet Cosmos Satellites
150.1100		NFM	Nationwide		Oil Slick Markers
150.1850		NFM	Nationwide		Oil Slick Markers

152.0000 - 152.9875 MHz   Police PMR (England & Wales)   12.5 kHz

Base	Mobile	Mode	Location		User & Notes
152.0000	143.2625	AM	London	MP	Channel 11
152.0125		AM	Isle of Man		Channel 1
152.0125		NFM	Strathclyde		Data Channel
152.0250	143.0625	AM	London	MP	Channel 3
152.0250	143.2625	NFM	Strathclyde		Channel 49
152.0375	143.0750	AM	Leicester	NL	Channel 1
152.0500	143.1125	AM	London	MP	Channel 5
152.0625	143.3125	AM	North Yorkshire	XN	Channel 2
152.0750	143.4375	NFM	Strathclyde		Channel 32
152.0750		NFM	Strathclyde		Data Channel
152.0875	143.0750	AM	Northumbria	LB	Channel 1
152.1000	143.4375	NFM	Strathclyde		Channel 31
152.1000		AM	London	MP	
152.1000		NFM	Strathclyde		Data Channel
152.1250	143.1250	NFM	Strathclyde		Channel 35
152.1250	143.2000	NFM	Strathclyde		Channel 13
152.1250		NFM	Strathclyde		Data Channel
152.1375	143.3125	AM	London	MP	Channel 13
152.1500	143.2000	NFM	Strathclyde		Channel 11
152.1500	143.3625	AM	North Yorkshire	XN	Channel 4
152.1625	143.0625	NFM	Strathclyde Fire Brigade		Channel 8
152.1625	143.2875	NFM	Strathclyde		Channel 46
152.1625	143.3375	NFM	Strathclyde		Channel 28
152.1625	143.3375	NFM	Scottish Fire Brigades		Channel 28
152.1625	143.3375	NFM	Strathclyde Fire Brigade		Channel 3
152.1625	143.3750	NFM	Strathclyde		Channel 41
152.1625	143.3750	NFM	Scottish Fire Brigades		Channel 41
152.1625	143.3750	NFM	Strathclyde Fire Brigade		Channel 5
152.1625	143.5250	NFM	Strathclyde Fire Brigade		Channel 10
152.1750	143.2000	NFM	Strathclyde		Channel 12
152.1750	143.1000	AM	Leicester	NL	Channel 2
152.1875	143.1000	AM	Northumbria	LB	Channel 2

Base	Mobile	Mode	Location	User & Notes	
152.1875	143.0375	NFM	Strathclyde		Channel 22
152.1875	143.0375	NFM	Scottish Fire Brigades		Channel 22
152.1875	143.0375	NFM	Strathclyde Fire Brigade		Channel 2
152.1875	143.1000	NFM	Dumfries & Galloway Fire		Channel 15
152.1875	143.1375	NFM	Strathclyde Fire Brigade		Channel 11
152.1875	143.2875	NFM	Strathclyde Fire Brigade		Channel 6
152.1875	143.2875	NFM	Scottish Fire Brigades		Channel 46
152.1875	143.4625	NFM	Strathclyde Fire Brigade		Channel 7
152.2000	143.2000	NFM	Strathclyde		Channel 09
152.2000	143.2625	NFM	Strathclyde		Channel 47
152.2125	143.4125	NFM	Strathclyde		Channel 36
152.2250	143.1375	AM	London	MP	Channel 6
152.2250	143.2750	NFM	Strathclyde		Channel 18
152.2375	143.0375	NFM	Strathclyde		Channel 23
152.2375	143.0375	NFM	Scottish Fire Brigades		Channel 23
152.2375	143.0375	NFM	Strathclyde Fire Brigade		Channel 2
152.2375	143.4625	NFM	Strathclyde Fire Brigade		Channel 7
152.2375	143.6625	NFM	Strathclyde Fire Brigade		Channel 9
152.2375	143.7125	NFM	Strathclyde		Channel 35
152.2375	143.7125	NFM	Scottish Fire Brigades		Channel 35
152.2375	143.7125	NFM	Strathclyde Fire Brigade		Channel 4
152.2500	143.2750	NFM	Strathclyde		Channel 13
152.2500	143.7125	NFM	Strathclyde Fire Brigade		
152.2625	143.4125	NFM	Strathclyde		Channel 37
152.2625	143.1875	AM	London	MP	Channel 8
152.2625	143.3875	AM	North Yorkshire	XN	Channel 5
152.2625	143.2625	NFM	Strathclyde		Channel 45
152.2750	143.4500	NFM	Strathclyde		Channel 26
152.2750		NFM	Strathclyde		Data Channel
152.2875	143.0375	NFM	Strathclyde		Channel 56
152.2875	143.0375	NFM	Strathclyde Fire Brigade		Channel 2
152.2875	143.0375	NFM	Scottish Fire Brigades		Channel 24
152.2875	143.7250	NFM	Dumfries & Galloway Fire		Channel 16
152.2875	143.1375	NFM	Strathclyde Fire Brigade		Channel 11
152.2875	143.2875	NFM	Strathclyde		Channel 48
152.2875	143.2875	NFM	Scottish Fire Brigades		Channel 48
152.2875	143.2875	NFM	Strathclyde Fire Brigade		Channel 6
152.2875	143.4625	NFM	Strathclyde Fire Brigade		Channel 1
152.2875	143.4625	NFM	Strathclyde Fire Brigade		Channel 7
152.2875	143.6625	NFM	Strathclyde Fire Brigade		Channel 9
152.2875	143.7125	NFM	Strathclyde		Channel 36
152.2875	143.7125	NFM	Scottish Fire Brigades		Channel 36
152.2875	143.7125	NFM	Strathclyde Fire Brigade		Channel 4
152.3000	143.2000	NFM	Strathclyde		Channel 14
152.3125	143.1625	AM	Humberside	XH	Channel 1
152.3125	143.0375	NFM	Strathclyde		Channel 27
152.3125	143.0375	NFM	Scottish Fire Brigades		Channel 27
152.3125	143.0375	NFM	Strathclyde Fire Brigade		Channel 2
152.3125	143.1000	NFM	Dumfries & Galloway Fire		Channel 15
152.3125	143.1375	NFM	Strathclyde Fire Brigade		Channel 11

Base	Mobile	Mode	Location		User & Notes
152.3125	143.2875	NFM	Strathclyde		Channel 44
152.3125	143.2875	NFM	Scottish Fire Brigades		Channel 44
152.3125	143.2875	NFM	Strathclyde Fire Brigade		Channel 6
152.3125	143.4625	NFM	Strathclyde Fire Brigade		Channel 7
152.3250	143.1625	AM	London	MP	Channel 7
152.3250	143.2750	NFM	Strathclyde		Channel 17 -
152.3375	143.0375	AM	West Yorkshire	XW	Channel 1
152.3375	143.0375	NFM	Strathclyde		Channel 26
152.3375	143.0375	NFM	Scottish Fire Brigades		Channel 26
152.3375	143.0375	NFM	Strathclyde Fire Brigade		Channel 2
152.3375	143.0625	NFM	Strathclyde Fire Brigade		Channel 8
152.3375	143.3375	NFM	Strathclyde		Channel 30
152.3375	143.3375	NFM	Scottish Fire Brigades		Channel 30
152.3375	143.3375	NFM	Strathclyde Fire Brigade		Channel 3
152.3375	143.3750	NFM	Strathclyde		Channel 39
152.3375	143.3750	NFM	Strathclyde Fire Brigade		Channel 5
152.3375	143.4625	NFM	Strathclyde Fire Brigade		Channel 7
152.3500	143.1250	AM	Leicester	NL	Channel 3
152.3500	143.2000	NFM	Strathclyde		Channel 10
152.3500	143.3125	NFM	Strathclyde		Channel 50
152.3500	143.6375	NFM	Strathclyde		Channel 30
152.3625	143.0375	AM	London	MP	Channel 2
152.3625	143.3625	NFM	Strathclyde		Channel 20
152.3625	143.3625	NFM	Scottish Fire Brigades		Channel 20
152.3750	143.3875	NFM	Strathclyde		Channel 19
152.3750	143.4500	NFM	Strathclyde		Channel 27
152.3750		NFM	Strathclyde		Data Channel
152.3875	143.0125	NFM	London	MP	Channel 1
152.3875	143.2375	AM	Cleveland	CZ	Channel 1
152.3875	143.3875	NFM	Scottish Fire Brigades		Channel 19
152.4000	143.2000	AM	Nottinghamshire	NH	Channel 1
152.4000	143.2750	NFM	Strathclyde		Channel 16
152.4125	143.1250	AM	Northumbria	LB	Channel 3
152.4125	143.1250	NFM	Strathclyde		Channel 33
152.4125	143.3875	AM	London	OJ	Channel 1
152.4125		NFM	Strathclyde		Data Channel
152.4250	143.2250	AM	Nottinghamshire	NH	Channel 2
152.4250	143.2625	NFM	Strathclyde		Channel 48
152.4250	143.6500	NFM	Strathclyde		Channel 20
152.4375	143.4500	NFM	Strathclyde		Channel 29
152.4375	143.4125	AM	London	OJ	Channel 2
152.4375		NFM	Strathclyde		Data Channel
152.4500	143.1875	AM	Humberside	XH	Channel 2
152.4500	143.4500	NFM	Strathclyde		Channel 28
152.4500		NFM	Strathclyde		Data Channel
152.4625	143.0875	AM	London	MN	Channel 4
152.4625	143.2875	NFM	Strathclyde		Channel 21
152.4625	143.2875	NFM	Scottish Fire Brigades		Channel 21
152.4625	143.3750	NFM	Strathclyde		Channel 40
152.4625	143.3750	NFM	Scottish Fire Brigades		Channel 40

Base	Mobile	Mode	Location	User & Notes	
152.4625	143.3750	NFM	Strathclyde Fire Brigade		Channel 5
152.4625	143.5250	NFM	Strathclyde Fire Brigade		Channel 10
152.4750	143.0625	NFM	Strathclyde Fire Brigade		Channel 8
152.4750	143.3375	NFM	Strathclyde		Channel 29
152.4750	143.3375	NFM	Scottish Fire Brigades		Channel 29
152.4750	143.3375	NFM	Strathclyde Fire Brigade		Channel 3
152.4750	143.7750	NFM	Dumfries & Galloway Fire		Channel 17
152.4875	143.2625	AM	Cleveland	CZ	Channel 4
152.4875	143.3125	NFM	Strathclyde		Channel 51
152.4875	143.6375	NFM	Strathclyde		Channel 30
152.4875	143.6500	NFM	Strathclyde		Channel 19
152.5000	143.2500	AM	Nottinghamshire	NH	Channel 3
152.5000	143.5375	NFM	Strathclyde		Channel 23
152.5125	143.0875	AM	West Yorkshire	XW	Channel 3
152.5125	143.6625	NFM	Strathclyde Fire Brigade		Channel 9
152.5125	143.7125	NFM	Strathclyde		Channel 33
152.5125	143.7125	NFM	Scottish Fire Brigades		Channel 33
152.5125	143.7125	NFM	Strathclyde Fire Brigade		Channel 4
152.5250	143.0375	NFM	Strathclyde		Channel 25
152.5250	143.0375	NFM	Scottish Fire Brigades		Channel 25
152.5250	143.0375	NFM	Strathclyde Fire Brigade		Channel 2
152.5250	143.4625	NFM	Strathclyde Fire Brigade		Channel 7
152.5375	143.5375	NFM	Strathclyde		Channel 22
152.5500	143.1500	AM	Northumbria	LB	Channel 4
152.5500	143.2875	AM	London	MN	Channel 12
152.5500	143.6625	NFM	Strathclyde Fire Brigade		Channel 9
152.5500	143.7125	NFM	Strathclyde		Channel 37
152.5500	143.7125	NFM	Scottish Fire Brigades		Channel 37
152.5500	143.7125	NFM	Strathclyde Fire Brigade		Channel 4
152.5625	143.6500	NFM	Strathclyde		Channel 21
152.5750	143.0250	AM	Northamptonshire	NG	Channel 1
152.5875	143.1125	AM	West Yorkshire	XW	Channel 4
152.5875	143.5375	NFM	Strathclyde		Channel 24
152.6000	143.0625	NFM	Strathclyde Fire Brigade		Channel 8
152.6000	143.1750	AM	Northumbria	LB	Channel 5
152.6000	143.2875	NFM	Strathclyde		Channel 46
152.6000	143.3375	NFM	Strathclyde		Channel 31
152.6000	143.3375	NFM	Scottish Fire Brigades		Channel 31
152.6000	143.3375	NFM	Strathclyde Fire Brigade		Channel 3
152.6000	143.3750	NFM	Strathclyde		Channel 43
152.6000	143.3750	NFM	Scottish Fire Brigades		Channel 43
152.6000	143.3750	NFM	Strathclyde Fire Brigade		Channel 5
152.6000	143.5250	NFM	Strathclyde Fire Brigade		Channel 10
152.6125	143.9000	NFM	Strathclyde		Channel 42
152.6250	143.2625	NFM	Strathclyde		Channel 46
152.6250	143.7000	NFM	Strathclyde		Channel 38
152.6375	143.4325	AM	London	MP	Channel 16
152.6375	143.6625	NFM	Strathclyde Fire Brigade		Channel 9
152.6375	143.7125	NFM	Strathclyde		Channel 32
152.6375	143.7125	NFM	Scottish Fire Brigades		Channel 32

Base	Mobile	Mode	Location	User & Notes	
152.6375	143.7125	NFM	Strathclyde Fire Brigade		Channel 4
152.6500	143.9000	NFM	Strathclyde		Channel 41
152.6625	143.9000	NFM	Strathclyde		Channel 44
152.6750	143.1250	NFM	Strathclyde		
152.6750		NFM	Strathclyde		Data Channel
152.6875	143.2125	AM	Humberside	XH	Channel 3
152.7000	143.0500	AM	Northamptonshire	NG	Channel 2
152.7000	143.7000	NFM	Strathclyde		Channel 40
152.7125	143.0625	AM	West Yorkshire	XW	Channel 2
152.7125	143.3750	NFM	Strathclyde		Channel 42
152.7125	143.3750	NFM	Scottish Fire Brigades		Channel 42
152.7125	143.3750	NFM	Strathclyde Fire Brigade		Channel 5
152.7125	143.5250	NFM	Strathclyde Fire Brigade		Channel 10
152.7250	143.3750	AM	Lincolnshire	NC	Channel 1
152.7500	143.5375	NFM	Strathclyde		Channel 25
152.7625	143.3125	NFM	Strathclyde		Channel 52
152.7625	143.6375	NFM	Strathclyde		Channel 30
152.7750	143.4625	AM	Lincolnshire	NC	Channel 2
152.7875	143.2750	AM	South Yorkshire	XS	Channel 1
152.7875	143.6625	NFM	Strathclyde Fire Brigade		Channel 9
152.7875	143.7125	NFM	Strathclyde		Channel 38
152.7875	143.7125	NFM	Scottish Fire Brigades		Channel 38
152.7875	143.7125	NFM	Strathclyde Fire Brigade		Channel 4
152.8000	143.0250	AM	County Durham	LA	Channel 1
152.8000		NFM	Strathclyde		Data Channel
152.8125	143.1375	AM	West Yorkshire	XW	Channel 5
152.8125	143.4375	NFM	Strathclyde		Channel 30
152.8125		NFM	Strathclyde		Data Channel
152.8150	143.4400	NFM	Strathclyde		
152.8250	143.7000	NFM	Strathclyde		Channel 39
152.8375	143.0500	AM	County Durham	LA	Channel 2
152.8375	143.6625	NFM	Strathclyde Fire Brigade		Channel 9
152.8375	143.7125	NFM	Strathclyde		Channel 34
152.8375	143.7125	NFM	Scottish Fire Brigades		Channel 34
152.8375	143.7125	NFM	Strathclyde Fire Brigade		Channel 4
152.8500	143.3250	AM	South Yorkshire	XS	Channel 3
152.8750	143.7750	NFM	Dumfries & Galloway Fire		Channel 17
152.8750	143.1375	NFM	Strathclyde Fire Brigade		Channel 11
152.8750	143.2875	NFM	Strathclyde		Channel 45
152.8750	143.2875	NFM	Scottish Fire Brigades		Channel 45
152.8750	143.2875	NFM	Strathclyde Fire Brigade		Channel 6
152.8875		AM	Isle of Man		Channel 2
152.9000	143.3000	AM	South Yorkshire	XS	Channel 2
152.9000	143.9000	NFM	Strathclyde		Channel 43
152.9250	143.3500	AM	South Yorkshire	XS	Channel 4
152.9250		NFM	Strathclyde		Data Channel
152.9500	143.1375	NFM	Strathclyde Fire Brigade		Channel 11
152.9500	143.2875	AM	North Yorkshire	XN	Channel 1
152.9500	143.2875	NFM	Strathclyde		Channel 47
152.9500	143.2875	NFM	Scottish Fire Brigades		Channel 47

Base	Mobile	Mode	Location		User & Notes
152.9500	143.2875	NFM	Strathclyde Fire Brigade		Channel 6
152.9500	143.3750	NFM	Scottish Fire Brigades		Channel 39
152.9500	143.5250	NFM	Strathclyde Fire Brigade		Channel 10
152.9625		AM	Isle of Man		Channel 3
152.9750	143.3375	AM	North Yorkshire	XN	Channel 3

153.0250 - 153.5000 MHz National Paging

Base	Mobile	Mode	Location	User & Notes
153.0750		NFM	Nationwide	Paging
153.1750		NFM	Nationwide	BT Pagers
153.2375		NFM	Nationwide	Paging
153.2500		NFM	Nationwide	Digital Mobile Comms
153.2750		NFM	Nationwide	Air Call Communications
153.3250		NFM	Nationwide	Air Call Communications
153.3375		NFM	Nationwide	Paging
153.3500		NFM	Nationwide	Inter City Pagers
153.3625		NFM	Nationwide	Paging
153.3750		NFM	Nationwide	Paging
153.4500		NFM	Nationwide	Paging

153.5000 - 154.0000 MHz MoD Tactical Channels 25 kHz NFM Simplex

154.0000 - 155.9875 MHz Police (England & Wales) Base Rpter 12.5 kHz

Base	Mobile	Mode	Location		User & Notes
154.0125	146.1250	FM	Lancashire	BD	Channel 3
154.0125		AM	M25 North	CH	Channel 5
154.0250	146.5125	AM	West Mercia	YK	Channel 3
154.0500	146.5375	AM	West Mercia	YK	Channel 4
154.0750	146.1750	AM	Lancashire	BD	Channel 4
154.0750		AM	M6/M56/M62		
154.0875	146.1625	AM	M25 South	CH	Channel 4
154.1000	146.1375	AM	Thames Valley	HB	Channel 2
154.1125	146.5625	AM	West Mercia	YK	Channel 5
154.1375		AM	Cambridge	VB	Channel 3
154.1375		AM	Thames Valley	HB	
154.1500	146.6750	AM	Gloucester	QL	Channel 1
154.2000	146.5875	AM	Staffordshire	YF	Channel 1
154.2250	146.7000	AM	Gloucester	QL	Channel 2
154.2625	146.4750	AM	West Midlands	YM	Channel 6
154.2750	146.5750	AM	Suffolk	VL	Channel 1
154.3125	146.4500	AM	West Midlands	YM	Channel 5
154.3375	146.7375	AM	Thames Valley	HB	Channel 1
154.3675	146.7250	AM	Gloucester	QL	Channel 3
154.4125	146.0875	AM	West Mercia	YK	Channel 2
154.4125	146.0875	AM	Warwickshire	YJ	Channel 1
154.4125	146.5750	AM	Suffolk	VL	Channel 2
154.4500	146.3500	AM	West Midlands	YM	Channel 2
154.5000	146.5000	AM	Jersey	M2JY	Channel 1
154.5500	146.5500	AM	Channel Islands		
154.5500	146.6125	AM	Staffordshire	YF	Channel 2
154.6000		AM	Guernsey Fire		
154.6000		AM	Jersey Fire		Channel 2

Base	Mobile	Mode	Location	User & Notes	
154.6125		AM	Melksham		
154.6250	146.3250	AM	West Midlands	YM	Channel 1
154.6250	146.6250	AM	Jersey	M2JY	Channel 2
154.6375	146.2875	AM	Kent	KA	Channel 5
154.6500	146.1125	AM	West Mercia	YK	Channel 2
154.6500	146.1125	AM	Warwickshire	YJ	Channel 2
154.7000	146.1000	AM	Suffolk	VL	Channel 3
154.7000	146.3750	AM	West Midlands	YM	Channel 3
154.7375	146.8625	AM	Bedfordshire	VA	Channel 2
154.7750	146.3000	AM	Thames Valley	HY	Channel 8
154.7875	146.0250	AM	Cambridge	VB	Channel 1
154.7875	146.8000	AM	Gwent	WE	Channel 2
154.8000	146.1875	AM	Dorset	QC	Channel 3
154.8000	146.7250	FM	Lancashire	BD	Channel 5
154.8125	146.8750	AM	Gwent	WE	Channel 3
154.8250	146.6875	AM	Cambridge	VB	Channel 2
154.8250	146.7000	AM	Cheshire	BA	Channel 2
154.8375	146.2750	AM	Thames Valley	HU	Channel 4
154.8625	146.1250	AM	South Wales	WL	Channel 1
154.8625	146.8250	AM	Merseyside	CH	Channel 1
154.8625	146.9000	AM	Suffolk	VL	Channel 1
154.8750	146.0375	AM	Thames Valley	HB	Channel 6
154.8875	146.1500	AM	South Wales	WL	Channel 2
154.8875	146.9250	AM	Suffolk	VL	Channel 2
154.9000	146.8375	AM	Bedfordshire	VA	Channel 1
154.9000	146.8500	AM	Merseyside	CH	Channel 2
154.9125	146.8250	AM	Norfolk	VK	Channel 1
154.9125	146.7750	AM	Gwent	WE	Channel 1
154.9250	146.0625	AM	Thames Valley	HB	Channel 5
154.9250	146.9000	AM	Merseyside	CH	Channel 3
154.9375	146.1750	AM	South Wales	WL	Channel 4
154.9375	146.6125	AM	Kent	KA	Channel 2
154.9500	146.0125	AM	Thames Valley	HB	Channel 7
154.9500	146.8500	AM	Birkenhead (M53)		
154.9625		AM	Isle of Man		Channel 3
154.9625	146.0750	AM	South Wales	WL	Channel 3
154.9625	146.6375	AM	Kent	KA	Channel 3
154.9750	146.4250	AM	West Midlands	YM	Channel 4
154.9750	146.6875	AM	Hampshire	HC	Channel 5
154.9875	146.8500	AM	Norfolk	VK	Channel 2
154.9875	146.8500	AM	Avon & Somerset	QP	Channel 3
155.0000	146.3125	AM	Sussex	KB	Channel 1
155.0125	146.2125	AM	Manchester	CK	Channel 1
155.0125	146.8125	AM	Avon & Somerset	QP	Channel 7
155.0250	147.1375	AM	Essex	VG	Channel 1
155.0250		AM	Manchester Traffic Police		
155.0375	146.7875	AM	Avon & Somerset	QP	Channel 6
155.0500	146.3625	AM	Sussex	KB	Channel 3
155.0500	146.6625	AM	North Wales	WA	Channel 1
155.0625	146.2375	AM	Manchester	CK	Channel 4

Base	Mobile	Mode	Location	User & Notes	
155.0625	146.7625	AM	Avon & Somerset	QP	Channel 5
155.0750	147.1875	AM	Essex	VG	Channel 2
155.0750		AM	Manchester Traffic Police		
155.0875	146.1000	AM	Manchester	CK	Channel 5
155.0875	146.4000	AM	Hampshire	HC	Channel 1
155.1000	146.4375	AM	North Wales	WA	Channel 2
155.1000	146.5875	AM	Kent	KA	Channel 1
155.1000		AM	Southport		
155.1125	146.4375	AM	Hampshire	HC	Channel 2
155.1250	146.6250	AM	Hertfordshire	VH	Channel 2
155.1375	146.6625	AM	Kent	KA	Channel 4
155.1500	146.0750	AM	Manchester	CK	Channel 6
155.1500	146.5750	AM	Avon & Somerset	QP	Channel 4
155.1625	146.4625	AM	Hampshire	HC	Channel 4
155.1625	147.6125	AM	Warwickshire	YJ	Channel 3
155.1750	146.1625	AM	Essex	VG	Channel 2
155.1750	146.9250	AM	Merseyside	CH	Channel 4
155.1875	146.3875	AM	Sussex	KB	Channel 4
155.1875		AM	Manchester Traffic Police		
155.2000	146.6000	AM	Hertfordshire	VH	Channel 1
155.2000		AM	Manchester		
155.2125	146.5000	AM	Avon & Somerset	QP	Channel 1
155.2250	146.3375	AM	Sussex	KB	Channel 2
155.2250	146.9500	AM	Merseyside	CH	Channel 5
155.2375	146.6500	AM	Hertfordshire	VH	Channel 3
155.2500	146.2750	FM	Lancashire	BD	Channel 2
155.2500	146.5250	AM	Avon & Somerset	QP	Channel 2
155.2500		AM	Manchester		
155.2625	146.3125	AM	Cheshire	BA	Channel 2
155.2625	146.4875	AM	Hampshire	HC	Channel 3
155.2750	146.9000	AM	Devon & Cornwall	BV	Channel 2
155.2750		AM	Powys	WH	Channel 4
155.2875	146.4125	AM	Sussex	KB	Channel 5
155.3000	146.9750	AM	Devon & Cornwall	QB	Channel 7
155.3250	146.2375	AM	Devon & Cornwall	CV	Channel 4
155.3250	146.9750	AM	Merseyside	CH	Channel 6
155.3625	146.2625	AM	Devon & Cornwall	FV	Channel 5
155.3875	146.3625	AM	Cheshire	BA	Channel 1
155.3875	146.9500	AM	Devon & Cornwall	DV	Channel 6
155.4125	146.6250	AM	Dyfed & Wales	WH	Channel 2
155.4250	146.3875	AM	Manchester	CK	Channel 2
155.4375	146.6000	AM	Dyfed & Wales	WH	Channel 1
155.4625	146.2125	AM	Devon & Cornwall	AV	Channel 1
155.4625	146.3000	FM	Lancashire	BD	Channel 1
155.4875	146.3375	AM	Cheshire	BA	Channel 3
155.5000	146.6500	AM	Dyfed & Wales	WH	Channel 3
155.5125	146.9250	AM	Devon & Cornwall	EV	Channel 3
155.5375	146.4625	AM	North Wales	WA	Channel 3
155.5750	146.2875	AM	Devon & Cornwall	QB	Channel 8
155.5875	146.2500	AM	Cumbria	BB	Channel 3

Base	Mobile	Mode	Location	User & Notes	
155.6000		AM	Derbyshire		Channel 1
155.6125	146.4125	AM	Greater Manchester	CK	Channel 3
155.6250	146.4125	AM	North Wales		Channel 4
155.6375	146.4500	AM	Cumbria	BB	Channel 2
155.6375	147.9375	AM	Jersey Fire		Channel 1
155.6250	146.4875	AM	North Wales	W A	Channel 4
155.6625	146.2000	AM	Cumbria	BB	Channel 1
155.6625	147.9500	AM	Guernsey Fire		Channel 1
155.6750		AM	Jersey Fire		
155.7000	155.7000	AM	RCS Handheld	CS	
155.7000		AM	Manchester		
155.7250	155.7250	AM	RCS Handheld	CS	
155.7500	155.7500	AM	RCS Handheld	CS	
155.7500		AM	Chester		
155.7500		AM	Manchester		
155.7750	155.7750	AM	RCS Handheld	CS	
155.7750		AM		KW	Link
155.8000		AM	Durham		Channel 1
155.8000	146.2000	AM	Surrey	HJ	Channel 1
155.8000	147.7625	AM	Guernsey	QY	Channel 1
155.8125	146.9125	AM	Wiltshire	QJ	Channel 1
155.8250	147.9750	AM	Guernsey Fire		Channel 2
155.8375	146.2875	AM	Derbyshire	NA	Channel 2
155.8375	146.2250	AM	Surrey	HJ	Channel 2
155.8500	146.9375	AM	Wiltshire	QJ	Channel 2
155.8625	146.2500	AM	Surrey	HJ	Channel 3
155.8625	147.7500	AM	Jersey	M2JY	Channel 3
155.8750	146.1000	AM	Dorset	QC	Channel 1
155.9000	146.9625	AM	Wiltshire	QJ	Channel 3
155.9000	147.8500	AM	Guernsey		Channel 2
155.9250	147.0125	AM	Dorset	QC	Channel 2
155.9375		AM	Thames Valley	SC	Channel 3
155.9500		AM	Dorset		Channel 3

154.0000 - 155.9875 MHz Central & N. Scottish & Irish Police and Fire Brigades Base Repeaters 12.5 kHz

Base	Mobile	Mode	Location	User & Notes	
154.0125		NFM	Fife		
154.0500		NFM	Edinburgh	T	
154.0625		NFM	Dumfermline	F	Fire
154.0625		NFM	Edinburgh	F	Fire
154.1125		NFM	Aberdeen	Fire Control	
154.1125		NFM	Dumfermline	F	Fire
154.1250		NFM	Edinbugh	T	
154.1250		NFM	Fife	T	
154.1500		NFM	Stirling	F	Fire
154.1500	146.4875	NFM	Central Region		
154.1625		NFM	Fife	B	
154.1875		NFM	Fife		
154.2250		NFM	Lanark	ZS	
154.2375		NFM	Edinburgh		

Base	Mobile	Mode	Location	User & Notes	
154.2375		NFM	West Lothian	F	
		NFM	Lanark	ZS	
154.2500		NFM	East Lothian	E	
154.3000		NFM	Fettes	ZH	Channel 1
154.3500		NFM	Fettes	ZH	Channel 2
154.4375		NFM	Edinburgh	T	Channel 3
154.4500	147.1875	NFM	Perth	W	Channel 2
		NFM	Pitlochry	WP	
154.4750	146.6250	NFM	Edinburgh	F	Fire
154.4875	146.4875	NFM	Central Region		
154.5250	146.4875	NFM	Central Region		
154.5375	146.5375	NFM	Edinburgh	F	Fire
154.5625	146.4875	NFM	Central Region		
154.6000	147.2375	NFM	Perth	W	Channel 5
154.6125		NFM	Edinburgh		
154.6125		NFM	Lanark	S	
154.6250	146.6250	NFM	Edinburgh	F	Fire
154.6375		NFM	Aberdeen		
154.6500		NFM	Edinburgh	F	Fire
154.6625	146.6625	NFM	Perth	Fire Control	Channel 2
154.7125	146.9625	NFM	Perth	W	Channel 4
154.7625	147.2375	NFM	Perth	W	Channel 5
154.7875		NFM	Aberdeen	Fire Control	
154.7875		NFM	Lanark	ZS	
154.7875	146.9500	NFM	Perth	Fire Control	Channel 3
154.8375	146.9625	NFM	Perth	E	Channel 4
154.8750	146.6625	NFM	Perth	Fire Control	
154.8875	147.2375	NFM	Perth	W	Channel 5
154.9125	147.0500	NFM	Perth	W	Channel 3
		NFM	Blairgowrie	WB	
154.9250		NFM	Grampian		
154.9375	146.9375	NFM	Perth	Fire Control	Channel 5
154.9500		NFM	Aberdeen	Fire Control	
154.9625	146.9625	NFM	Perth	E	Channel 4
154.9750	147.0500	NFM	Perth	W	Channel 3
		NFM	Blairgowrie	WB	
154.9750		NFM	Aberdeen	Fire Control	
154.9875		NFM	Grampian		
155.0250	147.0250	NFM	Dundee	ZS	Channel 6
155.0375	147.0625	NFM	Perth	Fire Control	Channel 4
155.0500	147.0500	NFM	Perth	W	Channel 3
		NFM	Blairgowrie	WB	
155.0625	147.0625	NFM	Perth	Fire Control	Channel 4
155.0750		NFM	Aberdeen	Fire Control	
155.1000	147.1875	NFM	Perth	W	Channel 2
		NFM	Pitlochry	WP	
155.1375	147.1375	NFM	Perth	Fire Control	Channel 1
155.1500	147.2250	NFM	Perth	W	Channel 1
155.1750	147.4875	NFM	Lothian & Borders		Fire
155.1875	147.1875	NFM	Perth	W	Channel 2

Base	Mobile	Mode	Location	User & Notes	
155.2000	147.0625	NFM	Perth	Fire Control Channel 4	
155.2125	147.2250	NFM	Perth	W	Channel 1
155.2250	147.2250	NFM	Perth	W	Channel 1
155.2375	147.2375	NFM	Perth	W	Channel 5
155.2500	147.4875	NFM	Lothian & Borders		Fire
155.2625		NFM	Fife		
155.2875		NFM	Aberdeen		
155.2875	147.2250	NFM	Perth	W	Channel 1
155.3000		NFM	Braemar	UBH	
		NFM	Banchory	UBK	
155.3125	147.2375	NFM	Perth	W	Channel 5
155.3375	147.1875	NFM	Perth	W	Channel 2
		NFM	Pitlochry	WP	
155.3625	147.1875	NFM	Perth	W	Channel 2
155.4000		NFM	Edinburgh	Control	Fire
		NFM	Perth	Fire Control	
155.4125	147.0375	NFM	Highlands & Islands		Fire
155.4250		NFM	Aberdeen		
155.4250	147.2375	NFM	Perth	W	Channel 5
155.4375		NFM	Perth	Fire Control	
155.4875	147.4875	NFM	Lothian & Borders		Fire
155.5000		NFM	Grampian		
155.5125	147.4875	NFM	Lothian & Borders		Fire
155.5375		NFM	Aberdeen		
155.5375		NFM	Grampian		
155.5500		NFM	Aberdeen		
155.5625		NFM	Edinburgh	ZS	
155.5625		NFM	Perth	W	
155.5750		NFM	Perth	W	
155.5875		NFM	Aberdeen		
155.5875		NFM	Grampian		
155.6125	147.0375	NFM	Highlands & Islands		Fire
155.6250		NFM	Aberdeen		
155.7375		NFM	Grampian	Fire Control	
155.7375	147.0375	NFM	Highlands & Islands		Fire
155.7875		NFM	Grampian	Fire Control	
155.8000		NFM	Aberdeen		
155.8000		NFM	Belfast/Larne		
155.8125		NFM	Grampian		
155.8250		NFM	Belfast/Larne		
155.8750		NFM	Dumfermline	F	Fire
155.9500		NFM	Belfast/Larne		

156.0000 - 162.5000 MHz   Maritime Band 25 kHz (Ship TX/Shore TX)

Base	Mobile	Mode	Location		User & Notes
156.0000	156.0000	NFM	Nationwide	Channel 00	Coastguard Primary
156.0250	160.6250	NFM	Islay	Channel 60	Working Chan
		NFM	Start Point	Channel 60	Working Chan
156.0500	160.6500	NFM	Nationwide	Channel 01	Port Ops
156.0750	160.6750	NFM	Anglesey	Channel 61	Working Chan

Base	Mobile	Mode	Location	User & Notes	
		NFM	Scillies	Channel 61	Working Chan
156.1000	160.7000	NFM	St. Malo	Channel 02	Working Chan
		NFM	Thames	Channel 02	Working Chan
156.1250	160.7250	NFM	Forth	Channel 62	Working Chan
		NFM	Orfordness	Channel 62	Working Chan
		NFM	Pendennis	Channel 62	Working Chan
		NFM	St Peter Port	Channel 62	Working Chan
156.1500	160.7500	NFM	Bacton	Channel 03	Working Chan
		NFM	Cardigan Bay	Channel 03	Working Chan
		NFM	Port en Bessin	Channel 03	Working Chan
156.1750	160.7750	NFM	Bacton	Channel 63	Working Chan
		NFM	Hastings	Channel 63	Working Chan
156.2000	160.7000	NFM	Grimsby	Channel 04	Working Chan
		NFM	Morecambe Bay	Channel 04	Working Chan
		NFM	Niton	Channel 04	Working Chan
156.2250	160.7250	NFM	Bacton	Channel 64	Working Chan
		NFM	Lands End	Channel 64	Working Chan
		NFM	Niton	Channel 64	Working Chan
		NFM	Rouen	Channel 64	Working Chan
156.2500	160.7500	NFM	Ilfracombe	Channel 05	Working Chan
		NFM	Lewis	Channel 05	Working Chan
		NFM	North Foreland	Channel 05	Working Chan
		NFM	Weymouth Bay	Channel 05	Working Chan
156.2750	160.7750	NFM	North Foreland	Channel 65	Working Chan
		NFM	Start Point	Channel 65	Working Chan
156.3000	156.3000	NFM	Nationwide	Channel 06	Intership Comms Primary. Used by Coastguard Vessels during SAR Situation.
156.3250	160.9250	NFM	North Foreland	Channel 66	Working Chan
		NFM	Pendennis	Channel 66	Working Chan
156.3500	160.9500	NFM	Bacton	Channel 07	Working Chan
		NFM	Hastings	Channel 07	Working Chan
		NFM	Ilfracombe	Channel 07	Working Chan
156.3750	156.3750	NFM	Nationwide	Channel 67	Coastguard Secondary
		NFM	Bantry	Channel 67	Working Chan
		NFM	Belmullet	Channel 67	Working Chan
		NFM	Cork	Channel 67	Working Chan
		NFM	Glen Head	Channel 67	Working Chan
		NFM	Jersey	Channel 67	Working Chan
		NFM	Malin Head	Channel 67	Working Chan
		NFM	Minehead	Channel 67	Working Chan
		NFM	Rosslare	Channel 67	Working Chan
		NFM	Shannon	Channel 67	Working Chan
		NFM	Valentia	Channel 67	Working Chan
156.4000	156.4000	NFM	Nationwide	Channel 08	Intership

Base	Mobile	Mode	Location	User & Notes	
					Comms
156.4250	156.4250	NFM	Nationwide	Channel 68	Intership Comms
		NFM	Ouistreham Marina	Channel 68	Working Chan
156.4500	156.4500	NFM	Solent	Channel 09	Pilot Office & Vessel
		NFM	French Marinas	Channel 09	Working Chan
		NFM	Brest	Channel 09	Working Chan
		NFM	Cherbourg	Channel 09	Working Chan
		NFM	Granville	Channel 09	Working Chan
156.4750	156.4750	NFM	Nationwide	Channel 69	Intership Comms
156.5000	156.5000	NFM	Nationwide	Channel 10	Pollution Frequency
156.5250	156.5250	NFM	Nationwide	Channel 70	Intership Comms Digital Selective Calling
156.5500	156.5500	NFM	Nationwide	Channel 11	Port Ops
		NFM	Liverpool	Channel 11	Pilot
156.5750	156.5750	NFM	Nationwide	Channel 71	Port Ops
		NFM	Portland	Channel 71	Royal Navy
156.6000	156.6000	NFM	Braye	Channel 12	Port
		NFM	Granville	Channel 12	Port
		NFM	Langstone Har.	Channel 12	Port
		NFM	St. Malo	Channel 12	Port
		NFM	St Peter Port	Channel 12	Working Chan
156.6250	156.6250	NFM	Nationwide	Channel 72	Intership Comms
156.6500	156.6500	NFM	Nationwide	Channel 13	Port Ops
156.6750	156.6750	NFM	Nationwide	Channel 73	Coastguard Tertiary
156.7000	156.7000	NFM	Chichester	Channel 14	Port
		NFM	Gorey Harbour	Channel 14	Port
		NFM	Humber	Channel 14	Pilot
		NFM	Shoreham	Channel 14	Port
		NFM	St Helier	Channel 14	Port & Marina
156.7250	156.7250	NFM	Nationwide	Channel 74	Ports, Lock Keepers & Swing Bridges
		NFM	Aldernay	Channel 74	Working Chan
		NFM	Gorey	Channel 74	Working Chan
156.7500	156.7500	NFM	Nationwide	Channel 15	Port Ops
156.7625	156.7625	NFM	Nationwide	Channel 75	Guard Band 156.7625 - 156.7875 MHz
156.8000	156.8000	NFM	Nationwide	Channel 16	Distress Safety

Base	Mobile	Mode	Location	User & Notes	
					And Calling Coastguard & Harbour Pilots
156.8125	156.8125	NFM	Nationwide	Channel 76	Guard Band 156.8125 - 156.8375 MHz
156.8250	156.8250	NFM	Nationwide		Direct Printing Telegraphy (DPT)
156.8500	156.8500	NFM	Nationwide	Channel 17	Port Ops
		NFM	BCIF		
156.8750	156.8750	NFM	Nationwide	Channel 77	Intership Comms
156.9000	156.9000	NFM	Nationwide	Channel 18	Port Ops
		NFM	Port en Bessin	Channel 18	Marina Chan
156.9250	161.5250	NFM	St Peter Port	Channel 78	Working Chan
156.9500	161.5500	NFM	Nationwide	Channel 19	Port Ops
156.9750	161.5750	NFM	Nationwide	Channel 79	Port Ops
157.0000	161.6000	NFM	Nationwide	Channel 20	Port Ops
157.0250	161.6250	NFM	Nationwide	Channel 80	Port Ops
		NFM	Dover	Channel 80	Coastguard
157.0500	161.6500	NFM	Nationwide	Channel	Shore TX also 156.0500 MHz
157.0750	161.6750	NFM	Anglesay	Channel 81	Working Chan
		NFM	Niton	Channel 81	Working Chan
157.1000	161.7000	NFM	Nationwide	Channel 22	Port Ops
157.1250	161.7250	NFM	Jersey	Channel 82	Working Chan
		NFM	Morecambe Bay	Channel 82	Working Chan
		NFM	Orfordness	Channel 82	Working Chan
157.1500	161.7500	NFM	Bantry	Channel 23	Working Chan
		NFM	Le Havre	Channel 23	Working Chan
		NFM	Malin Head	Channel 23	Working Chan
		NFM	Nationwide	Channel 23	Shore TX also 156.1500 MHz
		NFM	Rosslare	Channel 23	Working Chan
157.1750	161.7750	NFM	Belmullet	Channel 83	Working Chan
		NFM	Dublin	Channel 83	Working Chan
		NFM	Minehead	Channel 83	Working Chan
		NFM	Nationwide	Channel 83	Shore TX also 156.1750 MHz
		NFM	Plougasnou	Channel 83	Working Chan
		NFM	Thames	Channel 83	Working Chan
157.2000	161.8000	NFM	Celtic	Channel 24	Working Chan
		NFM	Collafirth	Channel 24	Working Chan
		NFM	Forth	Channel 24	Working Chan
		NFM	Glen Head	Channel 24	Working Chan
		NFM	Humber	Channel 24	Working Chan
		NFM	Ouessant	Channel 24	Working Chan
		NFM	Shannon	Channel 24	Working Chan
		NFM	Skye	Channel 24	Working Chan

Base	Mobile	Mode	Location	User & Notes	
		NFM	Valentia	Channel 24	Working Chan
157.2250	161.8250	NFM	Cromarty	Channel 84	Working Chan
		NFM	Paimpol	Channel 84	Working Chan
157.2500	161.8500	NFM	Buchan	Channel 25	Working Chan
		NFM	Islay	Channel 25	Working Chan
		NFM	Jersey	Channel 25	Working Chan
		NFM	Whitby	Channel 25	Working Chan
157.2750	161.8750	NFM	Bantry	Channel 85	Working Chan
		NFM	Humber	Channel 85	Working Chan
		NFM	Lands End	Channel 85	Working Chan
		NFM	Malin End	Channel 85	Working Chan
		NFM	Niton	Channel 85	Working Chan
157.3000	161.9000	NFM	Anglesey	Channel 26	Working Chan
		NFM	Brest	Channel 26	Working Chan
		NFM	Clyde	Channel 26	Working Chan
		NFM	Cork	Channel 26	Working Chan
		NFM	Cullercoats	Channel 26	Working Chan
		NFM	Hebrides	Channel 26	Working Chan
		NFM	Humber	Channel 26	Working Chan
		NFM	North Foreland	Channel 26	Working Chan
		NFM	Orkney	Channel 26	Working Chan
		NFM	Start Point	Channel 26	Working Chan
		NFM	Stonehaven	Channel 26	Working Chan
157.3250	161.9250	NFM	Nationwide	Channel 86	Link Calls
157.3500	161.9500	NFM	Cherbourg	Channel 27	Working Chan
		NFM	Grimsby	Channel 27	Working Chan
		NFM	Lands End	Channel 27	Working Chan
		NFM	Portpatrick	Channel 27	Working Chan
		NFM	Shetland	Channel 27	Working Chan
157.3750	161.9750	NFM	Buchan	Channel 87	Working Chan
		NFM	Niton	Channel 87	Working Chan
157.4000	162.0000	NFM	Anglesey	Channel 28	Working Chan
		NFM	Brest	Channel 28	Working Chan
		NFM	Cromarty	Channel 28	Working Chan
		NFM	Niton	Channel 28	Working Chan
		NFM	Shannon	Channel 28	Working Chan
		NFM	Valentia	Channel 28	Working Chan
		NFM	Whitby	Channel 28	Working Chan
157.4250	162.0250	NFM	Nationwide	Channel 88	Lighthouse Ch
		NFM	Anvil Point	Channel 88	Lighthouse
		NFM	Calais Main	Channel 88	Lighthouse
		NFM	Lands End	Channel 88	Working Chan
		NFM	North Foreland	Channel 88	Lighthouse
		NFM	Pillar Rock Pt	Channel 88	Lighthouse
157.4500	162.0500	NFM	Channel Island	Channel 29	BIFerries
		NFM	Torpoint Ferry	Channel 29	Operations
157.4750	162.0750	NFM	Nationwide	Channel 89	
157.5000	162.1000	NFM	English Channel	Channel 30	Herm Seaway
157.5250	162.1250	NFM	Nationwide	Channel 90	
157.5500	162.1500	NFM	Nationwide	Channel 31	Fisheries

Base	Mobile	Mode	Location	User & Notes	
					Patrol Boats
		NFM	Sark	Channel 31	Working Chan
157.5750	162.1750	NFM	Nationwide	Channel 91	
157.6000	162.2000	NFM	Nationwide	Channel 32	
157.6250	162.2250	NFM	Nationwide	Channel 92	
157.6500	162.2500	NFM	English Channel	Channel 33	Hovercraft Ch
		NFM	English Channel	Channel 33	Fishermens Cooperative
157.6750	162.2750	NFM	Nationwide	Channel 93	
157.7000	162.3000	NFM	English Channel	Channel 34	Herm
157.7250	162.3250	NFM	Nationwide	Channel 94	
157.7500	162.3500	NFM	Dover	Channel 35	Hovercraft
157.7750	162.3750	NFM	Nationwide	Channel 95	
157.8000	162.4000	NFM	Nationwide	Channel 36	
157.8250	162.4250	NFM	Nationwide	Channel 96	
157.8500	162.4500	NFM	Nationwide	Channel M1	Marinas
157.8750	162.4750	NFM	Nationwide	Channel 97	
157.9000	162.5000	NFM	Guernsey	Channel 38	Sealink
		NFM	English Channel	Channel 38	British Ferries
157.9250	162.5250	NFM	Nationwide	Channel 98	
157.9500	162.5500	NFM	Nationwide	Channel 39	
157.9750	162.5750	NFM	Nationwide	Channel 99	
158.0000	162.6000	NFM	Nationwide	Channel 40	
158.0250	162.6250	NFM	Nationwide	Channel 100	
158.0500	162.6500	NFM	English Channel	Channel 41	Battricks
158.0750	162.6750	NFM	Nationwide	Channel 101	
158.1000	162.7000	NFM	Nationwide	Channel 42	
158.1250	162.7250	NFM	Nationwide	Channel 102	
158.1500	162.7500	NFM	Nationwide	Channel 43	
158.1750	162.7750	NFM	Nationwide	Channel 103	
158.2000	162.8000	NFM	Moray Firth	Channel 44	Beatrice Alpha/ Bravo Platform
158.2125	162.8125	NFM	Nationwide	Channel 104	
158.2500	162.8500	NFM	Moray Firth	Channel 45	Beatrice Alpha/ Bravo Platform
158.2750	162.8750	NFM	Nationwide	Channel 105	
158.3000	162.9000	NFM	Moray Firth	Channel 46	Beatrice Alpha/ Bravo Platform
158.3125	162.9125	NFM	Nationwide	Channel 106	
158.3500	162.9250	NFM	Moray Firth	Channel 47	Beatrice Alpha/ Bravo Platform
158.3750	162.9750	NFM	Nationwide	Channel 107	
158.4000	163.0000	NFM	Aberdeen	Channel 48	Dockside
158.4250	163.0250	NFM	English Channel	Channel 108	Condor
158.4500	163.0500	NFM	Moray Firth	Channel 49	Beatrice Alpha/ Bravo Platform
		NFM	Nigg Bay	Channel 49	Oil Tanker Ldg
158.4750	163.0750	NFM	Aberdeen	Channel 109	Shipping Info
158.5000	163.1000	NFM	Nigg Bay	Channel 50	Oil Tanker Ldg
		NFM	English Channel	Channel 50	Emeraude Line

Base	Mobile	Mode	Location	User & Notes
158.5250	163.1250	NFM	Nationwide	Channel 110

158.5375 - 159.9125 MHz Pacnet Commercial X-25 Data Network

Base	Mobile	Mode	Location	User & Notes
158.5375	163.0375	NFM	Nationwide	Channel 1
158.6375	163.1375	NFM	Nationwide	Channel 2
158.7375	163.2375	NFM	Nationwide	Channel 3
158.8375	163.3375	NFM	Nationwide	Channel 4
158.8500	163.3500	NFM	Nationwide	British Telecom
158.9375	163.4375	NFM	Nationwide	Channel 5
159.0375	163.5375	NFM	Nationwide	Channel 6
159.1375	163.6375	NFM	Nationwide	Channel 7
159.2375	163.7375	NFM	Nationwide	Channel 8
159.3375	163.8375	NFM	Nationwide	Channel 9
159.4375	163.9375	NFM	Nationwide	Channel 10
159.4875	163.9875	NFM	Nationwide	Short Term Hire
159.5000	164.0000	NFM	Nationwide	Short Term Hire
159.5375	164.0375	NFM	Nationwide	Channel 11
159.5875	164.0875	NFM	Nationwide	Short Term Hire
159.6250	164.1250	NFM	Nationwide	Short Term Hire
159.6375	164.1375	NFM	Nationwide	Channel 12
159.6875	164.1875	NFM	Nationwide	Short Term Hire
159.7375	164.2375	NFM	Nationwide	Channel 13
159.8375	164.3375	NFM	Nationwide	Channel 14

159.9250 - 160.5500 MHz PMR Repeaters & Message Handling

Base	Mobile	Mode	Location	User & Notes
159.4875		NFM	Nationwide	New PMR Allocation
159.5000		NFM	Nationwide	New PMR Allocation
159.5875		NFM	Nationwide	New PMR Allocation
159.6250		NFM	Nationwide	New PMR Allocation
159.6875		NFM	Nationwide	New PMR Allocation
159.8750		NFM	Space	Myr
160.1250		NFM	Space	Soviet Mir Space Station

160.5500 - 161.0000 MHz International Maritime Shore Transmit

Base	Mobile	Mode	Location	User & Notes
160.5500		NFM	Nationwide	OAP Alarm System
160.5625		NFM	Nationwide	OAP Alarm System
160.5750		NFM	Nationwide	OAP Alarm System

161.0000 - 161.1000 MHz Paging Acknowledgement Channels

161.1000 - 161.5000 MHz Private Marine Allocation 25 kHz NFM Simplex

Base	Mobile	Mode	Location	User & Notes
161.2000	161.2000	NFM	Bacton	Philips Petroleum
161.3500	161.3500	NFM	BCIF 'Pride of Portsmouth'	
161.4250	161.4250	NFM	Marina & Yachts	Channel M2

161.5000 - 162.0500 MHz International Maritime Shore Transmit

Base	Mobile	Mode	Location	User & Notes
161.2750		NFM	Nationwide	Small Boats Alarms
162.0000		NFM	Various Satellites	

Base	Mobile	Mode	Location	User & Notes

162.0500 - 163.0000 MHz Private Marine Allocation 25 kHz NFM Simplex
Message Handling & NOAA Weather Sats

Base	Mobile	Mode	Location	User & Notes
162.4000		NFM	Space	NOAA Satellite
162.4250		NFM	Space	NOAA Satellite
162.4750		NFM	Space	NOAA Satellite
162.5500		NFM	Space	NOAA Satellite

163.0375 - 164.4125 MHz Pacnet Commercial X-25 Data Network WFM

Base	Mobile	Mode	Location	User & Notes
163.3000	158.7000	NFM	Lea Valley	Hoddesdon & Herts Buses

163.4375 - 165.0375 MHz Private Message Handling & CBS PMR Band

Base	Mobile	Mode	Location	User & Notes
163.4375	158.9375	NFM	Nationwide	New PMR Allocation
		NFM	Aberdeen	
163.5250	159.0250	NFM	Nationwide	New PMR Allocation
		NFM	Aberdeen	
163.7250	159.2250	NFM	Nationwide	New PMR Allocation
		NFM	Aberdeen	
163.9000	159.4000	NFM	Nationwide	New PMR Allocation
163.9250	159.4250	NFM	Nationwide	New PMR Allocation
163.9625	159.4625	NFM	Nationwide	New PMR Allocation
		NFM	Aberdeen	
163.9875	159.4875	NFM	Nationwide	New PMR Allocation
164.0000	159.5000	NFM	Nationwide	New PMR Allocation
164.0500	159.5500	NFM	Nationwide	National Scout Association
		NFM	Nationwide	St. John's Ambulance Ch 4
164.0625	159.5625	NFM	Nationwide	National Scout Association
164.0875	159.5875	NFM	Nationwide	New PMR Allocation
164.1250	159.6250	NFM	Nationwide	New PMR Allocation
164.1875	159.6875	NFM	Nationwide	New PMR Allocation
164.4000	159.9000	NFM	Nationwide	BT Selcal to Mobiles
164.4375	159.9375	NFM	Nationwide	Mobile Phone Link
164.4500	159.9500	NFM	Peterborough	Beeline
164.4750	159.9750	NFM	Nationwide	Aircall
164.5125	160.0125	NFM	Nationwide	Aircall
164.5500	160.0550	NFM	Nationwide	Aircall
164.5625	160.0625	NFM	Nationwide	Aircall
164.5750	160.0750	NFM	Nationwide	Aircall
164.6000	160.1000	NFM	Nationwide	Aircall
164.6250	160.1250	NFM	Nationwide	Aircall
		NFM	Plymouth	Emergency Doctors Service
164.7000	160.2000	NFM	Nationwide	Teleacoustic
164.7250	160.2250	NFM	Crewe	Taxis
164.7375	160.2375	NFM	Nationwide	Lodge Radio Service
164.7875	160.2875	NFM	Nationwide	Teleacoustic
164.8000	160.3000	NFM	Nationwide	Teleacoustic
164.8500	160.3500	NFM	Nationwide	Message Handling
		NFM	Colwyn	BR Station
164.8750	160.3750	NFM	Nationwide	Mobile Phone Link
164.8875	160.3375	NFM	Nationwide	Mobile Phone Link

Base	Mobile	Mode	Location	User & Notes
165.0125 - 168.2250 MHz VHF PMR High Band Base/Repeater				
Ambulance Service (England & Wales)				
165.0125	169.8125	NFM	Aberdeen	Northern Garage
165.0500	169.8500	NFM	Plymouth	AA Taxis
165.0625	169.8625	NFM	Aberdeen	ANC
		NFM	Crewe	Garage
		NFM	Perth	Community Repeater
165.0750	169.8650	NFM	Fife	Fife Regional Council
		NFM	M1	Associated Asphalt
		NFM	Perth	Community Repeater
165.0875	169.8875	NFM	Hatfield	Tarmac Construction
		NFM	Plymouth	University Security
165.1000	169.9000	NFM	Fordham	D. Jenkins TV
		NFM	Gorlestone	Ace Day & Night Taxis
		NFM	Kings Lynn	Geoff's Taxis
		NFM	Little Downham	Mott Farmers
		NFM	Saxmundham	Fishwick
165.1125	169.9125	NFM	Cambridge	Inter-City Cabs
		NFM	Dymchurch	Dymchurch Light Railway
		NFM	Sandy	Ariston Group Service
165.1250	169.9250	AM	Aberdeen	Amtrak
		NFM	Hadleigh	Wilsons Corn & Milling
		NFM	Huntingdon	Mercury Bluebird Taxis
		NFM	Perth	Taxi
		NFM	Sudbury	Woods Taxis
		NFM	Waltham	Ariston Group Service
165.1375	169.9375	NFM	London	Royal Parks Police
		NFM	London	Rimmington Minicabs
165.1500	169.9500	NFM	Nationwide	Group Four Security Ch 1
165.1625	169.9625	NFM	Ipswich	Spotcheck Security
		NFM	Nottingham	Nottingham University
165.1750	169.9750	NFM	Nationwide	Group Four Security Ch 2
165.1875	169.9875	NFM	Aberdeen	Security
		NFM	E Anglia	Pritchard Security
165.2000	170.0000	NFM	Aberdeen	Aberdeen Vets
		NFM	Norwich	Esso Heating
165.2125	170.0125	NFM	Grafton	Grosvenor Estates
		NFM	Nationwide	British Coal Security
		NFM	US Embassy London	US Secret Service Ch Mike
165.2250	170.0250	NFM	Jersey	Flying Dragon Cabs
165.2375	170.0375	NFM	Crewe	Taxi
		NFM	Leighton Buzzard	Choake Billington
165.2500	170.0500	AM	Aberdeen	Lucas
		NFM	March	Middle Level Commissioner
165.2625	170.0625	NFM	Jersey	Community Repeater
		NFM	Lancashire	Barkley Council
165.2750	170.0750	NFM	Halstead	Gosling Bros.
		NFM	Mansfield	Ace Taxis
		NFM	Shireoaks Colliery	Pit Security
165.2875	170.0875	NFM	Chesterford	Park Research

Base	Mobile	Mode	Location	User & Notes
165.3000	170.1000	NFM	Nationwide	
165.3125	170.1125	NFM	Ely	Garrett
		NFM	Stansted	Aircars
165.3250	170.1250	NFM	London	Diplomatic Transport
		NFM	Plymouth	City Security C/S Papa Ctrl
165.3375	170.1375	NFM	Clacton	Bernies Taxis
		NFM	Kings Lynn	Simons
		NFM	Letchworth	John's Taxis
		NFM	London	Diplomatic Transport
		NFM	Stowmarket	ICI Paint Depot
165.3500	170.1500	NFM	Nottingham	Doctors Service
165.3625	170.1625	NFM	Leicester	Taxi
		NFM	Dundee	Car Hire Service
		NFM	Edinburgh	Community Repeater
		NFM	Liverpool	Taxis
		NFM	Perth	Community Repeater
165.3750	170.1750	NFM	Cambridge	Able Cars
		NFM	Harlow	Regency Cars
		NFM	US Embassy London	US Secret Serv. Ch Charlie
		NFM	Plymouth	Olympic Taxis
		NFM	Warrington	Warrington Borough County
165.3875	170.1875	NFM	Clacton	Apollo Taxis
		NFM	Fakenham	Selective Fertilisers
		NFM	Luton Airport	Lep Transport
		NFM	Morecambe	Joe's Taxis
		NFM	Retford	Malcolm's Taxis
		NFM	Soham	Tompsett
165.4000	170.2000	NFM	Aberdeen	Oil Fabricators
		NFM	Bedford	Community Repeater
		NFM	Lancashire	Andersons Pumps
		NFM	London	IBA Maintenance
		NFM	London Underground	Ch 2 Spare
		NFM	Baker Street	London Underground
		NFM	Tooting Bec	London Underground
		NFM	Clapham North	London Underground
		NFM	Oxford Circus	London Underground
		NFM	Tooting Broadway	London Underground
		NFM	Clapham South	London Underground
		NFM	Balham	London Underground
		NFM	Clapham Common	London Underground
		NFM	Monument	London Underground
165.4125	170.2125	NFM	Fife	Fife Regional Council
		NFM	London Underground	Bakerloo/District/Northern
		NFM	Perth	King Contractors Channel 1
165.4250	170.2250	NFM	Chatteris	W. Barnes
		NFM	Hitchin	DER Television
		NFM	Plymouth	Night Watch Security
165.4375	170.2375	NFM	Carnforth	Council Roads Department
		NFM	Nationwide	Community Repeater Chan
		NFM	Mansfield	Doctors Service

Base	Mobile	Mode	Location	User & Notes
		NFM	W Midlands	Castle Security
		NFM	London Underground	Ch 4 East London/ Hamersmith & City/ Metropolitan/Piccadilly
165.4500	170.2500	NFM	Nationwide	Community Repeater Chan
		NFM	Nationwide	IBA Maintenance
165.4625	170.2625	NFM	Nationwide	Community Repeater Chan
		NFM	London Underground	Ch8 Central/Jubilee/Victoria
165.4750	170.2750	NFM	Anglia	Anglia Carphones
		NFM	Jersey	Honoury Police Channel 1
		NFM	Plymouth	Red Lightning Dispatch
165.4875	170.2875	NFM	London	ODRATS Channel 3
165.5000	170.3000	NFM	Cambridge	DER Television
		NFM	Letchworth	G. Folly Builders
		NFM	Plymouth	Tower Cabs
165.5125	170.3125	NFM	Aberdeen	Taxi
		NFM	Cambridge	Browns Taxis
165.5250	170.3250	NFM	Peterborough	DER Television
165.5375	170.3375	NFM	Oxford	Community Repeater
165.5500	170.3500	NFM	Nationwide	Community Repeater Chan
		NFM	Edinburgh	Tarmac Roadstone
		NFM	Plymouth	Military Security C/S RM
165.5625	170.3625	NFM	Nationwide	Community Repeater Chan
		NFM	Edinburgh	Castle Security
165.5750	170.3750	NFM	Cambridge	H. Robinson
		NFM	Kings Lynn	DER Television
		NFM	Woodbridge	K. Tuckwell Engineers
165.5875	170.3875	NFM	Harpenden	DER Television
		NFM	Luton	DER Television
165.6000	170.4000	NFM	London	Carreras Rothams Displays
165.6125	170.4125	NFM	Lakenheath	Base Taxis
165.6250	170.4250	NFM	London	BTP Ch 3
		NFM	Paddington Station	BR Transport Police
		NFM	Leicester	City Buses
165.6375	170.4750	NFM	London	BTP Ch 2
		NFM	Crewe	BR Transport Police
		NFM	Chester	BR Transport Police
		NFM	Liverpool	BR Transport Police
		NFM	Manchester	BR Transport Police
		NFM	Birmingham	BR Transport Police
		NFM	Victoria	BR Transport Police
165.6500	170.8500	NFM	London	BTP Ch 1
		NFM	London Underground	BR Transport Police
165.6625	170.8625	NFM	Jersey	Public Works
165.6750	170.8750	NFM	Dunstable	E.J. Allan
165.6875	170.4875	NFM	London	Bell Fruit Machines
165.7000	170.5000	NFM	Chatteris	Catwood Potatoes
		NFM	Kings Lynn	Ambassador Taxis
		NFM	Saffron Walden	Crusader Cars
165.7125	170.5125	NFM	Folkestone	City Buses

Base	Mobile	Mode	Location	User & Notes
		NFM	Ipswich	E.H. Roberts
		NFM	Newcastle	Castle Cars
165.7250	170.5250	NFM	Hatfield Tunnel	Tarmac Construction
		NFM	Hockwold	Bob's Taxis
165.7375	170.5375	NFM	Danbury	Amey Roadstones
		NFM	Sudbury	Amey Roadstones
165.7500	170.5500	NFM	Cambridge	Camhus
		NFM	Luton	Luton Borough Council
		NFM	Norwich	ECOC
		NFM	Peterborough	Camhus
165.7625	170.5625	NFM	Aberdeen	Waste Masters
		NFM	Bedfordshire	Mid Bedfordshire Council
		NFM	Dundee	City Council Dog Catcher
		NFM	Edinburgh	Lothian Regional Council
		NFM	Norfolk	Norfolk County Council
165.7750	170.5750	NFM	Norwich	Norwich City Council
		NFM	Southend	District Council
165.7875	170.5875	NFM	Hyndburh	Accrington Bus Control
		NFM	Suffolk	Mid Suffolk District Council
		NFM	London	US Secret Serv. Chan Baker
165.8000	170.6000	NFM	Dundee	Taxis
		NFM	Felixstowe	Peewit Caravans
		NFM	Luton	C.J. Private Hire
165.8125	170.6125	NFM	Cambridge	Panther Cars
		NFM	London	ODRATS Channel 5
165.8250	170.6250	NFM	Ipswich	Wilding & Smith
		NFM	March	Worral Potatoes
		NFM	Oxford	ABC Taxis
165.8375	170.6375	NFM	Ely	A.E. Lee Farms
		NFM	London	ODRATS Channel 6
		NFM	St. Osyth	Tudor Taxis
165.8500	170.6500	NFM	Haddenham	A.F. Buck
		NFM	London	ODRATS Channel 4
		NFM	Three Holes	Hallsworth Framing Co.
165.8625	170.6625	NFM	Nationwide	Securicor Channel 6
165.8750	170.6750	NFM	Nationwide	Securicor Channel 2
165.8750		NFM	Space	Myr Space Station
165.8875	170.6875	NFM	Nationwide	Securicor Channel 3
165.9000	170.7000	NFM	Dundee	Taxi
		NFM	Hitchin	Duggan's Taxis
		NFM	US Embassy London	US Secret Serv. Ch Romeo
165.9125	170.7125	NFM	Nationwide	Securicor Channel 7
165.9250	170.7250	NFM	Nationwide	
		NFM	Aberdeen	Oil Industry
165.9375	170.7375	NFM	Nationwide	Securicor Channel 8
165.9500	170.7500	NFM	Carnforth	Council Bin Men
		NFM	March	Guy Morton
		NFM	Norfolk	Norfolk Farm Produce
165.9625	170.7625	NFM	Nationwide	Securicor Channel 4
165.9750	170.7750	NFM	Nationwide	Securicor Chan 1 Emergency

Base	Mobile	Mode	Location	User & Notes
165.9875	170.7875	NFM	Nationwide	Securicor Channel 5
166.0000	170.8000	NFM	Martham	Fleggmart
		NFM	Space	Soviet Satellite (Myr)
166.0125	170.8125	NFM	Aberdeen	Taxi
		NFM	Peterborough	On Site Tyres
166.0250	170.8250	AM	Aberdeen	Oil Industry
		NFM	Kenney Hill	J.A Butcher
		NFM	Peterborough	Hotpoint
166.0375	170.8375	AM	Aberdeen	Taxi
		NFM	Cromer	Biffa Ltd
		NFM	Jersey	Pentagon Ltd
166.0500	170.8500	NFM	Oxford	Oxford City Council
		NFM	Suffolk	Suffolk Coastal Council
166.0625	170.8625	NFM	Colchester	Borough Council
		NFM	Dundee	City Council Workshop
		NFM	Eakering	BP Depot
166.0750	170.8750	NFM	Ipswich	Borough Council (Parks)
		NFM	Jersey	Public Services
		NFM	Nationwide	Securicor
		NFM	N. Herefordshire	District Council Channel 1
166.0875	170.8875	NFM	Essex	Havering Council
		NFM	Perth	Council Roads Department
		NFM	Perth	Council Leisure Dept
166.1000	170.9000	NFM	Jersey	Turner/Bluebird Cabs
		AM	Norfolk	Ambulance
		AM	Nottinghamshire	Ambulance
		AM	Oxfordshire	Ambulance/Central
166.1125	170.9125	NFM	Nationwide	
166.1250	170.9250	NFM	Aylesbury	District Council
		NFM	Space	Soviet Mir Telemetry
166.1375	170.9375	NFM	Basildon	District Council
		NFM	N. Herefordshire	District Council
		NFM	Oxford	Bus Company
166.1500	170.9500	NFM	Nationwide	
		NFM	Aberdeen	Dee Van Hire
		NFM	Liverpool	City Council
166.1625	170.9625	NFM	Sheffield	Parks Security
166.1750	170.9750	NFM	Grays	District Council
		NFM	Welwyn & Hatfield	District Council
		NFM	Kings Lynn	District Council
		NFM	Thanet	District Council
		NFM	W. Norfolk	District Council
		NFM	S. Cambridgeshire	District Council
		NFM	Vale of White Horse	District Council
166.1875	170.9875	NFM	Aberdeen	Council Roads Dept
		NFM	Folkestone	District Council
166.2000	171.0000	AM	SE London	Ambulance Ch 13/Orange
		NFM	Perth	Community Data Repeater
166.2125	171.0125	NFM	London	US Secret Serv Chan Romeo
166.2250	171.0250	NFM	Aberdeen	Fish Market

Base	Mobile	Mode	Location	User & Notes
		NFM	Ashford	District Council
		NFM	Jersey	Gas Board
		AM	Plymouth	City Bus Company
		NFM	Nationwide	Local Authorities
166.2375	171.0375	NFM	Nationwide	
		NFM	Aberdeen	TV Repairs
166.2500	171.0500	NFM	Dundee	Electrical Repairs
166.2750	171.0750	AM	E Lancashire	Lancashire Ambulances
		AM	SW London	Ambulance Ch 11/Orange
166.2875	171.0875	AM	Buckinghamshire	Ambulance
		AM	Cornwall	Ambulance Service
		AM	E Kent	Ambulance
		AM	Powys	Ambulance
		AM	Withenshaw	Paramedics
166.3000	171.1000	AM	NE London	Ambulance Ch 9/Gold Base
		AM	Leicester	Ambulance
		AM	Manchester	Ambulance
166.3125	171.1125	AM	London	Ambulance Ch 1/Red Base
		AM	Plymouth	Devon Ambulance Service
		AM	Powys	Ambulance
166.3250	171.1250	AM	SW London	Ambulance Ch 12/Orange
166.3375	171.1375	NFM	Dundee	City Council Data Link
		AM	Kent	Ambulance
		AM	Suffolk	Ambulance Control
166.3500	171.1500	AM	East London	Ambulance Ch 2/Red Base
166.3625	171.1625	NFM	Channel Tunnel	Maintenance
		AM	Crewe	Ambulance
		AM	Essex	Ambulance
		AM	Hampshire	Ambulance
166.3750	171.1750	NFM	Dundee	Taxi
		AM	SE London	Ambulance Ch 14/Green
166.3875	171.1875	AM	Berkshire	Ambulance
		AM	N Staffordshire	N Staffs Infirmary
		AM	W Kent	Ambulance
166.4000	171.2000	AM	East Sussex	Ambulance
		NFM	US Embassy London	US Secret Serv Chan Golf
166.4125	171.2125	AM	South London	Ambulance Ch 3/Red Base
166.4250	171.2250	AM	Essex	Ambulance
		AM	NE London	Ambulance Ch 7/Gold Base
		AM	Stoke on Trent	Emergency Doctor
166.4375	171.2375	AM	London	Ambulance Ch 15 Emrgncy
		AM	Oxfordshire	Ambulance
		AM	Powys	Ambulance
166.4500	171.2500	NFM	Jersey	St Ouen Parish
		AM	NW London	Ambulance Ch 6/Blue Base
166.4625	171.2625	AM	East Sussex	Ambulance
		NFM	US Embassy London	US Secret Serv Chan Victor
166.4750	171.2750	NFM	Jersey	Honoury Police
		AM	NW London	Ambulance Ch 5/Blue Base
166.4875	171.2875	AM	Essex	Ambulance

The UK Scanning Directory

Base	Mobile	Mode	Location	User & Notes
		NFM	Jersey	Honoury Police Channel 3
		NFM	US Embassy London	US Secert Serv Chan Zulu
		AM	Oxfordshire	Ambulance
166.5000	171.3000	AM	Gt Manchester	Ambulance
		AM	NE London	Ambulance Ch 8/Gold Base
166.5125	171.3125	AM	Cambridgeshire	Ambulance
		AM	Lancashire	Ambulance
		NFM	US Embassy London	US Secret Serv Chan Sierra
		AM	Northumberland	Ambulance
		AM	Surrey	Ambulance
		AM	Warwickshire	Ambulance
166.5250	171.3250	AM	East Sussex	Ambulance
		AM	London	Ambulance Ch 4/Red Base
		AM	Suffolk	Ambulance
		AM	W Lancashire	Lancashire Ambulances
		AM	Warwickshire	Ambulance
166.5375	171.3375	AM	Leicestershire	Ambulance
		AM	Surrey	Ambulance
		AM	West Yorkshire	Ambulance
166.5500	171.3500	AM	Devon	Patient Ambulance Service
		AM	Essex	Ambulance Ch 3
		AM	Northamptonshire	Ambulance
166.5625	171.3625	AM	Buckinghamshire	Ambulance
		NFM	Dundee	National Carriers
		AM	Herefordshire	Ambulance
		NFM	Perth	Ready Mixed Concrete
		AM	West Sussex	Ambulance
		AM	West Yorkshire	Ambulance
166.5750	171.3750	AM	Inner London	Ambulance Ch 10/White
		AM	Northamptonshire	Ambulance
166.5875	171.3875	NFM	Aberdeen	Surveyors
		AM	Durham	Ambulance
		AM	Hampshire	Ambulance
		AM	Hertfordshire	Ambulance
		AM	Sefton	Ambulance
		AM	Staffordshire	Ambulance
166.6000	171.4000	NFM	Colchester	NEEDES
		AM	England & Wales	Doctors Special Services
		AM	Peterborough	Dr. Gray
		AM	Saffron Walden	Accident Group
		AM	Surrey	Ambulance
		AM	Swaffham	Dr. Pilkington
166.6125	171.4125	AM	Berkshire	Ambulance
		AM	Hertfordshire	Ambulance
		AM	Oxfordshire	Ambulance
		AM	Staffordshire	Ambulance
166.6250	171.4250	AM	Aberdeen	Retail Park
		NFM	Cambridgeshire	Todd
		AM	Nationwide	Ambulance-to-Hospital Link
166.6375	171.4375	NFM	Aberdeen	Radio Specialists

Base	Mobile	Mode	Location	User & Notes
		NFM	Jersey	Holiday Tours
		NFM	Nationwide	Rediffussion Comms
		NFM	Rosyton	Farmers Fertilisers
		NFM	Whittlesey	S & S Tractors
		NFM	Woodbridge	Greenwell Farms
166.6500	171.4500	NFM	Bedfordshire	Bedfordshire Growers
		NFM	Bury	Byford Taxis
		NFM	Huntingdon	R. O'Connell
		NFM	Ipswich	Taxi Association
		NFM	London	ODRATS Channel 1
166.6625	171.4625	NFM	Baldock	Butts Taxis
		NFM	Bury	British Sugar
		NFM	Cantley	British Sugar
		NFM	Dullingham	P.B. Taylor
		NFM	Nationwide	ICL Channel 1
		NFM	Perth	Tay Taxis
		NFM	Sudbury	A Line Taxis
166.6750	171.4750	NFM	Alconbury	Steve's Taxis
		NFM	Nationwide	ICL Channel 2
166.6875	171.4875	NFM	Bury	British Sugar
		NFM	Chatteris	Whitworth Produce
		NFM	Ipswich	Ransomes
		NFM	Kings Lynn	British Sugar
		NFM	Norfolk	Crane & Son
		NFM	Peterborough	Co-Op TV Services
166.7000	171.5000	NFM	US Embassy London	US Sec. Serv. Ch November
		NFM	Methwold	Darby Bros. Farms
		NFM	Norwich	Bestway Taxis
166.7125	171.5125	NFM	Wissington	British Sugar
166.7250	171.5250	NFM	Lowestoft	Birds Eye
		NFM	Sharnbrook	Unilever
166.7375	171.5375	NFM	Nationwide	ICL Channel 3
		NFM	Norwich	Knight Benjamin
166.7500	171.5500	NFM	Banchory	Taxi
		AM	Crewe	Ambulance
		NFM	Surrey	Health Service
166.7625	171.5625	NFM	Castle Donington	Race Control
166.7750	171.5750	NFM	Abingdon	Abingdon Hospital
		NFM	Bedfordshire	Midwives
		NFM	Dover	Council
		NFM	Hampshire	Health Service
		NFM	Ipswich	Midwives
166.7875	171.5875	AM	Aberdeen	Oil Servcing
		NFM	Nationwide	Community Repeater
		NFM	Plymouth	Co-Op Store Detectives
166.8000	171.8000	NFM	Nationwide	Community Repeater
		NFM	Mansfield	Kings Mill Hospital
166.8125	171.6125	NFM	Essex	Havering Council
		NFM	Herts	Doctors Channel
		NFM	Upwell	Health Centre

Base	Mobile	Mode	Location	User & Notes
166.8250	171.6250	NFM	Clare	Dr. Carter
		NFM	Jersey	Yellow Cabs
		NFM	Kent	Health Service
		AM	Thanet	Ambulance
166.8375	171.6375	NFM	Dorest	Health Service
		NFM	Castle Donington	Medic Control
166.8500	171.6500	NFM	Dundee	Christian Salvensen
166.8625	171.6625	AM	Aberdeen	Rig Servicing
		NFM	Crewe	City Council
		NFM	Edinburgh	Community Repeater
166.8750	171.6750	NFM	Aberdeen	Community Repeater
166.8875	171.6875	NFM	Nationwide	
166.9000	171.7000	NFM	Aberdeen	Aberdeen Vets
		NFM	Colchester	A.E. Arnold
		NFM	Ely	Vets
166.9125	171.7125	NFM	Ipswich	British Sugar
		NFM	Peterborough	British Sugar
		NFM	Stretham	N. Rose Builders
166.9250	171.7250	NFM	Edinburgh	Community Repeater
		NFM	Littleport	J.H. Martin
		NFM	Lowestoft	Birds Eye
166.9375	171.7375	NFM	Aberdeen	Reatil Park Security
166.9500	171.7500	NFM	Nationwide	BBC TV O/B Crews
166.9750	171.7750	NFM	Edinburgh	Community Repeater
		NFM	Harwich	Daves Taxis
		NFM	Oxford	001 Cars
166.9875	171.7875	NFM	Littleport	Sallis Bros
167.0000	171.8000	NFM	Nationwide	
167.0125	171.8125	NFM	Essex	Telecom Repeater
167.0250	171.8250	NFM	E. Dereham	Fransham Farm Co.
		NFM	Ely	Stopps Taxis
		NFM	US Embassy London	US Sec. Serv. Ch Whiskey
167.0375	171.8375	NFM	London	Burns Security
		NFM	Leicester	Security
		NFM	Lowestoft	Birds Eye
167.0500	171.8500	NFM	Nationwide	Community Repeater
167.0625	171.8625	NFM	Sharnbrook	Assoc. Asphalte
167.0750	171.8750	NFM	N. Walsham	Norfolk Cannaries
167.0875	171.8875	NFM	Gt. Yarmouth	Botton Bros.
167.1000	171.9000	NFM	Brandon	F. Hiam Farms
		NFM	Cambridge	Ace Taxis
		NFM	Hempstead	R.D. Haylock
		NFM	Perth	Taxi
		NFM	Thorney	M.S. Smith
167.1125	171.9125	NFM	Gt. Yarmouth	J & H Bunn
167.1250	171.9250	NFM	Edinburgh	Taxi
		NFM	Kings Lynn	Watlington Plant
		NFM	March	Ross-Produce
167.1325	171.9375	NFM	Edinburgh	Community Repeater
		NFM	Kings Lynn	Wheelers TV

Base	Mobile	Mode	Location	User & Notes
167.1500	171.9500	AM	Aberdeen	Taxi
		NFM	Crewe	Taxi
		NFM	St. Neots	T & R Taxis
		NFM	Worlington	Tuckwell
167.1625	171.9625	NFM	Jersey	Gorey Cabs
		NFM	Keswick	Ambulance
		NFM	Norfolk	ICL Channel 4
167.1750	171.9750	NFM	Norfolk	ICL Channel 5
167.1875	171.9875	NFM	Lancaster	Council Roads Dept
167.2000	172.0000	NFM	Nationwide	PMR Short Term Hire
167.2125	172.0125	NFM	Newcastle	Silver Cars
		NFM	Norfolk	ICL Channel 6
167.2250	172.0250	NFM	Royston	Meltax
		NFM	Worksop	Bee Line Taxis
167.2375	172.0375	NFM	Aberdeen	Taxi
		NFM	Haverhill	Chequer Cabs
		NFM	Huntingdon	Pete's Taxis
		NFM	Oxford	G.T. Taxis
167.2500	172.0500	NFM	Aberdeen	Taxi
		NFM	Plymouth	Armada Taxis
		NFM	Powys	Thomas Jones (Vet)
167.2625	172.0625	NFM	Haverhill	Havtax
		NFM	Stevenage	Freewheelers
167.2750	172.0750	NFM	Aberdeen	Office Security
		NFM	Hockwold	J. Denney Taxix
		NFM	Huntingdon	H. Raby
		NFM	Jersey	States Motor Traffic Dept
		NFM	Southend	Assoc. Radio Cars
167.2875	172.0875	NFM	Norfolk	ICL Channel 7
		AM	Plymouth	Chequars Cabs
167.3000	172.1000	NFM	Kings Lynn	R.D. Carter
		NFM	Norfolk	ICL Channel 8
		NFM	Oxford	Radiotaxis
167.3125	172.1125	NFM	Aberdeen	Office Security
		NFM	Nationwide	
167.3250	172.1250	NFM	Bedford	Bedfordia Farms
		NFM	Chatteris	Graves & Graves
167.3375	172.1375	NFM	Biggleswade	Whitbread Farms
167.3500	172.1500	AM	Aberdeen	Taxi
		NFM	Fordham	Allen Newport
		NFM	Whittlesey	Luxicabs
167.3625	172.1625	NFM	Soham	Greens of Soham
167.3750	172.1750	NFM	Aberdeen	Tyre Service
		NFM	Nationwide	Comet Television
167.3875	172.1875	NFM	Aberdeen	Taxi
		NFM	Norfolk	ICL Channel 9
167.4000	172.2000	NFM	Colchester	Paxmans Diesels
		NFM	Leigh on Sea	Kellys Radio
		NFM	Norfolk	Assoc. Leisure
		NFM	Perth	Taxi

Base	Mobile	Mode	Location	User & Notes
167.4125	172.2125	NFM	Aberdeen	Estate Security
		NFM	Norfolk	ICL Channel 10
		NFM	Oxford	City Taxis
		NFM	Stoke on Trent	Haulage Company
167.4250	172.2250	NFM	Aberdeen	Taxi
		NFM	Downham Market	B.W. Mack
		NFM	Huntingdon	R. Brading
		NFM	Jersey	Jersey States Repeater
		NFM	Norwich	J.B. Green
167.4375	172.2375	NFM	Norwich	C. Wace
		NFM	S. Bumpstead	Shore Hall Estates
167.4500	172.2500	NFM	Knebworth	Vendustrial Ltd
		NFM	Newmarket	Chilcotts Taxis
		NFM	Norfolk	ICL Channel 11
167.4625	172.2625	NFM	Aberdeen	Crane Hire
		NFM	Lowestoft	Birds Eye
167.4875	172.2875	NFM	Aberdeen	Taxi
		NFM	Edinburgh	Taxi
167.5000	172.3000	NFM	Aberdeen	Farm
		NFM	Barway	Shropshire Produce
		NFM	Shipham	William Moorfoot
167.5125	172.3125	NFM	Norfolk	ICL Channel 12
167.5250	172.3250	NFM	Huntingdon	A.E. Abraham
		NFM	Norwich	Five Star Taxis
167.5375	172.3375	NFM	Aberdeen	Farm
167.5500	172.3500	NFM	Colchester	Abbeygate Taxis
		NFM	Jersey	Fort Regent Leisure Centre
167.5625	172.3625	NFM	Aberdeen	Estate
		NFM	March	G.E. Tribe
		NFM	Plymouth	Key Cab Taxis
167.5750	172.3750	NFM	Jersey	Pioneer Coaches
		NFM	March	David Johnson Farms
167.5875	172.3875	NFM	Bedford	Riverside taxis
167.6000	172.4000	NFM	Ipswich	Peter Green
167.6125	172.4125	NFM	Bedford	Windshield Ent.
167.6250	172.4250	NFM	Aberdeen	Deeside Shop Fitters
		NFM	Barway	Shropshire Produce
		NFM	Perth	Tayside Shopper Fitters
		NFM	Rochford	Andrews Taxis
167.6375	172.4375	NFM	Suffolk	Rumbolows Television
167.6500	172.4500	NFM	Wickford	Carter & Ward
167.6625	172.4625	NFM	Colchester	Smythe Motors
		NFM	Leicester	Taxi
167.6750	172.4750	NFM	Perth	Tayside Regional Council
167.6875	172.4875	NFM	Aberdeen	Slatters
		NFM	Nationwide	
167.7000	172.5000	AM	Aberdeen	Security
		NFM	Littleport	H. Thompson
167.7125	172.5125	NFM	Cambridge	Jakubowski Builders
167.7250	172.5250	NFM	Lowestoft	Birds Eye

Base	Mobile	Mode	Location	User & Notes
167.7500	172.5500	NFM	Lowestoft	Oulton Radio Taxis
167.7625	172.5625	NFM	Nationwide	
167.7750	172.5750	NFM	Welwyn Graden City	Industrial Services
		NFM	Woburn	Speedwell Farms
167.7875	172.5875	NFM	Basildon	Ace Taxi Group
		NFM	Edinburgh	Falcon Delivery
167.8000	172.6000	AM	Aberdeen	Security
		NFM	Cambridge	United Taxis
		NFM	Dovercourt	Starling Taxis
167.8125	172.6125	NFM	Bedford	M.W. Ward
		NFM	Jersey	Westmount Repeater
		NFM	Watlington	Watlington Plant Hire
167.8250	172.6250	NFM	Ipswich	Robin Hood Taxis
167.8375	172.6375	NFM	Nationwide	
167.8500	172.6500	NFM	Cambridge	Clearaway
167.8625	172.6625	NFM	Nationwide	
167.8750	172.6750	NFM	Nationwide	
167.8875	172.6875	NFM	Nationwide	
167.9000	172.7000	NFM	Jersey	Hire Channel
167.9125	172.7125	NFM	Royston	B & B Taxis
		NFM	Woodbridge	Normans Transport
167.9250	172.7250	NFM	Newmarket	Sound City Cars
167.9375	172.7375	NFM	Luton	Harvey Plant Hire
167.9500	172.7500	NFM	Jersey	Clarendon Cabs
167.9625	172.7625	NFM	Jersey	Dail-a-Cab
		NFM	Peterborough	Horrells Dairies
167.9750	172.7750	NFM	March	Coy & Manchett
167.9875	172.7875	NFM	Thanet	B.C. Taxis
168.0000	172.8000	NFM	Cambridge	Camtax
168.0125	172.8125	NFM	Nationwide	
		NFM	Stoke on Trent	Fourstar Taxis
168.0250	172.8250	NFM	Hitchin	Swan Garage
168.0375	172.8375	NFM	March	Rowe, Manchett & Till
168.0500	172.8500	NFM	Cambridge	A1 Taxis
		NFM	Cowley	Rover Plant Ambulance
		NFM	Worksop	J.J & J.R. Jacksons
168.0625	172.8625	NFM	Nationwide	
168.0750	172.8750	NFM	West Bergholt	John Willsher
168.0875	172.8875	NFM	Stevenage	W.G. Silverton
168.1000	172.9000	NFM	Welwyn Garden City	752 Taxis
168.1125	172.9125	NFM	Caister	Avenue Taxis
168.1375	172.9375	NFM	Aberdeen	Plant Hire
		NFM	Stoke on Trent	Sid's Taxis
168.1500	172.9500	NFM	Benwick	Bank Farms
		NFM	Ely	Evans Taxis
168.1625	172.9625	NFM	Aberdeen	Taxi
168.1750	172.9750	NFM	Bedford	Key Cars
168.1875	172.9875	NFM	Sutton	Salisbury Bros.
168.2000	173.0000	NFM	Nationwide	
168.2125	173.0125	NFM	Cambridge	Bettacars

Base	Mobile	Mode	Location	User & Notes
		NFM	Norfolk	ICL Channel 13
168.2250	173.0250	NFM	Jersey	Interlink Delivery
		NFM	March	Central Security
		NFM	Nationwide	Radio Investigation Service
168.2375	173.0375	NFM	Great Yarmouth	Halcyon Shipping
		NFM	Jersey	States Housing Dept
		NFM	Nottingham	Holme Pier Water Sports

168.2500 - 168.9375 MHz   Government Agencies/British Telecom

Base	Mobile	Mode	Location	User & Notes
168.2500	173.0500	NFM	Nationwide	DTI Radio Investigations
		NFM	Nationwide	BT Linesmen
168.2625	173.0625	NFM	Nationwide	BT Cable Laying
168.2750	173.0750	NFM	Nationwide	DTI Radio Investigations
168.2750	173.0750	NFM	Nationwide	Radiocommunications Agny
		NFM	North West England	RCA
168.2875	173.0875	NFM	Nationwide	
168.3000	173.1000	NFM	Nationwide	BT Linesmen
		NFM	Nationwide	DTI Radio Investigation
168.3125	173.1125	NFM	Nationwide	
168.3250	173.1250	NFM	Nationwide	
168.3375	173.1375	NFM	Nationwide	
168.3500	173.1500	NFM	Nationwide	
168.3625	173.1625	NFM	Nationwide	
168.3750	173.1750	NFM	Nationwide	
168.3875	173.1875	NFM	Aberdeen	
168.3875		NFM	Formula One Racing	Williams Voice (Patrese)
168.4000		NFM	Formula One Racing	Williams Team Voice Link
168.4375	173.2375	NFM	Aberdeen	Security
168.8625	173.6625	NFM	Nationwide	National Seismic Studies
168.8875	173.6875	NFM	Nationwide	BT Video Set Up Link
168.9000	173.7000	NFM	Jersey	Honoury Police Channel 7
168.9250	173.7250	NFM	London	Bullion Movement Security

168.9500 - 169.8375 MHz   PMR High Band Simplex 12.5 kHz

Base	Mobile	Mode	Location	User & Notes
168.9500	168.9500	NFM	Nationwide	British Telecom
168.9625	168.9625	NFM	Aberdeen	Docks
		NFM	Bury St. Edmunds	Rushbrooke Farms
		NFM	Jersey	Industrial (Motors) Ltd
168.9750	168.9750	NFM	Nationwide	BBC Engineering
		NFM	Leighton Buzzard	Joseph Arnold
168.9875	168.9875	NFM	Aberdeen	Docks
		NFM	Cambridge	Fitzwilliam College
		NFM	Jersey	Harbour
		NFM	London	Wembley Stadium Stewards
		NFM	Nationwide	Ordnance Survey
		NFM	Nationwide	IBA Riggers
169.0000	169.0000	NFM	Bournemouth	Synagogue Security
		NFM	Cambridge	Trinity College
		NFM	Crewe	Oakley Centre

Base	Mobile	Mode	Location	User & Notes
		NFM	Tilbury	Docks
		NFM	Southampton	Docks
		NFM	Nationwide	RAC Rally
169.0125	169.0125	NFM	Dundee	Tay Bridge Maintenance
		NFM	Nationwide	Short Term Lease PMR
		NFM	Parkeston Quay	Harwich Transport
169.0250	169.0250	NFM	Bury St. Edmunds	District Council
		NFM	Cambridge	Posthouse Forte Hotel
		NFM	Nationwide	St Johns Ambulance Chan B
		NFM	Powys	Powys County Council
		NFM	Nationwide	RAC Rally
169.0375	169.0375	NFM	Bournemouth	Internat. Centre Security
		NFM	Clacton	Pier Company
		NFM	Tendring	Tendring Hundreds Water
169.0500	169.0500	NFM	England	CEGB
		NFM	Nationwide	RAC Rally
169.0625	169.0625	NFM	Jersey	Docks
		NFM	Lowestoft	Christian Salvesen
		NFM	Nationwide	ICL Computers
169.0750	169.0750	NFM	Poole	Valiant Security
169.0875	169.0875	NFM	Nationwide	Limited DHSS Use
		NFM	Nationwide	Red Cross
		NFM	Plymouth	Plymouth Market Security
		NFM	Southampton	Docks
		NFM	Sutton Coldfield	Belfry Golf Course
169.1000	169.1000	NFM	Cambridge	Medical Research Council
169.1125	169.1125	NFM	Bournemouth	Internat Centre Security
		NFM	Halesworth	K.W. Thomas
		NFM	Worksop	Fox Covert Scrap Yard
169.1250	169.1250	NFM	Sheffield	Radio Sheffield Beacon
169.1300		NFM	Formula One Racing	Williams Team Voice Link
169.1375		NFM	Formula One Racing	Williams Voice (Mansell)
169.1375	169.1375	NFM	Aberdeen	BonAccord Centre Security
		NFM	Nationwide	Short Term Lease PMR
169.1500	169.1500	NFM	Nationwide	National Trust
		NFM	Worksop	Bassetlaw Hospital
169.1625	169.1625	NFM	Nationwide	Short Term Lease PMR
169.1750	169.1750	NFM	Martlesham Heath	Department of Energy
		NFM	Three Holes	Frank Hartley
169.1875		NFM	Formula One Racing	Lotus Team Ch 3
169.1875	169.1875	NFM	Jersey	Thomson Hire
		NFM	Nationwide	Short Term Lease PMR
		NFM	Nationwide	NCB Emergencies
169.2000	169.2000	NFM	Aberdeen	Aberdeen Ice Rink
		NFM	Jersey	British Airways
		NFM	Ware	Glaxo Operations
169.2125	169.2125	NFM	Bournemouth	M+J Security
		NFM	E Midlands	Airport Security
169.2250	169.2250	NFM	Jersey	Hurricaine Despatch

Base	Mobile	Mode	Location	User & Notes
		NFM	Nationwide	RAC Rallies
169.2375	169.2375	NFM	England	Water Baliffs
169.2500	169.2500	NFM	Southampton	University Security
169.2625	169.2625	NFM	Felixstowe	Repcon
		NFM	Great Yarmouth	St.Nicholas Hospital
169.2750	169.2750	NFM	Woodbridge	Kemball
169.3000	169.3000	NFM	Jersey	British Airways Loading
		NFM	West Bromwich	West Brom. Albion FC
169.3125	169.3125	NFM	Jersey	Commodore Shipping
169.3375	169.3375	NFM	Cambridge	Kings College
		NFM	Jersey	Island Sports Officials
		NFM	Perth	St John's Centre Security
		NFM	Devon	Tamar Bridge Security
		NFM	Silverstone	RAC General Use
169.3500	169.3500	NFM	Powys	South Wales Electricity
169.3625	169.3625	NFM	Leighton Buzzard	George Garside Sand
		NFM	Plymouth	Plymouth Argyle FC Ch 1
		NFM	Nationwide	St Johns Ambulance Ch A/1
169.3750	169.3750	NFM	South Walden	Bell College
169.3875	169.3875	NFM	Nationwide	St Johns Ambulance Chan 2
		NFM	Kent	Prismo Road Surfacing
		NFM	Nottingham	Technical Services

169.4000 - 169.8000 MHz   New European Messaging Service (ERMES)

Base	Mobile	Mode	Location	User & Notes
169.4875		NFM	Formula One Racing	Benetton Voice (Brundle)
169.7250		NFM	Plymouth	St John's Ambulance
		NFM	Plymouth	Plymouth Argyle Ch 2
169.8125		NFM	Birmingham	Aston Villa FC Stewards
169.8250		NFM	Folkestone	Folkestone Hotel
		NFM	Jersey	Honoury Police Channel 5

169.8500 - 173.0500 MHz PMR High Band Mobiles (Base Split - 5.8 MHz)

Base	Mobile	Mode	Location	User & Notes
173.1875		NFM	Nationwide	Mobile Alarm Paging

173.2000 - 173.3500 MHz   Telemetry Allocations & Garage Door Openers

Base	Mobile	Mode	Location	User & Notes
173.2250		NFM	Nationwide	Building Site Alarms
		NFM	Nationwide	Radio Ctrl'd Garage Doors

173.3500 - 174.0875 MHz Radio Deaf Aids & Biological Telemetry

173.3500 - 175.0200 MHz Cordless Microphones Simplex

Base	Mobile	Mode	Location	User & Notes
173.8000		NFM	Nationwide	Radio Mic. (Yellow)
		NFM	Nationwide	Theatre Radio Microphone
174.1000		NFM	Nationwide	Radio Microphone (Red)
		NFM	Nationwide	Theatre Radio Microphone
174.5000		NFM	Nationwide	Radio Microphone (Blue)
174.6625		NFM	Aberdeen	ASD
174.8000		NFM	Nationwide	Radio Microphone (Green)
		NFM	Nationwide	Theatre Radio Microphone

Base	Mobile	Mode	Location	User & Notes
175.0000		NFM	Nationwide	Radio Microphone (White)
		NFM	Nationwide	Theatre Radio Microphone

175.0200 - 176.5000 MHz IBA Broadcast Links 12.5 kHz Radio Mics

176·270

Base	Mobile	Mode	Location	User & Notes
175.5200 *PHONES*		NFM	Nationwide	ITV Radio Microphone
176.4000		NFM	Nationwide	Theatre Radio Microphone

176.5000 - 183.5000 MHz PMR Base Repeaters 12.5 kHz (Split + 8.0 MHz)
Radio Microphones

Base	Mobile	Mode	Location	User & Notes
176.6000		NFM	Nationwide	BBC Radio Microphone
176.8000		NFM	Nationwide	BBC News Radio Mics
177.0000		NFM	Nationwide	Theatre Radio Microphone
178.7250		NFM	Tayside	Data Repeater
179.0250		NFM	Tayside	Data Repeater
180.0000		NFM	Space	Cosmos 1870 Satellite
180.1250		NFM	Space	Cosmos Satellite
180.9500		NFM	Tayside	Data Repeater
182.5000		WFM	France	French TV Channel 5 Sound

183.5000 - 184.5000 MHz VHF PMR 12.5 kHz NFM Simplex

184.5000 - 191.5000 MHz VHF M1 PMR Mobiles 12.5 kHz (Split - 8.0 MHz)
Radio Microphones

Base	Mobile	Mode	Location	User & Notes
184.6000	NFM		Nationwide	BBC News Radio Mics
184.8000	NFM		Nationwide	ITV Radio Microphone
185.0000	NFM		Nationwide	ITV Radio Microphone
186.9850	NFM		Cumbria	Cumbria Fire Brigade
190.5000	WFM		France	French TV Channel 6 Sound

191.5000 - 192.5000 MHz VHF PMR 12.5 NFM Simplex

192.5000 - 199.5000 MHz PMR Base Repeaters 12.5 kHz (Split + 8.0 MHz)
Radio Microphones

Base	Mobile	Mode	Location	User & Notes
192.6000	192.6000	NFM	Nationwide	BBC News Radio Mics
192.8000	192.8000	NFM	Nationwide	ITV Radio Microphone
193.0000	193.0000	NFM	Nationwide	BBC Radio Microphone
195.2000		NFM	Walney Island	Mobile Coastguard
196.0250		NFM	Aberdeen	
196.6500		NFM	Aberdeen	

196.8500 - 198.3000 MHz British Rail Base to Cab Channels 50 kHz

Base	Mobile	Mode	Location	User & Notes
196.9000	204.9000	NFM	Nationwide	Channel 353
		NFM	Caersws	MWL RETB
196.9500	204.9500	NFM	Nationwide	Channel 357
197.0000	205.0000	NFM	Nationwide	Channel 361
197.0500	205.0500	NFM	Nationwide	Channel 365
197.1000	205.1000	NFM	Nationwide	Channel 369
197.1500	205.1500	NFM	Nationwide	Channel 373
		NFM	Nationwide	BR Cab Channel

Base	Mobile	Mode	Location	User & Notes
		NFM	Newtown	MWL RETD
		NFM	Welshpool	MWL RETD
		NFM	Westbury LC	MWL RETD
		NFM	Sutton Bridge Junction	MWL RETD
		NFM	Shrewsbury	MWL RETD
197.2000	205.2000	NFM	Nationwide	Channel 377
197.2500	205.2500	NFM	Nationwide	Channel 381
		NFM	Nationwide	BR Cab Channel
197.3000	205.3000	NFM	Nationwide	Channel 385
197.3500	205.3500	NFM	Nationwide	Channel 389
		NFM	Nationwide	BR Cab Channel
		NFM	Aberyswyth	MWL RETD
		NFM	Borth	MWL RETD
		NFM	Dovey Junction	MWL RETD
197.4000	205.4000	NFM	Nationwide	Channel 393
		NFM	Machynlleth	MWL RETD
		NFM	Newtown	MWL RETD
		NFM	Welshpool	MWL RETD
		NFM	Westbury LC	MWL RETD
		NFM	Sutton Bridge Junction	MWL RETD
		NFM	Shrewsbury	MWL RETD
197.4500	205.4500	NFM	Nationwide	Channel 397
197.5000	205.5000	NFM	Nationwide	Channel 401
197.5500	205.5500	NFM	Nationwide	Channel 405
197.6000	205.6000	NFM	Nationwide	Channel 409
		NFM	Ntionwide	BR Cab Channel
197.6500	205.6500	NFM	Nationwide	Channel 413
197.6750	205.6750	NFM	Nationwide	Channel 415
		NFM	Nationwide	BR Cab Channel
197.7000	205.7000	NFM	Nationwide	Channel 417
		NFM	Nationwide	BR Cab Channel
197.7250	205.7250	NFM	Nationwide	Channel 419
197.7500	205.7500	NFM	Nationwide	Channel 421
197.8000	205.8000	NFM	Nationwide	Channel 425
		NFM	Nationwide	BR Cab Channel
197.8375	205.8375	NFM	Nationwide	Channel 428
197.9000	205.9000	NFM	Nationwide	Channel 433
		NFM	Nationwide	BR Cab Channel
197.9500	205.9500	NFM	Nationwide	Channel 437
		NFM	Nationwide	BR Cab Channel
198.0000	206.0000	NFM	Nationwide	Channel 441
		NFM	Nationwide	BR Cab Channel
198.1000	206.1000	NFM	Nationwide	Channel 449
198.1500	206.1500	NFM	Nationwide	Channel 453
		NFM	Nationwide	BR Cab Channel
198.2500	206.2500	NFM	Nationwide	Channel 461
198.3000	206.3000	NFM	Nationwide	Channel 465
		NFM	Nationwide	BR Cab Channel
198.5000		WFM	France	French TV Channel 7 Sound

Base	Mobile	Mode	Location	User & Notes
199.5000 - 200.5000 MHz VHF PMR 12.5 NFM Simplex				
200.6000	200.6000	NFM	Nationwide	BBC Radio Microphone
200.8000	200.8000	NFM	Nationwide	ITV News Radio Mics
201.0000	201.0000	NFM	Nationwide	ITV Radio Microphone
201.3125 - 206.2625 MHz PMR London Transport 12.5 kHz (Split - 8.0 MHz)				
201.4625	193.4625	NFM	London	London Transport
201.5125	193.5125	NFM	London	London Transport
201.5500		NFM	East Scotland	Data Link
201.6125	193.6125	NFM	London	London Transport
201.6625	193.6625	NFM	West Midlands	West Midlands Bus Travel
		NFM	London	London Transport
201.7000		NFM	East Scotland	Data Link
201.7125	193.7125	NFM	London	London Transport
201.7625	193.7625	NFM	London	London Transport
201.8500		NFM	East Scotland	Data Link
201.8625	193.8625	NFM	London	London Transport
201.9125	193.9125	NFM	London	London Transport
201.9625	193.9625	NFM	London	London Transport
202.0000		NFM	East Scotland	Data Link
202.0125	194.0125	NFM	London	London Transport
202.0625	194.0625	NFM	London	London Transport
202.1500		NFM	East Scotland	Data Link
202.1625	194.1625	NFM	West Midlands	West Midlands Bus Travel
202.2125	194.2125	NFM	West Midlands	West Midlands Bus Travel
202.3125	194.3125	NFM	London	London Transport
202.4625	194.4625	NFM	London	London Transport
203.0125	195.0125	NFM	London	London Transport
203.1625	195.1625	NFM	Plymouth	Western National Buses
203.2625	195.2625	NFM	London	London Trans. Emerg Chan
203.3125	195.3125	NFM	London	London Transport
203.3625	195.3625	NFM	London	London Transport
203.6625	195.6625	NFM	London	London Transport
203.7125	195.7125	NFM	London	London Transport
204.5875		NFM	East Scotland	Data Link
204.7125	196.7125	NFM	West Midlands	West Midlands Bus Travel
204.7625	196.7625	NFM	West Midlands	West Midlands Bus Travel
204.8125	196.8125	NFM	London	London Transport
204.8625	196.8625	NFM	London	London Transport
204.9125	196.9125	NFM	London	London Transport
204.9625	196.9625	NFM	West Midlands	West Midlands Bus Travel
		NFM	London	London Transport
205.0125	197.0125	NFM	West Midlands	West Midlands Bus Travel
205.0375		NFM	East Scotland	Data Link
205.0625	197.0625	NFM	Gt Manchester	City Bus Inspectors
		NFM	Plymouth	City Buses
205.1875		NFM	East Scotland	Data Link
205.8125	197.8125	NFM	West Midlands	West Midlands Bus Travel
205.8375		NFM	East Scotland	Data Link
205.9375		NFM	East Scotland	Data Link

Base	Mobile	Mode	Location	User & Notes
206.0125	198.0125	NFM	West Midlands	West Midlands Bus Travel
206.0625	198.0625	NFM	West Midlands	West Midlands Bus Travel
206.1500		NFM	East Scotland	Data Link
206.1625	198.1625	NFM	Gt Manchester	City Bus Inspectors
		NFM	West Midlands	West Midlands Bus Travel
		NFM	London	London Trans. Talking Clock
206.2125	198.2125	NFM	West Midlands	West Midlands Bus Travel
		NFM	London Stanford Hill	London Transport
206.2625	198.2625	NFM	West Midlands	West Midlands Bus Travel
		NFM	London Stockwell	London Transport

206.5000 - 208.5000 MHz BBC & IBA Simplex/Duplex Allocations

Base	Mobile	Mode	Location	User & Notes
206.5000		WFM	France	French TV Channel 8 Sound

208.5000 - 215.5000 MHz VHF B3 PMR Base Repeaters 12.5 kHz
(Split + 8.0 MHz)O/B Radio Microphones

Base	Mobile	Mode	Location	User & Notes
208.6000	208.6000	NFM	Nationwide	Radio 1 Roadshow Mics
208.8000	208.8000	NFM	Nationwide	ITV Radio Microphone
209.0000	209.0000	NFM	Nationwide	BBC Radio Microphone
212.2000		WFM	Wolverhampton	Beacon Radio O/B
214.5000		WFM	France	French TV Channel 9 Sound

215.0000 - 216.0000 VHF PMR 12.5 NFM Simplex

216.5000 - 223.5000 MHz VHF M3 PMR Mobiles 12.5 kHz (Split - 8.0 MHz)
Radio Microphones

Base	Mobile	Mode	Location	User & Notes
224.23125		NFM	Bristol	BBC O/B Talkback

225.0000 - 399.9750 MHz Military Aeronautical Communications 25 kHz

Base	Mobile	Mode	Location	User & Notes
226.6000	226.6000	AM	RNAS Portland	Royal Navy
230.0500	230.0500	AM	RAF Buchan	Air Defence Region Ops
230.1500	230.1500	AM	RAF Boulmer	Air Defence Region Ops
		AM	RAF Staxton Wold	Air Defence Region Ops
230.6000	230.6000	AM	Nationwide	Air-Air
		AM	Nationwide	Air Defence Region
		AM	Watton	Eastern Radar
231.2500	231.2500	AM	RAF Staxton Wold	Air Defence Region Ops
231.5500	231.5500	AM	RAF Buchan	Air Defence Region Ops
231.6250	231.6250	AM	RAF West Drayton	London Mil
232.3500	232.3500	AM	RAF Neatishead	Air Defence Region Ops
232.5500	232.5500	AM	RAF Neatishead	Air Defence Region Ops
232.7000	232.7000	AM	RAF Neatishead	Air Defence Region Ops
233.0000	233.0000	AM	Royal Navy	Ship-Air
233.1500	233.1500	AM	RAF Portreath	Air Defence Region Ops
233.2000	233.2000	AM	Royal Navy	Ship-Air
233.9250	233.9250	AM	RNAS Portland	Royal Navy
234.6500	234.6500	AM	Cumbria	RAF Low Flying Air-Air
234.9000	234.9000	AM	Nationwide	RAF Volmet
235.0500	235.0500	AM	RAF West Drayton	London Mil
235.2500	235.2500	AM	Nationwide	USAF Displays

Base	Mobile	Mode	Location	User & Notes
237.5000		NFM	Nationwide	Dynamic Sciences Surv.
240.3000	240.3000	AM	Nationwide	Air-Air Refuelling
		AM	RAF Neatishead	Radar
240.4000	240.4000	AM	ARA 1, 7 & 10	Refuelling Primary
241.8250	241.8250	AM	RAF Aldergrove	RAF Ops
		AM	RAF St Mawgan	Tower
241.8500	241.8500	AM	RAF Neatishead	Radar Ops
241.9500	241.9500	AM	RNAS Culdrose	Approach
241.9750	241.9750	AM	RAF Honington	Tower
242.1500	242.1500	AM	RAF Upper Heyford	ATIS
243.0000	243.0000	AM	International	Air Distress
243.4500	243.4500	AM	Nationwide	Red Arrows Display
243.6000	243.6000	AM	RAF Lakenheath	Radar
243.8000	243.8000	AM	RAF Aerodrome	Radio Failure Frequency

243.9450 - 244.2500 MHz US AFSATCOM Down Links

Base	Mobile	Mode	Location	User & Notes
243.9450		NFM	AFSATCOM F2	NB Channel 11
243.9550		NFM	AFSATCOM F2	NB Channel 12
243.9600		NFM	AFSATCOM F2	NB Channel 13
243.9650		NFM	AFSATCOM F2	NB Channel 14
243.9700		NFM	AFSATCOM F2	NB Channel 15
243.9750		NFM	AFSATCOM F2	NB Channel 16
243.9800		NFM	AFSATCOM F2	NB Channel 17
243.9850		NFM	AFSATCOM F2	NB Channel 18
243.9900		NFM	AFSATCOM F2	NB Channel 19
243.9950		NFM	AFSATCOM F2	NB Channel 20
244.0000		NFM	AFSATCOM F2	NB Channel 21
244.0100		NFM	AFSATCOM F2	NB Channel 22
244.0450		NFM	AFSATCOM F3	NB Channel 11
244.0550		NFM	AFSATCOM F3	NB Channel 12
244.0600		NFM	AFSATCOM F3	NB Channel 13
244.0650		NFM	AFSATCOM F3	NB Channel 14
244.0700		NFM	AFSATCOM F3	NB Channel 15
244.0750		NFM	AFSATCOM F3	NB Channel 16
244.0800		NFM	AFSATCOM F3	NB Channel 17
244.0850		NFM	AFSATCOM F3	NB Channel 18
244.0900		NFM	AFSATCOM F3	NB Channel 19
244.0950		NFM	AFSATCOM F3	NB Channel 20
244.1000		NFM	AFSATCOM F3	NB Channel 21
244.1100		NFM	AFSATCOM F3	NB Channel 22
244.1450		NFM	AFSATCOM F1	NB Channel 11
244.1550		NFM	AFSATCOM F1	NB Channel 12
244.1600		NFM	AFSATCOM F1	NB Channel 13
244.1650		NFM	AFSATCOM F1	NB Channel 14
244.1700		NFM	AFSATCOM F1	NB Channel 15
244.1750		NFM	AFSATCOM F1	NB Channel 16
244.1800		NFM	AFSATCOM F1	NB Channel 17
244.1850		NFM	AFSATCOM F1	NB Channel 18
244.1900		NFM	AFSATCOM F1	NB Channel 19
244.1950		NFM	AFSATCOM F1	NB Channel 20

Base	Mobile	Mode	Location	User & Notes
244.2000		NFM	AFSATCOM F1	NB Channel 21
244.2100		NFM	AFSATCOM F1	NB Channel 22
244.2750	244.2750	AM	RAF Waddington	Ops
244.3000	244.3000	AM	RAF Valley	Air-to-Air
244.4250	244.4250	AM	RAF Northolt	Ops
244.6000	244.6000	AM	Nationwide	UK Distress
		AM	Plymouth Rescue	
244.6500	244.6500	AM	ARA 6	Air-Air Refuelling
		AM	RAF Buchan	Air Defence Region Ops
		AM	RAF Neatishead	Air Defence Region Ops
244.7750	244.7750	AM	RAF Bentwaters	Ground
244.8750	244.8750	AM	RAF Leconfield	Leconfield Rescue
245.0500	245.0500	AM	RAF Boulmer	Air Defence Region Ops
246.7000	246.7000	AM	RAF Sculthorpe	Approach
247.0000	247.0000	AM	RAF Boulmer	Air Defence Region
247.2500	247.2500	AM	RAF Staxton Wold	Air Defence Region Ops
247.2750	247.2750	AM	Nationwide	Air Defence Region
248.1000	248.1000	AM	RAF Neatishead	Air Defence Region Ops
248.3000	248.3000	AM	Nationwide	Airborne Intercept Cmd
248.8000	248.8000	AM	Nationwide	Sharks Helicopter Display
248.8500	302.4500	NFM	MARISAT	Channel 1
248.8750	302.4750	NFM	MARISAT	Channel 2
248.9000	302.5000	NFM	MARISAT	Channel 3
248.9250	302.5250	NFM	MARISAT	Channel 4
248.9500	302.5500	NFM	MARISAT	Channel 5
248.9750	302.5750	NFM	MARISAT	Channel 6
249.0000	302.7000	NFM	MARISAT	Channel 7
249.0250	302.7250	NFM	MARISAT	Channel 8
249.0500	302.7500	NFM	MARISAT	Channel 9
249.0750	302.7750	NFM	MARISAT	Channel 10
249.1000	302.8000	NFM	MARISAT	Channel 11
249.1250	302.8250	NFM	MARISAT	Channel 12
249.1500	302.8500	NFM	MARISAT	Channel 13
249.1750	302.8750	NFM	MARISAT	Channel 14
249.0000		NFM	Nationwide	Dynamic Sciences Surv.
249.2000	302.9000	NFM	MARISAT	Channel 15
249.2250	302.9250	NFM	MARISAT	Channel 16
249.2500	302.9500	NFM	MARISAT	Channel 17
249.2750	302.9750	NFM	MARISAT	Channel 18
249.3000	303.0000	NFM	MARISAT	Channel 19
249.3250	303.0250	NFM	MARISAT	Channel 20
249.3500	303.0500	NFM	MARISAT	Channel 21
249.4750	249.4750	AM	Prestwick	Scottish Mil
249.5500	249.5500	AM	RAF Wyton	Director
249.6250	249.6250	AM	RAF West Drayton	London Mil
249.7000		NFM	Nationwide	Dynamic Sciences Surv.
249.8000		NFM	Nationwide	Dynamic Sciences Surv.
249.8500	249.8500	AM	RAF Waddington	Departures
250.0500	250.0500	AM	RAF Cranwell	Zone
		AM	RAF Lossiemouth	Talkdown

Base	Mobile	Mode	Location	User & Notes
250.2750	250.2750	AM	Watton	Eastern Radar

250.3500 - 250.6500 MHz US FLTSATCOM Fleet Broadcast Down Links

Base	Mobile	Mode	Location	User & Notes
250.3500	291.3500	NFM	FLTSATCOM	Channel W1
250.4500	291.4500	NFM	FLTSATCOM F1	Channel X1
250.5500	291.5500	NFM	FLTSATCOM F3	Channel Y1
250.6500	291.6500	NFM	FLTSATCOM F2	Channel Z1
250.7000	250.7000	AM	Nationwide	Air Defence Region
250.9000		NFM	Nationwide	Dynamic Sciences Surv
261.5000	261.5000	AM	Nationwide	USAF Displays
251.6250	251.6250	AM	Watton	Eastern Radar
251.6000		NFM	Nationwide	Dynamic Sciences Surv
251.6500	251.6500	AM	Nationwide	Air Defence Region
251.7250	251.7250	AM	RAF Newton	Approach
251.7500	251.7500	AM	Nationwide	Air Defence Region
251.8000		NFM	Nationwide	Dynamic Sciences Surv
251.8500	292.8500	NFM	FLTSATCOM	Channel W 2
251.9000		NFM	Nationwide	Dynamic Sciences Surv
251.9500	292.9500	NFM	FLTSATCOM F1	Channel X 2
252.0500	293.0500	NFM	FLTSATCOM F3	Channel Y 2
252.1500	293.1500	NFM	FLTSATCOM F2	Channel Z 2
253.0000		NFM	Nationwide	Dynamic Sciences Surv
253.1000	253.1000	AM	RAF Lakenheath	Rapcon
253.5000	253.5000	AM	Netheravon	Salisbury Plain
253.5500	294.5500	NFM	FLTSATCOM	Channel W 3
253.6500	294.6500	NFM	FLTSATCOM F1	Channel X 3
253.7500	294.7500	NFM	FLTSATCOM F3	Channel Y 3
253.8000	253.8000	AM	Nationwide	NATO SAR Training
253.8500	294.8500	NFM	FLTSATCOM F2	Channel Z 3
253.9000		NFM	Nationwide	Dynamic Sciences Surv
254.2000	254.2000	AM	RAF Shawbury	Zone
254.2500	254.2500	AM	RAF Coltishall	Talkdown
254.4250	254.4250	AM	Nationwide	Air Defence Region
254.4750	254.4750	AM	RAF Brize Norton	ATIS
254.6500	254.6500	AM	RAF Lyneham	Ops
254.7500	254.7500	AM	RAF Brawdy	Zone Departures
254.8250	254.8250	AM	Watton	Eastern Radar
254.8750	254.8750	AM	RAF Mildenhall	Approach
255.1000	255.1000	AM	Nationwide	RAF Falcons Parachutists
255.2500	296.2500	NFM	FLTSATCOM	Channel W 4
255.3500	296.3500	AM	FLTSATCOM F1	Channel X 4
255.4000	255.4000	AM	RAF West Drayton	London Mil
255.4500	296.4500	AM	FLTSATCOM F3	Channel Y 4
255.5000	255.5000	AM	RAF Scampton	Approach
255.5500	296.5500	AM	FLTSATCOM F2	Channel Z 4
256.0000	256.0000	AM	Royal Navy	Ship-Air
256.1000	256.1000	AM	Royal Navy	Ship-Air
256.1250	256.1250	AM	MoD Filton	Approach
256.6000		NFM	Nationwide	Dynamic Sciences Surv
256.8500	297.8500	NFM	FLTSATCOM	Channel W 5

Base	Mobile	Mode	Location	User & Notes
256.9500	297.9500	AM	FLTSATCOM F1	Channel X 5
257.0500	298.0500	AM	FLTSATCOM F3	Channel Y 5
257.1000	257.1000	AM	RAF Brize Norton	Brize Radar
257.1500	298.1500	AM	FLTSATCOM F2	Channel Z 5
257.2250	257.2250	AM	RAF West Drayton	London Mil
257.8000	257.8000	AM	RAF Bentwaters	Tower
		AM	RAF Brawdy	Tower
		AM	RAF Brize Norton	Tower
		AM	RAF Church Fenton	Tower
		AM	RAF Cottesmore	Tower
		AM	RAF Cranwell	Tower
		AM	RAF Fairford	Tower
		AM	RAF Greenham Common	Tower
		AM	RAF Honington	Tower
		AM	RAF Kinloss	Tower
		AM	RAF Leeming	Tower
		AM	RAF Linton-on-Ouse	Tower
		AM	RAF Machrihanish	Tower
		AM	RAF Manston	Tower
		AM	RAF Marham	Tower
		AM	RAF Newton	Tower
		AM	RAF Northolt	Tower
		AM	RAF Odiham	Tower
		AM	RAF St Athan	Tower
		AM	RAF Scampton	Tower
		AM	RAF Sculthorpe	Tower
		AM	RAF Shawbury	Tower
		AM	RAF Topcliffe	Tower
		AM	RAF Upper Heyford	Tower
		AM	RAF Valley	Tower
		AM	RAF Waddington	Tower
		AM	RAF Wittering	Tower
		AM	RAF Woodbridge	Tower
		AM	RAF Wyton	Tower
258.0500	258.0500	AM	Nationwide	Air-Air Tanker Ops
258.3500	299.3500	NFM	FLTSATCOM	Channel W 6
258.4000	258.4000	AM	ARA 10	Air-Air Refuelling
258.4500	299.4500	NFM	FLTSATCOM F1	Channel X 6
258.5000	258.5000	AM	RAF Staxton Wold	Air Defence Region Ops
258.5500	299.5500	NFM	FLTSATCOM F3	Channel Y 6
258.6500	299.6500	NFM	FLTSATCOM F2	Channel Z 6
258.8000	258.8000	AM	Nationwide	British Army Air-Air
258.8250	258.8250	AM	RAF Mildenhall	Tower
258.9250	258.9250	AM	RAF Leuchars	Tower
258.9750	258.9750	AM	RAF Woodbridge	Dep Con
259.0000	259.0000	AM	RNAS Aberporth	Aberporth Information
		AM	MoD Farnborough	PAR
		AM	MoD West Freugh	Radar
259.1000	259.1000	AM	RAF Neatishead	Air Defence Region Ops
259.4000	259.4000	AM	RAF Bentwaters	Bentwaters Metro

Base	Mobile	Mode	Location	User & Notes
		AM	RAF Fairford	Fairford Metro
		AM	RAF Mildenhall	Mildenhall Metro
259.6000	259.6000	AM	RAF Neatishead	Air Defence Region Ops
259.7000	259.7000	NFM	International	Space Shuttle Down Link
259.7500	259.7500	AM	RNAS Culdrose	Talkdown
259.8250	259.8250	AM	RAF Dishforth	Tower
		AM	RAF Kinloss	Ops
259.8500	259.8500	AM	RAF Leuchars	Ground
		AM	RAF Scampton	Ground
259.8750	259.8750	AM	RAF Linton-on-Ouse	Talkdown
259.9250	259.9250	AM	RAF Greenham Common	Ground
259.9500	259.9500	AM	RAF Brawdy	Ground
		AM	RAF Woodvale	Tower
259.9750	259.9750	AM	RAF Fairford	Ground
		AM	RAF Kinloss	Director
		AM	RAF Lossiemouth	Director
		AM	RAF Machrihanish	Director
260.0000	260.0000	AM	RAF St Mawgan	Ops
260.0250	260.0250	AM	RAF West Drayton	London Mil
260.1500	260.1500	AM	Nationwide	Air Defence Region

260.3000 - 262.5500 MHz US FLTSATCOM Wideband Down Link

Base	Mobile	Mode	Location	User & Notes
260.3500	293.9500	NFM	FLTSATCOM F1	WB Channel A/X 1
260.3750	293.9750	NFM	FLTSATCOM F1	WB Channel A/X 2
260.4000	294.0000	NFM	FLTSATCOM F1	WB Channel A/X 3
260.4250	294.0250	NFM	FLTSATCOM F1	WB Channel A/X 4
260.4500	294.0500	NFM	FLTSATCOM F1	WB Channel A/X 5
260.4750	294.0750	NFM	FLTSATCOM F1	WB Channel A/X 6
260.5000	294.1000	NFM	FLTSATCOM F1	WB Channel A/X 7
260.5250	294.1250	NFM	FLTSATCOM F1	WB Channel A/X 8
260.5500	294.1500	NFM	FLTSATCOM F1	WB Channel A/X 9
260.5750	294.1750	NFM	FLTSATCOM F1	WB Channel A/X 10
260.6000	294.2000	NFM	FLTSATCOM F1	WB Channel A/X 11
260.6250	294.2250	NFM	FLTSATCOM F1	WB Channel A/X 12
260.6500	294.2500	NFM	FLTSATCOM F1	WB Channel A/X 13
260.6750	294.2750	NFM	FLTSATCOM F1	WB Channel A/X 14
260.7000	294.3000	NFM	FLTSATCOM F1	WB Channel A/X 15
260.7250	294.3250	NFM	FLTSATCOM F1	WB Channel A/X 16
260.7500	294.3500	NFM	FLTSATCOM F1	WB Channel A/X 17
260.7750	294.3750	NFM	FLTSATCOM F1	WB Channel A/X 18
260.8000	294.4000	NFM	FLTSATCOM F1	WB Channel A/X 19
260.8250	294.4250	NFM	FLTSATCOM F1	WB Channel A/X 20
260.8500	294.4500	NFM	FLTSATCOM F1	WB Channel A/X 21
261.4500	295.0500	NFM	FLTSATCOM F3	WB Channel B/Y 1
261.4750	295.0750	NFM	FLTSATCOM F3	WB Channel B/Y 2
261.5000	295.1000	NFM	FLTSATCOM F3	WB Channel B/Y 3
261.5250	295.1250	NFM	FLTSATCOM F3	WB Channel B/Y 4
261.5500	295.1500	NFM	FLTSATCOM F3	WB Channel B/Y 5
261.5750	295.1750	NFM	FLTSATCOM F3	WB Channel B/Y 6
261.6000	295.2000	NFM	FLTSATCOM F3	WB Channel B/Y 7

Base	Mobile	Mode	Location	User & Notes
261.6250	295.2250	NFM	FLTSATCOM F3	WB Channel B/Y 8
261.6500	295.2500	NFM	FLTSATCOM F3	WB Channel B/Y 9
261.6750	295.2750	NFM	FLTSATCOM F3	WB Channel B/Y 10
261.7000	295.3000	NFM	FLTSATCOM F3	WB Channel B/Y 11
261.7250	295.3250	NFM	FLTSATCOM F3	WB Channel B/Y 12
261.7500	295.3500	NFM	FLTSATCOM F3	WB Channel B/Y 13
261.7750	295.3750	NFM	FLTSATCOM F3	WB Channel B/Y 14
261.8000	295.4000	NFM	FLTSATCOM F3	WB Channel B/Y 15
261.8250	295.4250	NFM	FLTSATCOM F3	WB Channel B/Y 16
261.8500	295.4500	NFM	FLTSATCOM F3	WB Channel B/Y 17
261.8750	295.4750	NFM	FLTSATCOM F3	WB Channel B/Y 18
261.9000	295.5000	NFM	FLTSATCOM F3	WB Channel B/Y 19
261.9250	295.5250	NFM	FLTSATCOM F3	WB Channel B/Y 20
261.9500	295.5500	NFM	FLTSATCOM F3	WB Channel B/Y 21
262.0500	295.6500	NFM	FLTSATCOM F2	WB Channel C/Z 1
262.0750	295.6750	NFM	FLTSATCOM F2	WB Channel C/Z 2
262.1000	295.7000	NFM	FLTSATCOM F2	WB Channel C/Z 3
262.1250	295.7250	NFM	FLTSATCOM F2	WB Channel C/Z 4
262.1500	295.7500	NFM	FLTSATCOM F2	WB Channel C/Z 5
262.1750	295.7750	NFM	FLTSATCOM F2	WB Channel C/Z 6
262.2000	295.8000	NFM	FLTSATCOM F2	WB Channel C/Z 7
262.2250	295.8250	NFM	FLTSATCOM F2	WB Channel C/Z 8
262.2500	295.8500	NFM	FLTSATCOM F2	WB Channel C/Z 9
262.2750	295.8750	NFM	FLTSATCOM F2	WB Channel C/Z 10
262.3000	295.9000	NFM	FLTSATCOM F2	WB Channel C/Z 11
262.3250	295.9250	NFM	FLTSATCOM F2	WB Channel C/Z 12
262.3500	295.9500	NFM	FLTSATCOM F2	WB Channel C/Z 13
262.3750	295.9750	NFM	FLTSATCOM F2	WB Channel C/Z 14
262.4000	296.0000	NFM	FLTSATCOM F2	WB Channel C/Z 15
262.4250	296.0250	NFM	FLTSATCOM F2	WB Channel C/Z 16
262.4500	296.0500	NFM	FLTSATCOM F2	WB Channel C/Z 17
262.4750	296.0750	NFM	FLTSATCOM F2	WB Channel C/Z 18
262.5000	296.1000	NFM	FLTSATCOM F2	WB Channel C/Z 19
262.5250	296.1250	NFM	FLTSATCOM F2	WB Channel C/Z 20
262.5500	296.1500	NFM	FLTSATCOM F2	WB Channel C/Z 21
262.7000	262.7000	AM	RAF Church Fenton	Tower
		AM	RAF Leuchars	Approach
262.9000	262.9000	AM	RAF Cottesmore	Talkdown
262.9250	262.9250	AM	RAF Lakenheath	Radar
262.9500	262.9500	AM	RAF Coningsby	Director
263.0750	263.0750	AM	Watton	Eastern Radar
263.1500	263.1500	AM	Nationwide	Air Defence Region
263.5500	297.1500	NFM	FLTSATCOM	WB Channel W 1
263.5750	297.1750	NFM	FLTSATCOM	WB Channel W 2
263.6000	297.2000	NFM	FLTSATCOM	WB Channel W 3
263.6250	297.2250	NFM	FLTSATCOM	WB Channel W 4
263.6500	297.2500	NFM	FLTSATCOM	WB Channel W 5
263.6750	297.2750	NFM	FLTSATCOM	WB Channel W 6
263.7000	297.3000	NFM	FLTSATCOM	WB Channel W 7
263.7250	297.3250	NFM	FLTSATCOM	WB Channel W 8

Base	Mobile	Mode	Location	User & Notes
263.7500	297.3500	NFM	FLTSATCOM	WB Channel W 9
263.7750	297.3750	NFM	FLTSATCOM	WB Channel W 10
263.8000	297.4000	NFM	FLTSATCOM	WB Channel W 11
263.8250	297.4250	NFM	FLTSATCOM	WB Channel W 12
263.8500	297.4500	NFM	FLTSATCOM	WB Channel W 13
263.8750	297.4750	NFM	FLTSATCOM	WB Channel W 14
263.9000	297.5000	NFM	FLTSATCOM	WB Channel W 15
263.9250	297.5250	NFM	FLTSATCOM	WB Channel W 16
263.9500	297.5500	NFM	FLTSATCOM	WB Channel W 17
263.9750	297.5750	NFM	FLTSATCOM	WB Channel W 18
264.0000	297.6000	NFM	FLTSATCOM	WB Channel W 19
264.0250	297.6250	NFM	FLTSATCOM	WB Channel W 20
264.0500	297.6500	NFM	FLTSATCOM	WB Channel W 21
264.1250	264.1250	AM	RAF Upper Heyford	PAR
264.4750	264.4750	AM	RAF West Drayton	London Mil
264.9250	264.9250	AM	RAF Bentwaters	Tower
265.2500	306.2500	NFM	FLTSATCOM	Channel W 7
265.3500	306.3500	NFM	FLTSATCOM F1	Channel X 7
265.4500	306.4500	NFM	FLTSATCOM F3	Channel Y 7
265.5500	306.5500	NFM	FLTSATCOM F2	Channel Z 7
265.8500	265.8500	AM	Nationwide	Air-Air Refuelling
		AM	RAF Staxton Wold	Air Defence Region Ops
265.9000	265.9000	AM	Nationwide	Air Defence Region
266.4500	266.4500	AM	RAF Neatishead	Air Defence Region Ops
266.5000	266.5000	AM	Nationwide	USAF Air-Air Refuelling
266.5500	266.5500	AM	RAF Neatishead	Air Defence Region Ops
266.7500	307.7500	NFM	FLTSATCOM	Channel W 8
266.8000		NFM	Nationwide	Dynamic Sciences Surv
266.8500	307.8500	NFM	FLTSATCOM F1	Channel X 8
266.9500	307.9500	NFM	FLTSATCOM F3	Channel Y 8
267.0500	308.0500	NFM	FLTSATCOM F2	Channel Z 8
267.4000		NFM	Nationwide	Dynamic Sciences Surv
267.5500	267.5500	AM	RAF Boulmer	Air Defence Region Ops
		AM	RAF Neatishead	Air Defence Region Ops
267.9000	267.9000	AM	Nationwide	NATO Air-to-Air
268.1500	309.1500	NFM	FLTSATCOM	Channel W 9
268.2500	309.2500	NFM	FLTSATCOM F1	Channel X 9
268.3500	309.3500	NFM	FLTSATCOM F3	Channel Y 9
268.4000		NFM	Nationwide	Dynamic Sciences Surv
268.4000	268.4000	AM	RAF Brawdy	Talkdown
268.4500	309.4500	NFM	FLTSATCOM F2	Channel Z 9
268.7750	268.7750	AM	RAF Leuchars	Finals
		AM	RAF Valley	Valley Radar
268.8000		NFM	Nationwide	Dynamic Sciences Surv
268.8250	268.8250	AM	RAF Benson	Approach
269.0000		NFM	Nationwide	Dynamic Sciences Surv
269.1000	269.1000	AM	RAF Shawbury	Tower
269.1250	269.1250	AM	RAF Woodford	Approach
269.6500	310.6500	NFM	FLTSATCOM	Channel W 10
269.7500	310.7500	NFM	FLTSATCOM F1	Channel X 10

The UK Scanning Directory

Base	Mobile	Mode	Location	User & Notes
269.8500	310.8500	NFM	FLTSATCOM F3	Channel Y 10
269.9000		NFM	Nationwide	Dynamic Sciences Surv
269.9500	310.9500	NFM	FLTSATCOM F2	Channel Z 10
270.9000		NFM	Nationwide	Dynamic Sciences Surv
271.8000		NFM	Nationwide	Dynamic Sciences Surv
272.0750	272.0750	AM	RNAS Portland	Royal Navy Exercises
273.5250	273.5250	AM	RNAS Portland	Royal Navy
273.9000		NFM	Nationwide	Dynamic Sciences Surv
273.9000	273.9000	AM	Nationwide	NATO Low-Level Flying
274.4000		NFM	Nationwide	Dynamic Sciences Surv
275.3500	275.3500	AM	Nationwide	USAF Displays
275.4750	275.4750	AM	RAF West Drayton	London Mil
275.6500	275.6500	AM	RAF Sculthorpe	PAR
275.6750	275.6750	AM	Watton	Eastern Radar
275.7500	275.7500	AM	Nationwide	Air Defence Region
275.8000		NFM	Nationwide	Dynamic Sciences Surv
275.8000	275.8000	AM	RAF Upavon	Tower
275.8750	275.8750	AM	RAF Coningsby	Tower
275.9750	275.9750	AM	RAF Coltishall	Talkdown
276.0000	276.0000	AM	RAF Greenham Common	Ground
276.0750	276.0750	AM	RAF Chetwynd	Shawbury Approach
		AM	RAF Shawbury	Approach
276.1000	276.1000	AM	RAF Cosford	Approach
276.1750	276.1750	AM	RAF Odiham	ATIS
276.6000		NFM	Nationwide	Dynamic Sciences Surv
276.6500	276.6500	AM	RAF Neatishead	Air Defence Region Ops
276.8250	276.8250	AM	RAF Ternhill	Approach
276.8500	276.8500	AM	MoD Boscombe Down	PAR
277.0000	277.0000	AM	Nationwide	NATO Magic Surveillance
		AM	Royal Navy	Ship-Air
277.0750	277.0750	AM	RAF Sculthorpe	PAR
277.1250	277.1250	AM	RAF West Drayton	London Mil
277.1750	277.1750	AM	RAF Upper Heyford	Dispatcher
277.2250	277.2250	AM	RAF St Athan	Approach
277.4000	277.4000	AM	Nationwide	Air Defence Region
277.4750	277.4750	AM	RAF Wattisham	Director
		AM	RAF Waddington	Director
277.7500	277.7500	AM	Nationwide	Air Defence Region
277.9500	277.9500	AM	RAF West Drayton	London Mil
278.0250	278.0250	AM	RAF West Drayton	London Mil
278.9000		NFM	Nationwide	Dynamic Sciences Surv
279.0000		NFM	Space	Space Shuttle Down Link
279.3250	279.3250	AM	MoD Boscombe Down	ATIS
279.3500	279.3500	AM	RAF Benson	Tower
279.4750	279.4750	AM	RAF West Drayton	London Mil
279.5250	279.5250	AM	Nationwide	Air Defence Region
279.7250	279.7250	AM	Nationwide	Air Defence Region
280.5000	280.5000	AM	RAF Scampton	Departures
280.6000		NFM	Nationwide	Dynamic Sciences Surv
281.1000	281.1000	AM	Nationwide	Air Defence Region

Base	Mobile	Mode	Location	User & Notes
281.2000	281.2000	AM	Royal Navy	Ship-Air
281.7250	281.7250	AM	RNAS Portland	Naval Exercises
282.0000	282.0000	AM	RAF Cranwell	Director
282.0750	282.0750	AM	RAF Church Fenton	Fenton Radar
282.1250	282.1250	AM	RAF West Drayton	London Mil
282.1500	282.1500	AM	RAF Bentwaters	Command Post
282.2750	282.2750	AM	RAF Honington	Tower
282.4000	282.4000	AM	RAF Scampton	Tower
282.8000	282.8000	AM	RAF Boulmer	Boulmer Rescue
		AM	RAF Leconfield	Leconfield Rescue
		AM	RAF Valley	SAR Approach
		AM	RNAS Portland	Portland Radio
		AM	Nationwide	NATO SAR
283.4250	283.4250	AM	RAF Swinderby	Approach
283.5250	283.5250	AM	RAF West Drayton	London Mil
283.5750	283.5750	AM	RAF Waddington	Zone
283.6000	283.6000	AM	RAF Alconbury	Command Post
283.6250	283.6250	AM	RAF Upper Heyford	PAR
283.6500	283.6500	AM	Nationwide	Air Defence Region
284.8750	284.8750	AM	RAF West Drayton	London Mil
284.9250	284.9250	AM	RAF Alconbury	Alconbury Metro
284.9750	284.9750	AM	Nationwide	Air Defence Region
285.0250	285.0250	AM	RAF Leuchars	Ops
285.0500	285.0500	AM	RAF Waddington	Tower
285.1250	285.1250	AM	RAF Finningley	Radar
285.1500	285.1500	AM	RAF Cranwell	Talkdown
285.1750	285.1750	AM	RAF West Drayton	London Mil
285.6500	285.6500	AM	RAF Neatishead	Air Defence Region Ops
285.8500	285.8500	AM	RAF Boulmer	Boulmer Rescue
286.9000	286.9000	AM	RAF Buchan	Air Defence Region Ops
287.2500	287.2500	AM	Nationwide	RAF Air-Air Tanker Ops
287.6500	287.6500	AM	RAF Neatishead	Air Defence Region Ops
		AM	Royal Navy	Ship-Air
287.7000	287.7000	AM	Nationwide	Air Defence Region
287.9500	287.9500	AM	RAF Staxton Wold	Air Defence Region Ops
288.4000	288.4000	AM	RAF Boulmer	Air Defence Region Ops
		AM	RAF Neatishead	Air Defence Region Ops
289.0500	289.0500	AM	RAF Boulmer	Air Defence Region Ops
289.3500	289.3500	AM	RAF Neatishead	Air Defence Region Ops
290.0500	290.0500	AM	RAF Neatishead	Air Defence Region Ops
290.3750	290.3750	AM	Nationwide	Air Defence Region
290.8250	290.8250	AM	RAF Lakenheath	Radar
290.9500	290.9500	AM	Netheravon	Tower
291.0750	291.0750	AM	Nationwide	Air Defence Region
291.1750	291.1750	AM	Watton	Eastern Radar
291.3500	291.3500	AM	RAF Woodbridge	Tower
291.6500	291.6500	AM	MoD Boscombe Down	Approach
291.7750	291.7750	AM	Watton	Eastern Radar
291.8000	291.8000	AM	West Drayton	London Mil
291.9500	291.9500	AM	RAF Marham	Approach

Base	Mobile	Mode	Location	User & Notes
292.4500	292.4500	AM	RAF Boulmer	Air Defence Region Ops
292.4750	292.4750	AM	RAF Leuchars	Director
292.5250	292.5250	AM	West Drayton	London Mil
292.7000	292.7000	AM	RAF Leeming	Zone Radar
292.8000	292.8000	AM	RAF Church Fenton	Departures
		AM	RAF Linton-on-Ouse	Departures
292.9000	292.9000	AM	RAF Wyton	Talkdown
293.4250	293.4250	AM	RAF Coltishall	Zone
293.4750	293.4750	AM	Watton	Eastern Radar
293.6500	293.6500	AM	RAF Wyton	Ground
293.7750	293.7750	AM	RAF Marham	Radar
296.4000	296.4000	AM	Nationwide	Air Defence Region
296.7250	296.7250	AM	RAF Coltishall	Ground
296.7500	296.7500	AM	RAF Waddington	Radar
296.8000	296.8000	AM	International	Space Shuttle Down Link
296.9000	296.9000	AM	RAF Neatishead	Air Defence Region Ops
298.6500	298.6500	AM	Nationwide	Air Defence Region
299.1000	299.1000	AM	RAF Boulmer	Boulmer Rescue
299.4000	299.4000	AM	RAF Brawdy	Tower
		AM	RNAS Culdrose	Ground
		AM	RAF Lossiemouth	Ground
299.5000	299.5000	AM	ARA 3	Air-Air Refuelling
299.7000	299.7000	AM	Nationwide	Air Defence Region
299.9750	299.9750	AM	Watton	Eastern Radar
		AM	RAF West Drayton	London Mil
300.1500	300.1500	AM	RAF Benbecula	Air Defence Region Ops
300.1750	300.1750	AM	RNAS Portland	Radar
300.3500	300.3500	AM	RAF Northolt	ATIS
300.4250	300.4250	AM	RAF Linton-on-Ouse	Tower
300.4500	300.4500	AM	RAF OdihAM	Talkdown
300.4750	300.4750	AM	RAF LynehAM	Director
300.5750	300.5750	AM	RAF Waddington	Director
300.6500	300.6500	AM	RAF Neatishead	Air Defence Region Ops
300.7750	300.7750	AM	RAF Church Fenton	Approach
300.8250	300.8250	AM	RAF Lakenheath	Dispatcher
300.9250	300.9250	AM	RAF Coningsby	Talkdown
300.9500	300.9500	AM	Nationwide	Air Defence Region
304.0000		NFM	Worldwide	USAF Satcom Downlink
306.5000	306.5000	AM	ARA 6	Air-Air Refuelling
307.0000	307.0000	AM	RAF Neatishead	Air Defence Region Ops
307.6000	307.6000	AM	RAF Neatishead	Air Defence Region Ops
307.8000	307.8000	AM	RAF Fairford	Command Post
		AM	RAF Mildenhall	Command Post
		AM	Stanford	Stanford Ops
308.0000	308.0000	AM	ARA 3	Air-Air Refuelling
309.0750	309.0750	AM	RAF Lakenheath	Radar
309.5000	309.5000	AM	RAF Sculthorpe	Ground
309.5500	309.5500	AM	RAF Chetwynd	Ternhill Tower
309.6250	309.6250	AM	RAF OdihAM	Tower
309.6750	309.6750	AM	RAF Waddington	Talkdown

Base	Mobile	Mode	Location	User & Notes
309.7250	309.7250	AM	RAF Topcliffe	Tower
309.8750	309.8750	AM	RAF Leeming	Talkdown
309.9500	309.9500	AM	RAF Honington	Approach
310.0000	310.0000	AM	RAF Aldergrove	Approach
311.0000	311.0000	AM	ARA 2	Air-Air Refuelling
311.3000	311.3000	AM	Warton	Tower
311.3250	311.3250	AM	RAF Kinloss	Director
		AM	RAF Lossiemouth	Lossie Director
		AM	RNAS Yeovilton	Yeovil Ground
311.4000	311.4000	AM	RAF Binbrook	Tower
311.4750	311.4750	AM	Nationwide	USAF Air-Air
311.9500	311.9500	AM	RAF Wittering	Ground
311.9750	311.9750	AM	Nationwide	USAF General Air-Air
312.0000	312.0000	AM	Middle Wallop	Wallop Approach
312.0750	312.0750	AM	RAF Cottesmore	Approach
312.2250	312.2250	AM	RAF Coningsby	Approach
312.2750	312.2750	AM	RAF Wyton	Tower
312.3250	312.3250	AM	RAF Manston	Talkdown
312.3500	312.3500	AM	RAF Northolt	Tower
312.4000	312.4000	AM	RAF Lossiemouth	Talkdown
		AM	RNAS Portland	Talkdown
312.4250	312.4500	AM	RAF Chivenor	Director
312.5000	312.5000	AM	RAF Waddington	Approach
312.5500	312.5500	AM	RAF MarhAM	Ops
312.6500	312.6500	AM	RAF Staxton Wold	Air Defence Region Ops
312.6750	312.6750	AM	Middle Wallop	Director
312.7000	312.7000	AM	RNAS Merryfield	Tower
312.8000	312.8000	AM	RAF Woodvale	Approach
314.3500	314.3500	AM	RAF Bentwaters	Have Quick
314.4750	314.4750	AM	Nationwide	Air Defence Region
315.0000	315.0000	AM	RAF Neatishead	NATO AWACS Coord
315.2000	315.2000	AM	RAF Woodbridge	Dispatcher
315.5000	315.5000	AM	RAF Finningley	Director
315.5250	315.5250	AM	MoD Farnborough	Radar
315.5750	315.5750	AM	RAF Honington	Departures
		AM	RAF Lakenheath	Dep Con
315.6500	315.6500	AM	RNAS Lee-on-Solent	Tower
315.7500	315.7500	AM	RAF Abingdon	SRE
315.8500	315.8500	AM	RAF Neatishead	Air Defence Region Ops
315.9750	315.9750	AM	RAF Odiham	Approach
316.0000	316.0000	AM	RAF Upper Heyford	Tower
316.3500	316.3500	AM	ARA 8	Air-Air Refuelling
316.6000	316.6000	AM	ARA 4	Air-Air Refuelling
316.7500	316.7500	AM	Nationwide	USAF Air-Air
317.2000	317.2000	AM	ARA 4	Air-Air Refuelling
317.5000	317.5000	AM	Nationwide	Air Defence Region
317.8500	317.8500	AM	Nationwide	Air Defence Region
318.5500	318.5500	AM	Nationwide	Air-Air Refuelling
		AM	RAF Boulmer	Air Defence Region Ops
318.7500	318.7500	AM	RAF Neatishead	Air Defence Region Ops

Base	Mobile	Mode	Location	User & Notes
319.1500	319.1500	AM	RAF Staxton Wold	Air Defence Region Ops
319.4000	319.4000	AM	RAF Neatishead	Air Defence Region Ops
319.6000	319.6000	AM	Royal Navy	Ship-Air
322.2000	322.2000	AM	Nationwide	USAF Air-Air
322.4000	322.4000	NFM	Nationwide	TADIL-A Data Link
322.9500	322.9500	AM	Nationwide	USAF Displays
323.2000	323.2000	AM	Nationwide	USAF Air-Air

326.5000 - 328.6000 MHz Radio Astronomy

326.9000	326.500	AM	ARA 5	Air-Air Refuelling

329.1500 - 335.0000 MHz Aeronautical ILS (Glideslope Component)

Base		Mode	Location	User & Notes
329.1500		AM	Nationwide	Glideslope (Localiser 108.95 MHz)
		AM	Woodford	Runway 25
329.3000		AM	Nationwide	Glideslope (Localiser 108.90 MHz)
		AM	Cranfield	Runway 22
		AM	Edinburgh	Runway 07/25
		AM	Kerry	Runway 07/25
		AM	RAF Woodbridge	Runway 27
329.4500		AM	Nationwide	Glideslope (Localiser 110.55 MHz)
		AM	Filton	Runway 10/28
329.6000		AM	Nationwide	Glideslope (Localiser 110.50 MHz)
		AM	Bournemouth	Runway 08/26
		AM	London/Stansted	Runway 05/23
		AM	RAF Scampton	Runway 05/23
329.7500		AM	Nationwide	Glideslope (Localiser 108.55 MHz)
329.9000		AM	Nationwide	Glideslope (Localiser 108.50 MHz)
		AM	RAF Benson	Runway 19
330.0500		AM	Nationwide	Glideslope (Localiser 110.75 MHz)
330.2000		AM	Nationwide	Glideslope (Localiser 110.70 MHz)
		AM	Cardiff	Runway 12/30
		AM	Connaught	Runway 27
		AM	London/Heathrow	Runway 23
		AM	RAF Coningsby	Runway 26
330.3500		AM	Nationwide	Glideslope (Localiser 108.75 MHz)
		AM	Humberside	Runway 21
330.5000		AM	Nationwide	Glideslope (Localiser 108.70 MHz)
		AM	RAF Alconbury	Runway 30
		AM	RAF Leuchars	Runway 27
		AM	RAF St Mawgan	Runway 31
		AM	RAF Shawbury	Runway 19
330.6500		AM	Nationwide	Glideslope (Localiser 110.95 MHz)
330.8000		AM	Nationwide	Glideslope (Localiser 110.90 MHz)
		AM	Belfast/Aldergrove	Runway 17
		AM	Jersey	Runway 09
		AM	Leeds & Bradford	Runway 32/14
		AM	London/Gatwick	Runway 08R/26L
		AM	Norwich	Runway 27
		AM	Ronaldsway	Runway 27
330.9500		AM	Nationwide	Glideslope (Localiser 111.95 MHz)

Base	Mobile	Mode	Location	User & Notes
331.2500		AM	Nationwide	Glideslope (Localiser 109.15 MHz)
		AM	Luton	Runway 08/28
331.3000		AM	Nationwide	Glideslope (Localiser 111.90 MHz)
		AM	RAF Brize Norton	Runway 08/26
		AM	RAF Honington	Runway 27
331.4000		AM	Nationwide	Glideslope (Localiser 109.10 MHz)
331.5500		AM	Nationwide	Glideslope (Localiser 111.15 MHz)
331.7000		AM	Nationwide	Glideslope (Localiser 111.10 MHz)
		AM	RAF Fairford	Runway 09/27
		AM	RAF Lossiemouth	Runway 23
		AM	RAF Waddington	Runway 21
		AM	RAF Wattisham	Runway 23
331.8500		AM	Nationwide	Glideslope (Localiser 109.35 MHz)
332.0000		AM	Nationwide	Glideslope (Localiser 109.30 MHz)
		AM	Glasgow	Runway 23
		AM	RAF Church Fenton	Runway 24
		AM	RAF Wyton	Runway 27
332.1500		AM	Nationwide	Glideslope (Localiser 111.35 MHz)
332.3000		AM	Nationwide	Glideslope (Localiser 111.30 MHz)
		AM	Hatfield	Runway 24
		AM	Perth	Runway 21
		AM	Teesside	Runway 23
332.4500		AM	Nationwide	Glideslope (Localiser 109.55 MHz)
332.6000		AM	Nationwide	Glideslope (Localiser 109.50 MHz)
		AM	London/Heathrow	Runway 09R/27L
		AM	Manchester	Runway 06/24
		AM	Plymouth	Runway 31
		AM	Shannon	Runway 24
332.7500		AM	Nationwide	Glideslope (Localiser 111.55 MHz)
		AM	Newcastle	Runway 07/25
		AM	RAF Coltishall	Runway 22
		AM	RAF Finningley	Runway 20
		AM	RAF Upper Heyford	Runway 27
332.9000		AM	Nationwide	Glideslope (Localiser 111.50 MHz)
333.0500		AM	Nationwide	Glideslope (Localiser 109.75 MHz)
		AM	Coventry	Runway 23
333.2000		AM	Nationwide	Glideslope (Localiser 109.70 MHz)
		AM	Belfast/Aldergrove	Runway 25
		AM	Dinard	Runway 36
		AM	RAF Cranwell	Runway 27
		AM	RAF Kinloss	Runway 26
		AM	RAF Lyneham	Runway 25
		AM	RAF Valley	Runway 14
333.3500		AM	Nationwide	Glideslope (Localiser 111.75 MHz)
		AM	Liverpool	Runway 09/27
333.5000		AM	Nationwide	Glideslope (Localiser 111.70 MHz)
		AM	RAE Boscombe Dn	Runway 24
333.6500		AM	Nationwide	Glideslope (Localiser 109.95 MHz)
333.8000		AM	Nationwide	Glideslope (Localiser 109.90 MHz)
		AM	Aberdeen/Dyce	Runway 16/34

Base	Mobile	Mode	Location	User & Notes
		AM	Cherbourg	Runway 29
		AM	Cork	Runway 17/35
		AM	East Midlands	Runway 09/27
		AM	Exeter	Runway 26
		AM	RAF Bentwaters	Runway 25
		AM	Stornoway	Runway 18
		AM	Warton	Runway 26
333.9500		AM	Nationwide	Glideslope (Localiser 108.35 MHz)
334.1000		AM	Nationwide	Glideslope (Localiser 108.30 MHz)
		AM	Bedford	Runway 27
		AM	RAF Lakenheath	Runway 24
334.2500		AM	Nationwide	Glideslope (Localiser 110.15 MHz)
		AM	Bristol	Runway 09/27
334.4000		AM	Nationwide	Glideslope (Localiser 110.10 MHz)
		AM	Birmingham	Runway 15/33
		AM	Glasgow	Runway 05
		AM	RAF Marham	ILS Runway 24
		AM	Rennes	Runway 29
334.5500		AM	Nationwide	Glideslope (Localiser 108.15 MHz)
		AM	Blackpool	Runway 28
		AM	Lydd	Runway 22
334.7000		AM	Nationwide	Glideslope (Localiser 108.10 MHz)
		AM	Guernsey	Runway 09/27
		AM	RAF Abingdon	Runway 36
		AM	RAF Chivenor	Runway 28
		AM	RAF Mildenhall	Runway 11/29
334.7500	334.750	AM	RAF Neatishead	MRSA
334.8500		AM	Nationwide	Glideslope (Localiser 110.35 MHz)
335.0000		AM	Nationwide	Glideslope (Localiser 110.30 MHz)
		AM	Jersey	Runway 27
		AM	London/Heathrow	Runway 09L/27R
		AM	Prestwick	Runway 13/31
		AM	RAF Cottesmore	Runway 23
		AM	RAF Leeming	Runway 16

335.4000 - 399.9000 MHz UHF Military Aviation 25 kHz

Base	Mobile	Mode	Location	User & Notes
336.0000	336.0000	AM	RAF Woodbridge	ATIS
336.2750	336.2750	AM	MoD Farnborough	Approach
336.3500	336.3500	AM	RAF Kinloss	Tower
		AM	RAF Leeming	Talkdown
336.3750	336.3750	AM	RAF Cottesmore	Ground
336.4750	336.4750	AM	MoD Filton	Director
		AM	Warton	Approach
336.5250	336.5250	AM	RAF St Athan	Tower
336.5500	336.5500	AM	RAF St Mawgan	Talkdown
337.5750	337.5750	AM	RAF Fairford	Tower
337.7250	337.7250	AM	RAF Valley	Director
337.7500	337.7500	AM	RAF Cranwell	Ground
		AM	RAF Lossiemouth	Tower
		AM	RNAS Portland	Tower

Base	Mobile	Mode	Location	User & Notes
		AM	RNAS Prestwick	Navy Prestwick
337.8250	337.8250	AM	RAF Leeming	Approach
337.8750	337.8750	AM	RAF Cottesmore	Talkdown
337.9000	337.9000	AM	RAF Marham	Tower
		AM	RAF Shawbury	Ground
337.9250	337.9250	AM	MoD Bedford	Tower
		AM	MoD West Freugh	Tower
337.9500	337.9500	AM	RAF Wittering	Talkdown
337.9750	337.9750	AM	RAF Coningsby	Talkdown
		AM	RAF Machrihanish	Talkdown
338.2000	338.2000	NFM	Nationwide	TADIL-A Data Link
338.6250	338.6250	AM	RAF Manston	Manston Director
338.6500	338.6500	AM	RAF Brize Norton	Brize Talkdown
338.6750	338.6750	AM	RAF Lakenheath	Radar
338.8250	338.8250	AM	RAF Ternhill	Tower
338.8500	338.8500	AM	RAF Leeming	Ground
338.8750	338.8750	AM	RNAS Yeovilton	Yeovil Director
338.9750	338.9750	AM	RAF Honington	Zone
		AM	RNAS Predannack	Tower
339.9500	339.9500	AM	RAF Coltishall	Tower
		AM	RNAS Culdrose	Radar
339.9750	339.9750	AM	RNAS Yeovilton	Talkdown
340.0000	340.0000	AM	RAF Chivenor	Zone
		AM	RAF Honington	Approach
340.0250	340.0250	AM	RAF Linton-on-Ouse	Ground
		AM	RAF Lyneham	Zone
340.1000	340.1000	AM	RAF St Athan	Talkdown
340.1250	340.1250	AM	RAF Mildenhall	Ground
340.1500	340.1500	AM	RAF Brawdy	Talkdown
340.1750	340.1750	AM	RAF Finningley	Ground
		AM	RAF Valley	Tower
340.2000	340.2000	AM	RAF Church Fenton	Ground
340.3250	340.3250	AM	RAF Greenham Common	Tower
340.4500	340.4500	AM	RAF Neatishead	MRSA
340.4750	340.4750	AM	RAF Cranwell	Cranwell Approach
340.5250	340.5250	AM	RAF Barkston Heath	Barkston Approach
340.5750	340.5750	AM	RAF Cottesmore	Approach
340.8250	340.8250	AM	RAF Sculthorpe	Tower
341.6500	341.6500	AM	RAF Bentwaters	ATIS
341.6750	341.6750	AM	Nationwide	Air Defence Region
341.9250	341.9250	AM	RAF Scampton	Talkdown
342.0250	342.0250	AM	MoD Filton	Tower
342.0750	342.0750	AM	RAF Barkston Heath	Barkston Tower
342.1250	342.1250	AM	RAF Waddington	Ground
342.2250	342.2250	AM	RAF Alconbury	Dispatcher
342.2500	342.2500	AM	RAF Coltishall	Director
342.4500	342.4500	AM	RAF Brize Norton	Brize Approach
342.6500	342.6500	AM	RAF Neatishead	Air Defence Region Ops
342.7000	342.7000	AM	RAF Scampton	Recoveries
343.3000	343.3000	AM	RAF Neatishead	Air Defence Region Ops

Base	Mobile	Mode	Location	User & Notes
343.4250	343.4250	AM	RAF Wattisham	Tower
343.6000	343.6000	AM	RAF Croughton	Croughton Radio
343.6250	343.6250	AM	RAF Upper Heyford	PAR
343.7000	343.7000	AM	Warton	Radar
344.0000	344.0000	AM	RAF Abingdon	SRE
		AM	RAF Brize Norton	Director
		AM	RAF Coningsby	Director
		AM	RAF Finningley	Director
		AM	RAF Leeming	Director
		AM	RAF Lyneham	Director
		AM	RAF Machrihanish	Director
		AM	RAF Manston	Director
		AM	RAF Marham	Director
		AM	RAF St Athan	Director
		AM	RAF St Mawgan	Director
		AM	RAF Scampton	Recoveries
		AM	RAF Valley	Director
		AM	RAF Waddington	Director
		AM	RAF Wittering	Director
		AM	RAF Wyton	Director
344.3500	344.3500	AM	RAF Manston	Tower
		AM	RAF Topcliffe	Director
		AM	RNAS Yeovilton	Talkdown
344.4750	344.4750	AM	RAF Linton-on-Ouse	Director
344.5250	344.5250	AM	RAF Machrihanish	Approach
344.5750	344.5750	AM	RAF Leeming	Tower
344.6000	344.6000	AM	RAF Abingon	Tower
344.6250	344.6250	AM	RAF Coningsby	Approach
344.8000	344.8000	AM	RAF Mildenhall	Command Post
344.9750	344.9750	AM	RAF Northolt	Approach
345.0000	345.0000	AM	RAF Neatishead	Air Defence Region Ops
353.0000	353.0000	AM	RAF Boulmer	Air Defence Region Ops
353.0500	353.0500	AM	RAF Buchan	Air Defence Region Ops
353.2000	353.2000	AM	RAF Ternhill	Approach
354.4500	354.4500	AM	RNAS Portland	Naval Exercises
356.1750	356.1750	AM	RAF Chivenor	Talkdown
		AM	RAF Wattisham	Talkdown
356.2750	356.2750	AM	RAF Halton	Halton Aero Club
356.7250	356.7250	AM	RAF Leeming	Ops
356.8250	356.8250	AM	RAF Bentwaters	Dispatcher
356.8750	356.8750	AM	RAF Brize Norton	Brize Director
356.9750	356.9750	AM	RAF Shawbury	Talkdown
357.1000	357.1000	AM	RAF Scampton	Talkdown
357.1250	357.1250	AM	RAF Cosford	Tower
357.1500	357.1500	AM	RAF Wittering	Tower
357.1750	357.1750	AM	RAF St Athan	Radar
357.2000	357.2000	AM	RAF St Mawgan	Approach
357.3750	357.3750	AM	RAF Topcliffe	Approach
357.4000	357.4000	AM	MoD Farnborough	Tower
357.4750	357.4750	AM	RAF Brize Norton	Brize Ops

Base	Mobile	Mode	Location	User & Notes
357.9000	357.9000	AM	RAF Upper Heyford	Command Post
358.5000	358.5000	AM	RAF Brawdy	Director
358.5250	358.5250	AM	RAF Linton-on-Ouse	Talkdown
358.5500	358.5500	AM	RAF Coningsby	Ground
358.6000	358.6000	AM	RAF Machrihanish	Tower
		AM	RAF Upper Heyford	Metro
358.6500	358.6500	AM	RAF Leeming	Director
358.6750	358.6750	AM	RAF Lakenheath	Tower
		AM	RAF Valley	Talkdown
358.7000	358.7000	AM	RNAS Culdrose	Talkdown
358.7250	358.7250	AM	RAF Cottesmore	Director
358.7500	358.7500	AM	RAF Honington	Talkdown
		AM	RAF Mona	Tower
358.7750	358.7750	AM	RAF Finningley	Approach
358.8000	358.8000	AM	RAF Abingdon	Benson Zone
358.8500	358.8500	AM	RAF Church Fenton	Director
359.4250	359.4250	AM	RNAS Portland	Naval Exercises
359.5000	359.5000	AM	RAF Lyneham	Approach
359.5250	359.5250	AM	RAF Sculthorpe	PAR
359.8250	359.8250	AM	RAF Wattisham	Talkdown
359.8750	359.8750	AM	RAF Wittering	Approach
360.5500	360.5500	AM	RAF St Mawgan	Director
362.0500	362.0500	AM	Salisbury Plain	Salisbury Ops
362.0750	362.0750	AM	RAF Bentwaters	Approach
362.2000	362.2000	AM	MoD Bedford	Approach
362.2250	362.2250	AM	Netheravon	Approach
362.3000	362.3000	AM	RAF Abingdon	Approach
		AM	RAF Alconbury	Wyton Zone Approach
		AM	RAF Benson	Approach
		AM	RAF Brawdy	Approach
		AM	RAF Chivenor	Approach
		AM	RAF Church Fenton	Fenton Approach/Radar
		AM	RAF Coningsby	Approach
		AM	RAF Cosford	Approach
		AM	RAF Cranwell	Approach
		AM	RAF Dishforth	Approach
		AM	RAF Fairford	Brize Radar
		AM	RAF Honington	Approach
		AM	RAF Kemble	Approach
		AM	RAF Kinloss	Approach
		AM	RAF Leeming	Approach
		AM	RAF Leuchars	Approach
		AM	RAF Linton-on-Ouse	Approach
		AM	RAF Lossiemouth	Approach
		AM	RAF Lyneham	Approach
		AM	RAF Machrihanish	Approach
		AM	RAF Manston	Approach
		AM	RAF Marham	Approach
		AM	RAF Mildenhall	Approach
		AM	RAF Newton	Approach

Base	Mobile	Mode	Location	User & Notes
		AM	RAF Northolt	Approach
		AM	RAF Odiham	Approach
		AM	RAF St Athan	Approach
		AM	RAF St Mawgan	Approach
		AM	RAF Scampton	Approach/Radar
		AM	RAF Shawbury	Approach
		AM	RAF Ternhill	Approach
		AM	RAF Topcliffe	Approach
		AM	RAF Upper Heyford	Approach
		AM	RAF Valley	Approach
		AM	RAF Wittering	Approach
		AM	RAF Wyton	Approach
		AM	RNAS Portland	Approach/Radar/Tower
		AM	RNAS Yeovilton	Approach/Director
362.3750	362.3750	AM	RAF Alconbury	Wyton Zone
		AM	RAF Wyton	Approach
362.4500	362.4500	AM	RAF Chivenor	Tower
362.6750	362.6750	AM	RAF Linton-on-Ouse	Approach
362.8250	362.8250	AM	Nationwide	Air Defence Region
364.2000	364.2000	AM	RAF Neatishead	Air Defence Region Ops
		AM	Nationwide	NATO Magic Surveillance
364.7750	364.7750	AM	RAF Chivenor	Approach
364.8000	364.8000	AM	RAF Coltishall	Ops
364.8250	364.8250	AM	Middle Wallop	Talkdown
364.8750	364.8750	AM	RAF Upper Heyford	Approach
365.1000	365.1000	AM	RAF Mildenhall	Dispatcher
367.9500	367.9500	AM	RAF Brawdy	Approach
368.3250	368.3250	AM	RAF Bentwaters	Final Controller
369.1250	369.1250	AM	Nationwide	Air Defence Region
369.1500	369.1500	AM	RAF Spadeadam	Ops
369.8750	369.8750	AM	RNAS Yeovilton	Radar
370.0000	370.0000	AM	RNAS Predannack	Tower
370.0250	370.0250	AM	RAF Leuchars	Finals
370.0500	370.0500	AM	RAF Cottesmore	Tower
		AM	RAF Kinloss	Talkdown
370.1000	370.1000	AM	MoD Boscombe Down	Tower
370.3000	370.3000	AM	MoD Llanbedr	PAR
		AM	RAF Brize Norton	Brize Ground
371.2000	371.2000	AM	RAF Fairford	Command Post
		AM	Nationwide	Air Defence Region
372.0500	372.0500	AM	RAF Neatishead	Air Defence Region Ops
372.3000	372.3000	AM	RNAS Culdrose	Ground
372.3250	372.3250	AM	RAF Valley	Valley Approach
372.4250	372.4250	AM	HMS Cambridge	Royal Navy
372.6250	372.6250	AM	Middle Wallop	Wallop Tower
372.6500	372.6500	AM	RNAS Yeovilton	Tower
373.1000	373.1000	AM	Nationwide	Air Defence Region
375.1500	375.1500	AM	RAF Upper Heyford	Ground
375.2000	375.2000	AM	RAF Lyneham	Talkdown
375.3000	375.3000	AM	RAF Swinderby	Tower

Base	Mobile	Mode	Location	User & Notes
375.4250	375.4250	AM	RAF Newton	Tower
375.5000	375.5000	AM	RAF Northolt	Talkdown
375.5250	375.5250	AM	RAF Alconbury	Wyton Departure
		AM	RAF Wyton	Wyton Departure
376.5250	376.5250	AM	RAF Kinloss	Talkdown
376.5750	376.5750	AM	RAF Cottesmore	Director
		AM	RAF Wittering	Dep Control
376.6250	376.6250	AM	RAF Brize Norton	Director
376.6500	376.6500	AM	RAF Lossiemouth	Approach
376.6750	376.6750	AM	RAF Chivenor	Director
378.2000	378.2000	AM	RAF Neatishead	MRSA
378.3000	378.3000	AM	RAF Sculthorpe	PAR
379.0250	379.0250	AM	RAF Manston	Approach
379.2750	379.2750	AM	RAF Coltishall	Approach
379.4250	379.4250	AM	RAF Northolt	Director
379.4750	379.4750	AM	RAF Fairford	Dispatcher
379.5000	379.5000	AM	RNAS Culdrose	Radar
379.5250	379.5250	AM	RAF Cranwell	Tower
379.5500	379.5500	AM	RAF Finningley	Tower
379.6500	379.6500	AM	RAF Marham	Talkdown
379.6750	379.6750	AM	RAF Dishforth	Approach
379.7000	379.7000	AM	RAF Mona	Approach
379.7500	379.7500	AM	RNAS Yeovilton	ATIS
379.9250	379.9250	AM	RAF Chivenor	Ground
379.9750	379.9750	AM	MoD Farnborough	Ops
380.0250	380.0250	AM	MoD Boscombe Down	Approach
380.1250	380.1250	AM	RAF St Athan	Director
380.1750	380.1750	AM	MoD Llanbedr	Tower
380.2250	380.2250	AM	RNAS Culdrose	Tower
380.8000	380.8000	AM	Nationwide	Air-Air Refuelling
381.0000	381.0000	AM	RAF Lyneham	ATIS
381.0750	381.0750	AM	RAF Church Fenton	Departures
		AM	RAF Linton-on-Ouse	Departures
381.1250	381.1250	AM	MoD Boscombe Down	PAR
381.2000	381.2000	AM	RAF Brize Norton	Tower
382.9000	382.9000	AM	HMS Drake (Plymouth)	Royal Navy
383.1500	383.1500	NFM	Nationwide	TADIL-A Data Link
383.2250	383.2250	AM	RAF Wittering	Talkdown
383.2750	383.2750	AM	RAF Woodbridge	Ground
383.4500	383.4500	AM	RAF Alconbury	Tower
383.4750	383.4750	AM	RAF Cranwell	Talkdown
383.5000	383.5000	AM	RAF Finningley	Talkdown
383.5250	383.5250	AM	MoD West Freugh	Approach
383.6250	383.6250	AM	MoD Bedford	Approach
385.1500	385.1500	AM	RAF Bentwaters	Have Quick
385.4000	385.4000	AM	RAF Brize Norton	Talkdown
		AM	RAF Church Fenton	Fenton Talkdown
		AM	RAF Finningley	Talkdown
		AM	RAF Honington	Talkdown
		AM	RAF Leeming	Talkdown

Base	Mobile	Mode	Location	User & Notes
		AM	RAF Lyneham	Talkdown
		AM	RAF Machrihanish	Talkdown
		AM	RAF Manston	Talkdown
		AM	RAF Marham	Talkdown
		AM	RAF Northolt	Talkdown
		AM	RAF Odiham	Talkdown
		AM	RAF St Athan	Talkdown
		AM	RAF St Mawgan	Talkdown
		AM	RAF Shawbury	Talkdown
		AM	RAF Topcliffe	Talkdown
		AM	RAF Valley	Talkdown
		AM	RAF Waddington	Talkdown
		AM	RAF Wyton	Talkdown
386.6750	386.6750	AM	MoD Llanbedr	Approach
386.7250	386.7250	AM	RAF Church Fenton	Talkdown
		AM	MoD Bedford	PAR
386.7750	386.7750	AM	MoD Farnborough	Radar
		AM	RAF Odiham	Radar
386.8250	386.8250	AM	RAF Lyneham	Tower
386.8750	386.8750	AM	RAF Shawbury	Radar
386.9000	386.9000	AM	RAF Valley	Ground
388.0000	388.0000	AM	Nationwide	Sharks Helicopter Displays
397.9750	397.9750	AM	RAF Lakenheath	Ground
398.1000	398.1000	AM	RAF Lossiemouth	Approach
398.2000	398.2000	AM	RAF Lakenheath	Command Post
398.8750	398.8750	AM	Nationwide	Air Defence Region

399.9000 - 400.0500 MHz Radio Navigation Satellite Down Links

400.0500 - 400.1500 MHz Satellite Time Signal Down Links

400.1500 - 401.0000 MHz Satellite Down Links

401.0000		NFM	Nationwide	RAF Target Telemetry
406.0000		AM	International	Distress Frequency

406.0000 - 406.1000 MHz Mobile Satellite Terminal Up Link

406.1000		AM	International	Distress Frequency
				Monitored by UK
				USA & JAPAN

406.1000 - 406.5000 MHz Radio Astronomy

406.5000 - 409.5000 MHz North Sea Radio Positions Beacons and US Embassy Close Protection Teams

409.5000 - 410.0000 MHz MOD & Government Use

410.0000 - 425.0000 MHz MOD, USAF & Mould Use 25 kHz Formula One Racing Team Links

410.0000	410.0000	NFM	Nationwide	USAF Base Common

Base	Mobile	Mode	Location	User & Notes
410.2750	410.2750	NFM	RAF Bentwaters	USAF Ground Maintenance
		NFM	RAF Woodbridge	USAF Ground Maintenance
410.4750	410.4750	NFM	RAF Bentwaters	USAF Ground Maintenance
		NFM	RAF Woodbridge	USAF Ground Maintenance
410.5000	410.5000	NFM	RAF Bentwaters	USAF Ground Maintenance
		NFM	RAF Woodbridge	USAF Ground Maintenance
410.6000	410.6000	NFM	RAF Upper Heyford	USAF Ground
		NFM	RAF Bentwaters	USAF Ground Maintenance
		NFM	RAF Woodbridge	USAF Ground Maintenance
410.7750	410.7750	NFM	RAF Bentwaters	USAF Ground Maintenance
		NFM	RAF Woodbridge	USAF Ground Maintenance
410.9000	410.9000	NFM	RAF Upper Heyford	USAF Ground
		NFM	RAF Bentwaters	USAF Ground Maintenance
		NFM	RAF Woodbridge	USAF Ground Maintenance
		NFM	Nationwide	MoD Transport
411.0000	411.0000	NFM	RAF Fairford	USAF Ground
411.1500	411.5000	NFM	RAF Upper Heyford	USAF Ground
		NFM	RAF Bentwaters	USAF Ground Maintenance
		NFM	RAF Woodbridge	USAF Ground Maintenance
411.1750	411.1750	NFM	Leicester	City Council
411.4250	411.4250	NFM	RAF Bentwaters	USAF Ground Maintenance
411.5750	411.5750	NFM	RAF Lakenheath	USAF Ground Maintenance
412.0500	412.0500	NFM	Worldwide	US ATS Satellites
412.1750	412.1750	NFM	RAF Woodbridge	USAF Ground Maintenance
412.2750	412.2750	NFM	RAF Fairford	USAF Ground
		NFM	RAF Woodbridge	USAF Ground Maintenance
412.5250		NFM	South Wales	Mould
412.8375		NFM	South Wales	Mould
413.0750	413.0750	NFM	RAF Upper Heyford	USAF Ground
		NFM	RAF Bentwaters	USAF Ground Maintenance
		NFM	RAF Woodbridge	USAF Ground Maintenance
413.1000	413.1000	NFM	Nationwide	USAF Displays
413.1250	413.1250	NFM	RAF Upper Heyford	USAF Ground
413.1500	413.1500	NFM	Nationwide	USAF Base Common
413.1750	413.1750	NFM	RAF Upper Heyford	USAF Ground
413.4250	413.4250	NFM	MoD Malvern	Security
413.4500	413.4500	NFM	Nationwide	MoD Transport
413.5000	413.5000	NFM	Nationwide	MoD Transport
413.8000	413.8000	NFM	Nationwide	Nuclear Security
		NFM	RAF Upper Heyford	USAF Ground
414.1500	414.1500	NFM	RAF Upper Heyford	USAF Ground
414.3000	414.3000	NFM	RAF Fairford	USAF Base Security
414.4875		NFM	Formula One Racing	McLaren Team Voice Link
415.9875		NFM	Formula One Racing	McLaren Team Voice Link
417.9375		NFM	Manchester	Mould
419.9875		NFM	Formula One Racing	McLaren Team Voice Link
420.0125		NFM	Manchester	Mould
421.5625		NFM	Manchester	Mould
421.7875		NFM	Manchester	Mould

Base	Mobile	Mode	Location	User & Notes
425.0000 - 425.5000 MHz PMR Mobile 12.5 kHz NFM (Split + 20.5 MHz)				
425.5000 - 429.0000 MHz PMR Mobile 12.5 kHz NFM (Split + 14.5 MHz)				
429.0000 - 430.0000 MHz MoD Radiolocation Beacons				
430.0000 - 431.0000 MHz MoD Allocation				
430.9000	430.9000	NFM	RAF Newton	Base Security
433.0000 - 434.0000 MHz MoD Mould Repeaters 12.5 kHz				
433.0125		NFM	West Midlands	Mould
433.1375		NFM	West Midlands	Mould
433.2875		NFM	West Midlands	Mould
433.0000 - 440.0000 MHz 70cm Amateur Radio Band				
433.0000	434.6000	NFM		Raynet Channel RB0
		NFM	GB3BN	Bracknell, Berks
		NFM	GB3CK	Charing, Kent
		NFM	GB3DT	Wimborne, Dorset
		NFM	GB3EX	Exeter
		NFM	GB3KB	Farnborough, Kent
		NFM	GB3LL	Llandudno
		NFM	GB3MK	Milton Keynes
		NFM	GB3MS	Malvern Hills
		NFM	GB3NR	Norwich
		NFM	GB3NT	Newcastle-upon-Tyne
		NFM	GB3NY	Scarborough
		NFM	GB3PF	Pendle Forest
		NFM	GB3PU	Perth
		NFM	GB3SO	Boston, Lincs
		NFM	GB3SV	Bishop Stortford
		NFM	GB3US	Sheffield
		NFM	GB3WN	Wolverhampton
433.0250	434.6250	NFM		Raynet Channel RB1
		NFM	GB3BV	Hemel Hempstead
433.0500	434.6500	NFM		Raynet Channel RB2
		NFM	GB3AV	Aylesbury
		NFM	GB3BA	Aberdeen
		NFM	GB3CH	Plymouth
		NFM	GB3CI	Corby
		NFM	GB3CF	Fylde Coast
		NFM	GB3EK	Margate
		NFM	GB3HD	Huddersfield
		NFM	GB3LS	Lincoln
		NFM	GB3LV	Enfield
		NFM	GB3NN	Wells, Norfolk
		NFM	GB3NX	Crawley
		NFM	GB3OS	Stourbridge
		NFM	GB3PH	Portsmouth
		NFM	GB3ST	Stoke-on-Trent

Base	Mobile	Mode	Location	User & Notes
		NFM	GB3UL	Belfast
		NFM	GB3YL	Yeovil
433.0750	434.6750	NFM		Raynet Channel RB3
		NFM	GB3HL	Hillingdon
		NFM	GB3HU	Hull
		NFM	GB3VS	Taunton
433.1000	434.7000	NFM		Raynet Channel RB4
		NFM	GB3AN	Anglesey
		NFM	GB3GC	Goole
		NFM	GB3HZ	High Wycombe
		NFM	GB3IH	Ipswich
		NFM	GB3IW	Isle Of Wight
		NFM	GB3KL	Kings Lynn
		NFM	GB3KM	Knockmore
		NFM	GB3LE	Leicester
		NFM	GB3MA	Central Manchester
		NFM	GB3NK	Wrotham
		NFM	GB3OH	Linlithgow
		NFM	GB3SP	Pembroke
		NFM	GB3UB	Bath
433.1250	434.7250	NFM		Raynet Channel RB5
		NFM	GB3GH	Gloucester
		NFM	GB3HY	Sussex
		NFM	GB3NW	Hendon
		NFM	GB3OV	Huntingdon
		NFM	GB3WJ	Scunthrope
433.1500	434.7500	NFM		Raynet Channel RB6
		NFM	GB3BD	Bedford
		NFM	GB3BR	Brighton
		NFM	GB3BE	Bury St. Edmunds
		NFM	GB3CR	Mold, Clwyd
		NFM	GB3CW	Newton, Powys
		NFM	GB3HA	Hornsea
		NFM	GB3LW	Central London
		NFM	GB3ME	Rugby
		NFM	GB3NM	Nottingham
		NFM	GB3SK	Canterbury
		NFM	GB3SW	Salisbury
		NFM	GB3SY	Barnsley
		NFM	GB3WG	Port Talbot
433.1750	434.7750	NFM		Raynet Channel RB7
433.2000	434.8000	NFM		Raynet Channel RB8
		NFM	GB3EH	Warwickshire
		NFM		Raynet Channel SU8
433.2250	434.8250	NFM		Raynet Channel RB9
		NFM	GB3SW	Salisbury
433.2500	434.8500	NFM		Raynet Channel RB10
		NFM	GB3AW	Ashmansworth
		NFM	GB3BS	Bristol
		NFM	GB3DD	Dundee

Base	Mobile	Mode	Location	User & Notes
		NFM	GB3DY	Wirksworth
		NFM	GB3ER	Danbury
		NFM	GB3HU	Little Weighton
		NFM	GB3LI	Liverpool
		NFM	GB3LT	Luton
		NFM	GB3ML	Blackhill
		NFM	GB3MW	Leamington Spa
		NFM	GB3NS	Banstead
		NFM	GB3NU	Benbecula
		NFM	GB3PB	Peterborough
		NFM	GB3PD	Peterhead
		NFM	GB3WY	Queensbury
433.2750	434.8750	NFM		Raynet Channel RB11
		NFM	GB3AH	Swaffham
		NFM	GB3BK	Reading
		NFM	GB3DC	Sunderland
		NFM	GB3GR	Grantham
		NFM	GB3GY	Grimsby
		NFM	GB3HN	Hitchin
		NFM	GB3HT	Hinckley
		NFM	GB3LA	Leeds
		NFM	GB3LR	Lewes
		NFM	GB3NF	Southampton
		NFM	GB3SH	Devon
		NFM	GB3WP	East Manchester
		NFM	GB3ZI	Stafford
433.3000	434.9000	NFM		Raynet Channel RB12
		NFM	GB3GF	Guildford
		NFM	GB3MT	Bolton
		NFM	GB3PT	Barkway
		NFM	GB3RY	Leicester
		NFM		Raynet Chan SU17 RTTY
433.3250	434.9250	NFM		Raynet Channel RB13
		NFM	GB3CA	Carlisle
		NFM	GB3CY	York
		NFM	GB3DS	Worksop
		NFM	GB3GU	Guernsey
		NFM	GB3HW	Gidea Park
		NFM	GB3LC	Louth
		NFM	GB3SM	Leek
		NFM	GB3TD	Swindon
		NFM	GB3TH	Tamworth
		NFM	GB3VH	Hatfield
		NFM	GB3XX	Daventry
433.3500	434.9500	NFM		Raynet Channel RB14
		NFM	GB3AB	Aberdeen
		NFM	GB3CB	Birmingham
		NFM	GB3CE	Colchester
		NFM	GB3ED	Edinburgh
		NFM	GB3GL	Glasgow

Base	Mobile	Mode	Location	User & Notes
		NFM	GB3HE	Hastings
		NFM	GB3HK	Hawick
		NFM	GB3HO	Horsham
		NFM	GB3HR	Stanmore
		NFM	GB3LF	Staveley
		NFM	GB3MR	Stockport
		NFM	GB3ND	Ilfracombe
		NFM	GB3PY	Cambridge
		NFM	GB3SD	Weymouth
		NFM	GB3TS	Middlesbrough
		NFM	GB3WF	Leeds
		NFM	GB3YL	Lowestoft
433.3750	434.9750	NFM		Raynet Channel RB15
		NFM	GB3BF	North Berks
		NFM	GB3FN	Farnham
		NFM	GB3HB	St Austell
		NFM	GB3LH	Shrewsbury
		NFM	GB3OM	Omagh
		NFM	GB3OX	Oxford
		NFM	GB3PP	Preston
		NFM	GB3SG	Cardiff
		NFM	GB3SU	Sudbury
		NFM	GB3SZ	Bournemouth
		NFM	GB3WI	Wisbech
		NFM	GB3WU	Wakefield
433.4000	435.0000	NFM		Raynet Channel RB16
		NFM	GB3HC	Hereford
433.5000		NFM		Raynet Chan SU20 Calling
435.0750		NFM	Space	Oscar 14
435.2500		NFM	Space	Oscar 14
435.9100		NFM	Space	Oscar 20 Beacon
435.9750		NFM	Space	Polar Bear 8688A
437.1500		NFM	Space	Oscar 19 Data
438.5500		NFM	Edinburgh	Sky TV Sound Relay

440.0125 - 443.4875 MHz PMR Base Repeaters 12.5 kHz (Split - 14.5 MHz)

Base	Mobile	Mode	Location	User & Notes
440.1000	425.6000	NFM	Barrow	VSEL Submarine Guards
		NFM	Wimbledon	Council
440.1750	425.6750	NFM	London	London Transport Buses
440.2250	425.7250	NFM	Birmingham NEC	Security & Maintenance
		NFM	London	London Transport Buses
440.3250	425.8250	NFM	London	London Transport Buses Victoria Stn
440.4000	425.9000	NFM	Surrey	Ambulance Incident Control
440.4250	425.9250	NFM	Aberdeen	
440.4750	425.9750	NFM	London	London Transport Buses
440.5000	426.0000	NFM	London	London Transport Buses
440.5250	426.0250	NFM	London	London Transport Buses
440.5500	426.0500	NFM	London	London Transport Buses
440.5750	426.0750	NFM	London	London Taxis North/Central

Base	Mobile	Mode	Location	User & Notes
440.5875	426.0875	NFM	London	London Taxis N1/N7
440.6000	426.1000	NFM	London	London Taxis WC/SW1
440.6250	426.1250	NFM	London	London Taxis Gt Lon Area
440.6500	426.1500	NFM	London	London Taxis Gt London
440.7000	426.2000	NFM	London	Centracom Ltd
440.7250	426.2250	NFM	London	London Taxis
		NFM	London	Central London Messanger
440.7750	426.2750	NFM	Nationwide	Customs & Excise Chan 4
440.7750	440.7750	NFM	Nationwide	Customs & Excise
440.8250	426.3250	NFM	London Heathrow	Customs Surveillance
		NFM	Nationwide	Customs & Excise Chan 5
		NFM	Shoreham	Customs & Excise
440.8500	426.3500	NFM	Nationwide	Customs & Excise Chan 2
		NFM	Newhaven	Customs & Excise
440.8750	426.3750	NFM	Nationwide	Customs & Excise Chan 3
441.0000	426.5000	NFM	Nationwide	Xerox Copiers
441.0250	426.5250	NFM	Leicester	Leicester Buses
441.0500	426.5500	NFM	Aberdeen	
441.1000	426.6000	NFM	Nationwide	MediCall
441.1500	426.6500	NFM	Birmingham Airport	AirCall
441.2000	426.7000	NFM	Cambridge	Community Repeater
441.3500	426.8500	NFM	Crewe	Taxis
441.4750	426.9750	NFM	Aberdeen	
441.6250	427.1250	NFM	Aberdeen	
441.6500	427.1500	NFM	Aberdeen	
442.2875	427.7875	NFM	London	BBC TV Talkback
442.3250	427.8250	NFM	London	BBC TV Talckback
442.3375	427.8375	NFM	London	BBC TV Talkback
442.43125		NFM	Birmingham	ITN Talkback
442.4375	427.9375	NFM	London	ITV Talkback
442.4500	427.9500	NFM	London	ITV Talkback
442.4750	427.9750	NFM	Grays Inn Road	ITN MCR Channel 442
442.5000	428.0000	NFM	London	ITN News Talkback
442.6750	428.1750	NFM	Merseyside	Inshore Lifeboats

443.5000 - 445.3000 MHz MoD Radiolocation & Base Comms 25 kHz

Base	Mobile	Mode	Location	User & Notes
443.7500		NFM	Colchester	36 DWS REME Channel 1
444.0250		NFM	Colchester	36 DWS REME Channel 3
444.3250		NFM	Colchester	36 DWS REME Channel 4
444.6500		NFM	Colchester	36 DWS REME Channel 2

445.0000 - 447.9875 MHz Shop & Post Office Security 25 kHz

Base	Mobile	Mode	Location	User & Notes
445.1500		NFM	BR Birmingham	Post Office
445.2500		NFM	Aberdeen	BBC Film Crews
445.5500	445.5500	NFM	Plymouth	Post Office Security
445.7000		NFM	Nationwide	MediCall
445.7250	425.2250	NFM	London Underground	Station Radio Ch 1
		NFM	Aldgate East	London Underground
		NFM	Bayswater	London Underground
		NFM	Cannon street	London Underground

Base	Mobile	Mode	Location	User & Notes
		NFM	Chancery lane	London Underground
		NFM	Earls' Court	London Underground
		NFM	Embankment	London Underground
		NFM	Euston	London Underground
		NFM	Glouster Road	London Underground
		NFM	Heathrow Central	London Underground
		NFM	Holburn	London Underground
		NFM	Kings Cross	London Underground
		NFM	Leicster Square	London Underground
		NFM	Liverpool Street	London Underground
		NFM	Mile End	London Underground
		NFM	Moorgate	London Underground
		NFM	Hamersmith	London Underground
		NFM	Sloance Square	London Underground
		NFM	South Kensington	London Underground
		NFM	St James's Park	London Underground
		NFM	Temple	London Underground
		NFM	Tottenham Court Road	London Underground
		NFM	Tower Hill	London Underground
		NFM	Victoria	London Underground
		NFM	Whitechapel	London Underground
		NFM	Crystal Palace	DER Repeater
445.7750	425.2750	NFM	London Underground	Station Radio Ch 2
		NFM	Gloucester Road	London Underground
		NFM	Paddington	London Underground
		NFM	Westminster	London Underground
445.7750	445.7750	NFM	Nationwide	Customs & Excise
445.8000	425.3000	NFM	London Underground	Station Radio Ch 3
		NFM	Ladbrooke Grove	London Underground
		NFM	Mansion House	London Underground
445.8250	445.8250	NFM	Nationwide	Customs & Excise
445.8500	445.8500	NFM	Nationwide	Customs & Excise
446.1250		NFM	Guildford	Debenhams Security
		NFM	Nottingham	Debenhams Security
		NFM	Stapeley	Stapeley Water Gardens
446.2500		NFM	Gatwick	Thomas Cook
446.3000		NFM	London	Capital Radio's Flying Eye
446.4000		NFM	Nationwide	Radio Investigations
446.4500		NFM	Nationwide	Radio Investigations
446.4750	452.2500	NFM	Nationwide	Fire Brigade Channel 2
446.5625		NFM	Essex	BBC Radio Essex Links
446.6375		NFM	Stoke	Radio Stoke O/B
446.6875		NFM	Nationwide	BBC Local Radio Talkback
		NFM	Radio Nottingham	Outside Broadcast
446.7375		NFM	Nationwide	BBC Local Radio Talkback
446 8375		NFM	Nationwide	BBC Local Radio Talkback
		NFM	Radio Derby	Outside Broadcast
446.9000		NFM	BR Euston	Post Office
446.9375		NFM	Nationwide	BBC Local Radio Talkback
446.9375	141.3000	NFM	Leicester	BBC Radio Leicester O/B

Base	Mobile	Mode	Location	User & Notes

446.5000 - 447.5000 MHz IBA, BBC and MoD Allocations

Base	Mobile	Mode	Location	User & Notes
447.0000		NFM	Silverstone	Japanese TV Talkback
447.0875		NFM	Birmingham	BRMB/Xtra O/B
		NFM	Stoke-on-Trent	Signal Radio OB
447.4000		NFM	Silverstone	US TV Talkback
447.5000		NFM	Silverstone	US TV Talkback
448.2250	451.2250	NFM	Birmingham (Euro Summit)	French Diplomatic Service
448.25625		NFM	London	London Hilton Trunking Sys.

449.0000 - 450.0000 MHz MoD Radiolocation Beacons

450.0000 - 452.9750 MHz Police Mobile PMR System (England & Wales)

Base	Mobile	Mode	Location	User & Notes
450.0125	464.0125	NFM	Needles Lighthouse Phone Line	
450.0250	464.0250	NFM	Wilmslow	
450.0500	464.0500	NFM	Arsenal FC Security	Channel 60
		NFM	Luton Town FC Security	
		NFM	Wembley FC Security	
		NFM	Wolverhampton FC Security	
450.0500	450.0500	NFM	Chelsea FC Security	Channel 77
450.0750	464.0750	NFM	Millwall FC Security	Channel 61
		NFM	Port Vale FC Security	
450.0750	450.0750	NFM		Channel 78
450.1000	464.1000	NFM	Nationwide	Channel 62
450.1250	464.1250	NFM	Thames Valley Scrambled	Channel 63
450.1250	450.1250	NFM		Channel 79
450.1500	464.1500	NFM	Brentford FC Security	Channel 64
		NFM	Crystal Palace FC Security	Selhurst Control
		NFM	Thames Valley	
450.1500	450.1500	NFM		Channel 80
450.1750	464.1750	NFM	Arsenal FC Security	Channel 65
		NFM	Port Vale FC Security	
		NFM	Southend Utd FC Security	
		NFM	Tenterden	
450.1750	450.1750	NFM		Channel 81
450.2000	464.2000	NFM	Gillingham FC Security	Channel 66
		NFM	Millwall FC Security	
		NFM	Portsmouth FC Security	
450.2000	450.2000	NFM	Nationwide	Channel 82
450.2250	464.2250	NFM	Brands Hatch	Channel 67
		NFM	Crystal Palace FC Security	
		NFM	West Ham FC Security	
450.2250	450.2250	NFM	Nationwide	Channel 83
450.2500	450.2500	NFM	Maidstone Utd FC Security	Channel 68
450.2750	464.2750	NFM	Nationwide	Channel 69
450.3000	450.3000	NFM	Fire Command	Channel 70
450.6250	450.6250	NFM	Police Air to Ground	Channel 88
450.6750	450.6750	NFM	Police Air to Ground	Channel 89
451.0000	464.9000	NFM	Various Areas	

Base	Mobile	Mode	Location	User & Notes
451.0250	464.9250	NFM	Nationwide	Channel T1
451.0500	464.9500	NFM	Nationwide	Channel T2
451.0750	464.9750	NFM	Nationwide	Channel T3
451.1000	465.0000	NFM	Nationwide	Channel T4
451.1250	465.0125	NFM	Nationwide	Channel T5
451.1500	451.1500	NFM	Nationwide	Channel 00
451.1750	465.0750	NFM	Nationwide	Channel 01
451.2000	465.1000	NFM	Manchester, Moss Side	Channel 02
		NFM	Warrington	
451.2250	465.1250	NFM	Nationwide	Channel 03
		NFM	Gorton	
451.2500	465.1500	NFM	London Diplomatic Protection	Channel 04
451.2750	465.1750	NFM	Stockport	Channel 05
451.2750	465.8750	NFM	Cheadle	Channel 58
		NFM	Thames Valley	
		NFM	Nationwide	Reserve Channel B
451.3000	451.3000	NFM	Thames Valley	Channel 71
		NFM	Optica Surveillance Air-Ground	
451.3000	465.2000	NFM	Gatwick Immigration	Channel 06
		NFM	Police Airborne	
451.3250	465.2250	NFM	CID Covert	Channel 07
451.3250	451.3250	NFM	Nationwide	Channel 07
451.3250	465.3250	NFM	Thames Valley Traffic	Channel 07
		NFM	Airport Tactical Fire Liaison	Channel 1
451.3500	465.2500	NFM	Gatwick Immigration	Channel 08
		NFM	HM Prisons	
		NFM	Newhaven Immigration	
		NFM	Port of London Authority	
451.3750	465.2750	NFM	Basildon	Channel 09
		NFM	Canvey Island	
		NFM	Cardiff	
		NFM	Farnborough	
		NFM	Gloucester	
		NFM	Horley	
		NFM	High Wycombe	AE
		NFM	HMP Littlehey	
		NFM	HMP Pentonville	
		NFM	HMP Feltham	
		NFM	HMP Winson Green	
		NFM	Leeds Holbeck	
		NFM	Liverpool	
		NFM	Manchester Bootle St.	
		NFM	Northampton	
		NFM	Nottingham City Centre	
		NFM	Redhill	ER
		NFM	Reigate	ER
		NFM	Rennishaw	
		NFM	Rye	
451.4000	451.4000	NFM	Fire Services	Channel 01
451.4000	465.4000	NFM	London Fire Brigade	Channel 10

Base	Mobile	Mode	Location	User & Notes	
451.4250	465.3250	NFM	Acocks Green		Channel 11
		NFM	Aylesbury	AA	
		NFM	Bristol		
		NFM	Caterton		
		NFM	Collyhurst		
		NFM	Connah's Quay		
		NFM	Ecclesfield, S Yorkshire	F2	
		NFM	Essex		
		NFM	Faringdon		
		NFM	Grays		
		NFM	Leeds Pudsey		
		NFM	Liverpool St Helens		
		NFM	Long Eaton		
		NFM	Manchester Beswick		
		NFM	Newbury	FA	
		NFM	Nottingham Eastwood		
		NFM	Reading	EA	
		NFM	Reading East	EX	
		NFM	Reading West		
		NFM	Stansted Airport	GF/GM	
		NFM	Thames Valley	AB	
		NFM	Wantage		
		NFM	Wendover		
		NFM	West Midlands, Dunstall Rd		
		NFM	Witney		
		NFM	Wolverhampton North	M2YMG1	
451.4500	465.4500	NFM	Fire Service England		Channel 02
451.4500	451.4500	NFM	Fire Services		Channel 3
451.4500	465.3500	NFM			Channel 12
451.4750	451.4750	NFM	General Fire Incidents		Channel 4
451.4750	465.3750	NFM	Basingstoke		Channel 13
		NFM	Billericay		
		NFM	Borough Green		
		NFM	Burnham	CC	
		NFM	Campsfield House Detention Centre		
		NFM	Harrogate		
		NFM	HMP Wandsworth		
		NFM	London Met	CB CD MD	
		NFM	Maidenhead	CG	
		NFM	Oldham		
		NFM	Ripley		
		NFM	Stretford		
		NFM	Thames Valley		
		NFM	Thorne, S. Yorkshire	A2	
		NFM	Tonbridge		
		NFM	Urmston		
		NFM	Walsall	M2YMH1	
		NFM	West Malling		
		NFM	Wickford		
451.5000	465.4000	NFM	Various Areas		Channel 14

Base	Mobile	Mode	Location	User & Notes	
		NFM	London Diplomatic	Ranger	
451.5250	465.4250	NFM	Birkenhead		Channel 15
		NFM	Bradford Laisterdyke		
		NFM	Carterton	FJ	
		NFM	Collyhurst		
		NFM	Godalming	WO	
		NFM	Grays		
		NFM	Essex		
		NFM	Hazelmere	WO	
		NFM	Liverpool Birkenhead		
		NFM	Long Eaton		
		NFM	Manchester Collyhurst		
		NFM	Mexborough	A3	
		NFM	Miles Platting		
		NFM	Plymouth		
		NFM	Shirley		
		NFM	Solihull	M2YML1	
		NFM	Stafford		
		NFM	Thames Valley		
		NFM	West Bridgeford		
		NFM	West Midlands, Birmingham Rd		
		NFM	Wolverhampton South	M2YMG3	
		NFM	Winchester		
		NFM	Wirley	FI	
451.5250	451.5250	NFM	Fire Breathing Apparatus		Channel 6
451.5500	465.4500	NFM	London Diplomatic		Channel 16
451.5500	451.5500	NFM	Nationwide		Channel 02
451.5750	465.4750	NFM	Accrington		Channel 17
		NFM	Bradford		
		NFM	Castelford		
		NFM	Dover Detention Centre		
		NFM	Dudley	M2YMJ1	
		NFM	Faringdon	FE	
		NFM	Garstang		
		NFM	Haslingden		
		NFM	HMP Walton		
		NFM	HMP Ranby		
		NFM	HMP Swinton		
		NFM	HMP Latchmere House		
		NFM	HMP Long Lartin		
		NFM	Lancaster		
		NFM	Rochester Borstal		
		NFM	Somerset & Avon		
		NFM	Southampton		
		NFM	Taunton		
		NFM	Thames Valley		
		NFM	Wantage	FF	
		NFM	Warrington		
451.6000	465.5000	NFM	Addlestone		Channel 18
		NFM	Birmingham Central	M2YMF1	

This illustrates the variety of antennas used in both VHF and UHF PMR bands. The tower also supported several microwave relay dishes, which are becoming more common on many such towers around the country.

Base	Mobile	Mode	Location	User & Notes
		NFM	Broadstairs	
		NFM	Folkestone	
		NFM	HMP North Sea Camp	
		NFM	Hythe	
		NFM	Lydd	
		NFM	Maidstone	
		NFM	Manchester Sale	
		NFM	Manchester Trafford	
		NFM	Oldham	
		NFM	Ramsgate	
		NFM	Reading	EX
		NFM	Southwood	F3
		NFM	Staplehurst	
		NFM	Walton-on-Thames	NA
		NFM	Wellingborough	
		NFM	West Midlands, Bradford St	
		NFM	West Midlands, Steelhouse Lane	
451.6125	451.6125	NFM	Fire Breathing Apparatus	Channel 2
451.6250	465.5250	NFM	Gatwick Aiport	Channel 19
		NFM	Blackpool	
		NFM	Essex	
		NFM	Leicester	
		NFM	Sussex Emergency Use	
		NFM	Thames Valley Special Events	
451.6500	465.5500	NFM	Thames Valley	Channel 20
		NFM	Ulverstone	
		NFM	Mobile Repeaters Nottingham	
		NFM	Mobile Repeaters South Yorkshire	
		NFM	Mobile Repeaters West Midlands	
		NFM	Mobile Repeaters West Mercia	
		NFM	Mobile Repeaters Staffordshire	
		NFM	Mobile Repeaters Devon	
		NFM	Mobile Repeaters Cornwall	
		NFM	Mobile Repeaters Leicester	
		NFM	Mobile Repeaters West Yorkshire	
		NFM	Mobile Repeaters Sussex	
		NFM	Mobile Repeaters Hertfordshire	
		NFM	Mobile Repeaters Cambridgeshire	
451.6750	465.5750	NFM	Barnsley	Channel 21
		NFM	Benfleet	
		NFM	Bishop Stortford	
		NFM	Bradford	
		NFM	Canvey Island	
		NFM	Cardiff Central	
		NFM	City Of London	
		NFM	Henley	EE
		NFM	Leicester	
		NFM	Matlock	
		NFM	Northwich	
		NFM	Reading	EG

Base	Mobile	Mode	Location	User & Notes	
		NFM	Shinfield		
		NFM	Stoke on Trent		
		NFM	Sutton		
		NFM	Thames Valley Woodley	EB	
		NFM	Twyford		
		NFM	Welwyn Garden City		
		NFM	Wokingham		
		NFM	Mobile Repeater Devon		
		NFM	Mobile Repeater Cornwall		
451.7000	465.6000	NFM	Arnold		Channel 22
		NFM	Ashford		
		NFM	Bourneville	M2YMB2	
		NFM	Burscough		
		NFM	Chandlers Ford		
		NFM	Chorley		
		NFM	Coppull		
		NFM	Dorking	ED	
		NFM	Gillingham		
		NFM	Leatherhead	EL	
		NFM	Margate		
		NFM	Rainham		
		NFM	Salford		
		NFM	Thames Valley		
451.7250	465.6250	NFM	Blackpool		Channel 23
		NFM	Leicester		
		NFM	Thames Valley		
451.7500	465.6500	NFM	Bramford		Channel 24
		NFM	Bury		
		NFM	Flint		
		NFM	Hinckley		
		NFM	HMP Salisbury		
		NFM	Liverpool Huyton		
		NFM	Manchester Bury		
		NFM	Market Harborough		
		NFM	Melksham		
		NFM	Middleton		
		NFM	Retford		
		NFM	Skegness		
		NFM	St Annes		
		NFM	Thames Valley		
		NFM	Ward End	M2YME2	
		NFM	Worksop		
451.7750	465.6750	NFM	Blackpool		Channel 25
		NFM	Brighton Crown Court		
		NFM	Leicester		
		NFM	Plymouth Argyle FC Security		
		NFM	Thames Valley Special Events		
451.7750	465.6750	NFM	Fire Breathing Apparatus		Channel 3
451.8000	465.7000	NFM	Acocks Green	M2YME3	Channel 26
		NFM	Burton		

Base	Mobile	Mode	Location	User & Notes
		NFM	Leeds Gipton	
		NFM	Gloucester	
		NFM	Liverpool Marsh Lane	
		NFM	Manchester Ringway	
		NFM	Ringwood	
		NFM	Southport	
		NFM	Thames Valley	
		NFM	Totton	
		NFM	Wallington	
451.8250	465.7250	NFM	Abingdon	Channel 27
		NFM	Bristol	
		NFM	Canning	
		NFM	Canterbury	
		NFM	Cowley	BC
		NFM	Eastleigh	
		NFM	Fleetwood	
		NFM	Harlow	
		NFM	HMP Coldingley	
		NFM	HMP Eastchurch	
		NFM	Langley	CE
		NFM	Manchester Middleton	
		NFM	Nelson Colne	
		NFM	Oxford City	BA
		NFM	Pendle	
		NFM	Rochdale	
		NFM	Slough HQ	CA
		NFM	Stratford-Upon-Avon	
		NFM	Thames Valley	
		NFM	Wigan	
		NFM	Woodseats	E2
451.8500	465.7500	NFM	Amersham	Channel 28
		NFM	Aylesbury	
		NFM	Beaconsfield	AC
		NFM	Bletchley	DG
		NFM	Buckingham	DB
		NFM	Congleton	
		NFM	Gerrards Cross	
		NFM	Grays	
		NFM	Harlow	
		NFM	HMP Armley	
		NFM	HMP Morton Hall	
		NFM	Lincoln	
		NFM	London Met	CB
		NFM	Macclesfield	
		NFM	Milton Keynes	
		NFM	Ormskirk	
		NFM	Thames Valley	
		NFM	Wednesbury	M2YMK2
451.8750	465.7750	NFM	Special Events	Channel 29
		NFM	Thames Valley	

Base	Mobile	Mode	Location	User & Notes	
451.9000	465.8000	NFM	Camberley	NC	Channel 30
		NFM	Coventry South	M2YMM3	
		NFM	Egham	NE	
		NFM	Fletchamstead		
		NFM	HMP Holloway		
		NFM	Leeds Horsforth		
		NFM	Liverpool Kirby		
		NFM	Manchester Bury		
		NFM	Portsmouth		
		NFM	Thames Valley		
		NFM	Wednesfield	M2YMG2	
451.9250	465.8250	NFM	Birmingham Airport	M2YMEA	Channel 31
		NFM	Blackpool Div HQ		
		NFM	Bracknell	CH	
		NFM	Buckingham		
		NFM	Buckley		
		NFM	Crowthorne	CF	
		NFM	Hackenthorpe	E1	
		NFM	Hatfield		
		NFM	HM Prisons		
		NFM	Hungerford	FB	
		NFM	Lincoln		
		NFM	Langley		
		NFM	Milton Keynes		
		NFM	Milton Keynes North		
		NFM	Milton Keynes South		
		NFM	Morley		
		NFM	Newbury		
		NFM	Newport Pagnell	DD	
		NFM	Redhill	ER	
		NFM	Reigate	ER	
		NFM	Thatcham		
		NFM	Welwyn Garden City		
		NFM	Wolverton		
451.9500	465.8500	NFM	Brownhill		Channel 32
		NFM	Farnham	WF	
		NFM	Kings Heath	M2YMB3	
		NFM	Leeds Garforth		
		NFM	Newark		
		NFM	Thames Valley		
		NFM	Widnes		
		NFM	Mobile Repeaters		
452.1500	466.0500	NFM	Thames Valley Special Use		Channel 33
452.2500	446.4750	NFM	Fire Service England		Channel 02
452.2500	452.2500	NFM	Powys Fire Services		Channel 73
452.2750	465.9250	NFM	Chadderton		Channel 57
		NFM	Failsworth		
		NFM	Oldham		
		NFM	Thames Valley		
		NFM	Nationwide		Reserve Ch A

Base	Mobile	Mode	Location	User & Notes	
452.3000	465.9000	NFM	Northwich		Channel 60
		NFM	Thames Valley		
		NFM	Nationwide		Reserve Ch C
452.3250	452.3250	NFM	Sale		Channel 74
		NFM	Thames Valley Support Units		
		NFM	Nationwide		Channel 06
453.3500	466.2500	NFM	Liverpool Huyton		Channel 59
		NFM	Manchester Hyde		
		NFM	Thames Valley		
		NFM	Thameside		
		NFM	Nationwide		Reserve Ch B
452.3625	452.3625	NFM	Fire Services		Channel 2
452.3750	452.3750	NFM			Channel 75
		NFM	Tactical Firearms Unit		
452.3750	466.2750	NFM			Channel 76
		NFM	Tactical Firearms Unit		
452.4000	466.3000	NFM	Attercliffe	D2	Channel 33
		NFM	Birmingham Central	M2YMF1	
		NFM	Burnley		
		NFM	Cranbrook		
		NFM	Derby		
		NFM	Devonport		
		NFM	Digbeth		
		NFM	Epping		
		NFM	Exeter		
		NFM	Faversham		
		NFM	Ford Open Prison		
		NFM	Great Yarmouth		
		NFM	Harpenden		
		NFM	HMP Canterbury		
		NFM	HMP Lewes		
		NFM	HMP Northeye		
		NFM	HMP Stafford		
		NFM	HMP Welford Rd, Leicester		
		NFM	Huddersfield		
		NFM	Hunstanton		
		NFM	Huntingdon		
		NFM	Leicester		
		NFM	Liverpool Marsh Lane		
		NFM	Manchester Bootle St.		
		NFM	Neots		
		NFM	Plymouth		
		NFM	St Albans		
		NFM	Shipley		
		NFM	Sittingbourne		
		NFM	Sussex Prisons		
		NFM	Southend on Sea		
		NFM	Thames Valley		
		NFM	Thetford		
		NFM	West Midlands, Bradford St		

Base	Mobile	Mode	Location	User & Notes	
		NFM	Woodbridge		
		NFM	York		
452.4250	466.3250	NFM	Bedworth		Channel 34
		NFM	Bradford Queenshouse		
		NFM	Channel Tunnel		
		NFM	Derby South West		
		NFM	Hadleigh		
		NFM	HMP Wormwood Scrubbs		
		NFM	Leyland		
		NFM	Manchester Trafford		
		NFM	Stretford		
		NFM	Thames Valley		
		NFM	Wombwell	B2	
452.4375		NFM	Orpington		Channel 35
452.4500	466.3500	NFM	Amersham	AD	Channel 35
		NFM	Basildon		
		NFM	Bradford Odsal		
		NFM	Chatham		
		NFM	Cheltenham		
		NFM	Chequers	AZ	
		NFM	Chesham	AH	
		NFM	Derby Central		
		NFM	Dorking	ED	
		NFM	Eastbourne		
		NFM	Egbaston	M2YMB1	
		NFM	Exmouth		
		NFM	Farnborough		
		NFM	Fleet		
		NFM	Gorleston-on-Sea		
		NFM	Hammerton Rd	F1	
		NFM	Hillsborough		
		NFM	Huntingdon		
		NFM	Isle Of Grain		
		NFM	Knutsford		
		NFM	Leatherhead	EL	
		NFM	Newhaven		
		NFM	Nottingham Radford Rd		
		NFM	Rochdale		
		NFM	Rochester		
		NFM	Speke		
		NFM	Stroud		
		NFM	Westbury		
		NFM	West Midlands, Belgrave Rd		
		NFM	Wilmslow		
452.4750	466.3750	NFM	Basingstoke		Channel 36
		NFM	Brierley Hill	M2YMJ2	
		NFM	Brighton		
		NFM	Broadmoor Secure Hospital		
		NFM	Caterham		
		NFM	Chelmsley Wood		

Base	Mobile	Mode	Location	User & Notes	
		NFM	London Met	CD/CM	
		NFM	Rawmarsh	C2	
		NFM	Oxhey		
		NFM	Oxted		
		NFM	Shrewsbury		
		NFM	Skelmersdale		
		NFM	Surrey		
		NFM	Thames Valley		
452.5000	466.4000	NFM	Aldershot		Channel 37
		NFM	Aylesham		
		NFM	Braintree		
		NFM	Bristol		
		NFM	Clacton-on-Sea		
		NFM	Deal		
		NFM	Dover		
		NFM	East Grinstead		
		NFM	Fleet		
		NFM	Hailsham		
		NFM	Harbourne	M2YMC3	
		NFM	Harpenden		
		NFM	Herne Bay		
		NFM	HMP Ashford		
		NFM	HMP Brixton		
		NFM	Hove		
		NFM	Ladywood		
		NFM	Leeds Weetwood		
		NFM	Letchwood		
		NFM	Liverpool City Centre		
		NFM	Longsight		
		NFM	Loughborough		
		NFM	Manchester Platt Lane		
		NFM	Northampton		
		NFM	Plymouth		
		NFM	St Albans		
		NFM	Sandwich		
		NFM	Swanley		
		NFM	Thames Valley		
		NFM	Whitstable		
452.5250	466.4250	NFM	Various Areas		Channel 38
		NFM	Special Use		Channel 88
452.5375	452.5375	NFM	Fire/Police Link Repeaters		Channel 8
452.5500	466.4500	NFM	Ashton-under-Lyme		Channel 39
		NFM	Banbury	BD	
		NFM	Bisceter	BE	
		NFM	Carlton		
		NFM	Grantham		
		NFM	Hanley		
		NFM	Kings Lynn		
		NFM	Leeds Millgarth		
		NFM	Luton		

Base	Mobile	Mode	Location	User & Notes	
		NFM	Luton Airport		
		NFM	Manchester Huyton		
		NFM	March		
		NFM	Newport		
		NFM	Newton Abbot		
		NFM	Rotherham	C1	
		NFM	Southend on Sea		
		NFM	Smethwick	M2YMK3	
		NFM	Surrey		
		NFM	Swindon		
		NFM	Thames Valley		
		NFM	Thameside		
		NFM	Uckfield		
		NFM	Wilmslow		
		NFM	Woking	NW	
452.5750	466.4750	NFM	Bacup		Channel 40
		NFM	Barnsley	B1	
		NFM	Basingstoke		
		NFM	Billericay		
		NFM	Bradford Toller Lane		
		NFM	Canterbury		
		NFM	Edenbridge		
		NFM	Herne Bay		
		NFM	Morecambe		
		NFM	Rawtenstall		
		NFM	Rossendale		
		NFM	Sevenoaks		
		NFM	Thames Valley		
		NFM	Waterfoot		
		NFM	Welwyn Garden City		
		NFM	Westerham		
		NFM	Whitstable		
		NFM	Wickford		
		NFM	Wigan		
		NFM	British Transport Police		
452.5875		NFM	Chislehurst		
452.6000	466.5000	NFM	Ascot	CB	Channel 41
		NFM	Berkhamstead		
		NFM	Brighton		
		NFM	Bristol		
		NFM	Canterbury		
		NFM	Chatham		
		NFM	Chichester		
		NFM	County Durham		
		NFM	Crewe		
		NFM	Derby		
		NFM	Edenbridge		
		NFM	Fallowfields		
		NFM	Felixstowe		
		NFM	Fulwood		

Base	Mobile	Mode	Location	User & Notes	
		NFM	Glenfield		
		NFM	Gravesend		
		NFM	Grimsby		
		NFM	Hastings		
		NFM	Haverhill		
		NFM	Hertford		
		NFM	Holmfirth		
		NFM	Leamington Spa		
		NFM	Leicester		
		NFM	Maidstone		
		NFM	Manchester Platt Field		
		NFM	Meopham		
		NFM	Midhurst		
		NFM	Mildenhall		
		NFM	Petworth		
		NFM	Plymouth		
		NFM	Preston		
		NFM	Scunthorpe		
		NFM	Seaham		
		NFM	Stow		
		NFM	Syston		
		NFM	Thames Valley		
		NFM	Wallasey		
		NFM	West Bromwich	M2YMK1	
		NFM	Winchester		
		NFM	Windsor	CD	
452.6250	466.5250	NFM	Barrow		Channel 42
		NFM	Chichester		
		NFM	Erdington	M2YMD3	
		NFM	Gatwick		
		NFM	Habrough		
		NFM	Leicester City		
		NFM	London Met	CD/DM	
		NFM	Mansfield		
		NFM	Stoke		
		NFM	Thames Valley		
452.6375	452.6375	NFM	General Fire Incidents		
452.6500	466.5500	NFM	Aldridge & Brownhills	M2YMH3	Channel 43
		NFM	Bedford		
		NFM	Bexhill		
		NFM	Blackburn		
		NFM	Brentwood		
		NFM	Broadstairs		
		NFM	Camden Town		
		NFM	Chapletwon		
		NFM	Chester		
		NFM	Colchester		
		NFM	Crowbrough		
		NFM	Derby City Centre		
		NFM	Ely		

Base	Mobile	Mode	Location	User & Notes
		NFM	Gosport	
		NFM	Hemel Hempstead	
		NFM	Hodderson	
		NFM	HMP Lincoln	
		NFM	HMP Strangeways	
		NFM	Leicester	
		NFM	Littlehampton	
		NFM	Longridge	
		NFM	Longton	
		NFM	Manchester Strangeways	
		NFM	Nuneaton	M2YJN
		NFM	Ramsgate	
		NFM	Sheffield City	
		NFM	Sussex Headquarters	
		NFM	Thames Valley	
		NFM	Ware	
		NFM	West Bar	E3
452.6750	466.5750	NFM	Aston	M2YMD1 Channel 44
		NFM	Coventry North	M2YMM2
		NFM	Grantham	
		NFM	Maltby	C3
		NFM	Rotherham	
		NFM	South Manchester	
		NFM	Spalding	
		NFM	Thames Valley	
		NFM	West Midlands, Queens Rd	
		NFM	West Midlands, Steelhouse Lane	
		NFM	West Midlands, Stoney Stanton	
452.7000	466.6000	NFM	Battle	Channel 45
		NFM	Bognor Regis	
		NFM	Boston	
		NFM	Bristol	
		NFM	Burgess Hill	
		NFM	Cambridge	
		NFM	Chipping Norton	BF
		NFM	Congleton	
		NFM	Doncaster	A1
		NFM	Essex	
		NFM	Grays	
		NFM	Guildford	WG
		NFM	Halesowen & Stourbridge	M2YMJ3
		NFM	Halifax	
		NFM	Harrogate	
		NFM	Haywards Heath	
		NFM	Hucknall	
		NFM	Kettering	
		NFM	Leighton Buzzard	
		NFM	Liverpool St Helens	
		NFM	Merseyside, St Helens	
		NFM	Newark	

Base	Mobile	Mode	Location	User & Notes	
		NFM	Peterborough		
		NFM	Seaford		
		NFM	Shoreham-by-Sea		
		NFM	Sutton Coldfield	M2YMD2	
		NFM	Thames Valley		
		NFM	West Midlands, Halsowen		
		NFM	Woodstock		
452.7250	466.6250	NFM	Bloxwich	M2YMH2	Channel 46
		NFM	Cleckheaton		
		NFM	Cleethorpes		
		NFM	Coventry Central	M2YMM1	
		NFM	Horsham		
		NFM	Liverpool Toxteth		
		NFM	Manchester Strangeways		
		NFM	Margate		
		NFM	Stechford		
		NFM	Stockport		
		NFM	Thames Valley		
		NFM	Tunstall		
		NFM	Washington		
		NFM	West Midlands, Littel Park St		
		NFM	Willenhall		
452.7500	466.6500	NFM	Baldock		Channel 47
		NFM	Birmingham		
		NFM	Bury St Edmonds		
		NFM	Camberley	NC	
		NFM	Chelmsford		
		NFM	Dartford		
		NFM	Egham	NE	
		NFM	Gatwick		
		NFM	Halifax		
		NFM	Hitchin		
		NFM	Horsham		
		NFM	Ipswich		
		NFM	Isle Of Sheppey		
		NFM	Leicester Wigston		
		NFM	Lewes		
		NFM	Mexborough		
		NFM	Newmarket		
		NFM	Notingham, West Bridgeford		
		NFM	Paddock Wood		
		NFM	Rickmansworth		
		NFM	Sevenoaks		
		NFM	Sheerness		
		NFM	Stevenage		
		NFM	Stockport		
		NFM	Sudbury		
		NFM	Sullbridge	F1	
		NFM	Thames Valley		
		NFM	Tonbridge		

Base	Mobile	Mode	Location	User & Notes	
		NFM	Tunbridge Wells		
		NFM	Wakefield		
		NFM	Watford		
		NFM	Worthing		
		NFM	Yardley	M2YME1	
452.7750	466.6750	NFM	Basildon		Channel 48
		NFM	Birkenhead		
		NFM	Burslem		
		NFM	Dewsbury		
		NFM	Ellesmere Port		
		NFM	Handsworth	M2YMC1	
		NFM	Manchester Ringway		
		NFM	Thames Valley		
		NFM	Walton-on-Thames	NA	
		NFM	West Midlands, Thornhill Rd		
452.8000	466.7000	NFM	Abingdon	FH	Channel 49
		NFM	Beeston		
		NFM	Chelmsford		
		NFM	City Of London		
		NFM	Cornwall		
		NFM	Crawley East		
		NFM	Crawley West		
		NFM	Didcot	FC	
		NFM	Dyfed/Powys		
		NFM	Harwich		
		NFM	HMP Hull		
		NFM	Ladywood	M2YMC2	
		NFM	Leicester Beumont Leys		
		NFM	Liverpool Walton Lane		
		NFM	Maidstone		
		NFM	Newbury		
		NFM	Newcastle		
		NFM	Newhaven		
		NFM	Pontefract		
		NFM	Salford		
		NFM	Skipton		
		NFM	Stradshall		
		NFM	Stevenage		
		NFM	Stoke		
		NFM	Thames Valley		
		NFM	Wallingford		
		NFM	Watford		
452.8250	466.7250	NFM	Special Use		Channel 50
452.8500	466.7500	NFM	Altringham		Channel 51
		NFM	Harwich		
		NFM	Witham		
		NFM	Special Use		
452.8750	466.7750	NFM	Bolton		Channel 52
		NFM	Chelmsley Wood	M2YML2	
		NFM	Rayleigh		

Base	Mobile	Mode	Location	User & Notes
		NFM	Southend	
		NFM	Thames Valley	
452.9000	466.8000	NFM	Leigh	Channel 53
		NFM	Thames Valley	
452.9250	466.7750	NFM	Jersey Traffic Wardens	
452.9250	466.8250	NFM	Leicester	Channel 54
		NFM	Mobile Repeater S Yorkshire	
		NFM	Thames Valley	
452.9250	452.9250	NFM	Nationwide	Channel 05
		NFM	Warrington	
452.9500	466.8500	NFM	Kidsgrove	Channel 55
		NFM	Runcorn	
		NFM	South Elmsall	
		NFM	Thames Valley	
		NFM	Mobile Repeater S Yorkshire	
452.9750	466.8750	NFM	Bolton	Channel 56
		NFM	Thames Valley	
452.9750	452.9750	NFM	Nationwide	Channel 01

453.0000 - 454.0000 MHz PMR Mobile Band 12.5 kHz (Split + 6.5 MHz)

Base	Mobile	Mode	Location	User & Notes
453.0000	466.9000	NFM	Nationwide	British Transport Police
453.0250	459.5250	NFM	Guernsey	Harbour Channel
		NFM	Felixstowe	Harbour Channel
		NFM	Harwich	Harbrour Channel
		NFM	Moray Firth	Beatrice Alpha Platform
		NFM	Nigg Bay	Oil Terminal Control
		NFM	Staverton	Staverton Airport Ground
453.0500	459.5500	NFM	Peterborough	Peter Brotherhood
		NFM	Nationwide	Shopping centre Security
		NFM	Nationwide	R.R. Security
		NFM	Nationwide	Air Rangers
453.0750	459.5750	NFM	Felixstowe	Quay Shipping
		NFM	Harwich	Quay Shipping
		NFM	Jersey	Harbour/Marina Channel 2
		NFM	Manch	Manch Ground Services
		NFM	Oxford University	Science Area
		NFM	Nationwide	British Rail
453.1000	459.6000	NFM	Blackpool	Local Health Authority
		NFM	Cowley	Rover Plant Channel 1
		NFM	Moray Firth	Beatrice Bravo Platform
		NFM	Peterborough	Shopping Centre Security
		NFM	Nottingham	Boots Beeston
		NFM	Malty	Buttlers Roadstone
		NFM	London	Docklands Security
		NFM	South Walden	Schering Agrochemicals
453.1250	459.6250	NFM	Dover	Western Docks Jetfoil
		NFM	Cowley	Rover Plant Channel 4
		NFM	Harwich	Carless Solvents
		NFM	Nigg Bay	Oil Terminal Fire Channel
		NFM	Norwich	Colmans Foods

Base	Mobile	Mode	Location	User & Notes
		NFM	Nationiwde	British Rail
453.1375	459.6375	NFM	Dartford Tunnel	Police
453.1500	459.6500	NFM	Orpington	Metro Bus
		NFM	Nottingham	Boots Broardmarsh
		NFM	Fleetwood	P & O Security
		NFM	S. London	City Security
		NFM	Whittlesey	McCain Intenational
453.1750	453.1750	NFM	Moray Firth	Beatrice Alpha Platform
453.1750	459.6750	NFM	Bacton	Shell UK Ltd.
		NFM	Bedford	3M UK
		NFM	Jersey	Jersey States
		NFM	Bristol	Bus Inspectors
		NFM	Cardiff	Docks Security
		NFM	Ipswich	Docks Security
		NFM	Suffolk	County Council
		NFM	Essex	County Council
		NFM	Sheffield	City Parks Security
		NFM	Nationiwde	British Rail
453.2000	459.7000	NFM	Ipswich	British Transport Police
		NFM	Plymouth	City Bus Inspectors
		NFM	Rotherham	Bus Inspectors
453.2125	459.7125	NFM	Nationwide	British Rail
453.2250	459.7250	NFM	Heathrow	Heathrow Tower
		NFM	Jersey	Harbour Channel 1
		NFM	Tilbury	Rover Security
453.2500	459.7500	NFM	Leeds	Traffic wardens
453.2750	459.7750	NFM	Nationwide	Anbulance Emergency
		NFM	Oxfordshire	Ambulance Service
453.3000	459.8000	NFM	Bagg	Company Security
453.3250	459.8250	NFM	Manch	St Mary's Hospital security
453.3500	459.8500	NFM	Nottingham	City Engineers Channel 2
453.3625	459.8625	NFM	Nationwide	DSS
453.3750	459.8750	NFM	Cowley	Rover Plant Channel 2
		NFM	Nationwide	DSS
453.3875	459.8875	NFM	Nationwide	DSS
453.4000	459.9000	NFM	Ipswich	Port Authority Channel 3
		NFM	Jersey	Airport Staff
		NFM	Nationwide	British Rail
		NFM	Nigg Bay	Oil Terminal Maintenance
		NFM	Nottingham	Formans
		NFM	Oxford University	Science Area
453.4500	459.9500	NFM	Cardiff	Inland Revenue Security
		NFM	Cowley	Rover Plant Channel 3
		NFM	Guernsey	Customs
453.4625	459.9625	NFM	Nationwide	Inland Revenue Security
453.4750	459.9750	NFM	Manchester	University Security
453.4875	459.9875	NFM	E Midlands	Aiport Fire Channel 1
453.5000	460.0000	NFM	Nationwide	Britsih Rail Security
		NFM	Jersey	Sealink
453.5250	460.0250	NFM	E Midlands	Airport Fire Channel 2

Base	Mobile	Mode	Location	User & Notes
		NFM	Ipswich	British Rail
		NFM	Jersey	Harbour Channel 3
		NFM	Nationwide	British Rail Transport Police
		NFM	Nigg Bay	Oil Terminal Security
453.5500	460.0500	NFM	Nationwide	British Rail Stations Chan 1
		NFM	Bedford	British Rail
		NFM	Birmingham New Street	British Rail
		NFM	Bletchley Yard	British Rail
		NFM	Brighton	British Rail
		NFM	Brighton Depot	British Rail
		NFM	Cannon Street	British Rail
		NFM	Cardiff	British Rail
		NFM	Cardiff DMU Shunters	British Rail
		NFM	Carlisle	British Rail
		NFM	Chester	British Rail
		NFM	Crewe	British Rail
		NFM	Doncaster	British Rail
		NFM	East Croydon	British Rail
		NFM	Edinburgh Waverly	British Rail
		NFM	Ely	British Rail
		NFM	Fenchurch Street	British Rail
		NFM	Glasgow Queen Street	British Rail
		NFM	Harwich Parkeston Yard	British Rail
		NFM	Hull Paragon	British Rail
		NFM	Ilford Car Sheds	British Rail
		NFM	Ipswich	British Rail
		NFM	Leeds	British Rail
		NFM	Leicester	British Rail
		NFM	Liverpool Street	British Rail
		NFM	Manchester Picadilly	British Rail
		NFM	Newcastle	British Rail
		NFM	Norwich	British Rail
		NFM	Norwich Crown Point	British Rail
		NFM	Paddington	British Rail
		NFM	Preston	British Rail
		NFM	Reading	British Rail
		NFM	Selhurst Junction	British Rail
		NFM	Slade Green Depot	British Rail
		NFM	Sheffield	British Rail
		NFM	Shenfield	British Rail
		NFM	Shrewsbury	British Rail
		NFM	Stansted Airport	British Rail
		NFM	Stratford	British Rail
		NFM	Watreloo	British Rail
		NFM	Wimbledon Park Depot	British Rail
		NFM	Wolverhampton	British Rail
		NFM	Woking	British Rail
		NFM	York	British Rail
453.5500	453.5500	NFM	Nottingham	City Council
453.5625	460.0625	NFM	Nationwide	British Rail

Base	Mobile	Mode	Location	User & Notes
453.5750	460.0750	NFM	Nationwide	British Airways
		NFM	Felixstowe	Freightliners
		NFM	Worksop	Tesco
453.6000	460.1000	NFM	Cowley	Rover Assembly
		NFM	Nationwide	Bus Inspectors
453.6500	460.1500	NFM	Edinburgh	Lothian Regional Transport
453.6750	460.1750	NFM	Edinburgh	Lothian Regional Transport
		NFM	London	Docklands Light Railway
453.7000	460.2000	NFM	Dover	Eastern Docks Port Ops
453.7500	460.2500	NFM	Aberdeen	Airport
453.8250	460.3250	NFM	Ipswich	Cranfield Bros.
453.8500	460.3500	NFM	Aberdeen	Aberdeen Hospital
		NFM	Norwich	Norwich & Norfolk Hospital
453.8750	460.3750	NFM	Aberdeen	Airport
		NFM	Edinburgh	Lothian Regional Transport
		NFM	Luton Airport	Monarch Airlines
453.9000	460.4000	NFM	Nationwide	British Rail Stations
		NFM	Ashford	British Rail
		NFM	Barking	British Rail
		NFM	Birmingham International	British Rail
		NFM	Charing Cross	British Rail
		NFM	Derby	British Rail
		NFM	Doncaster	British Rail
		NFM	Euston	British Rail
		NFM	Gatwick Airport	British Rail
		NFM	Glasgow Central	British Rail
		NFM	Guildford	British Rail
		NFM	Haymarket Depot	British Rail
		NFM	Heaton Depot	British Rail
		NFM	Hoo Junction Deport	British Rail
		NFM	Kings Cross	British Rail
		NFM	Liverpool Lime Street	British Rail
		NFM	Norwich Crown Point	British Rail
		NFM	Nottingham	British Rail
		NFM	Perth	British Rail
		NFM	Peterborough	British Rail
		NFM	Victoria	British Rail
		NFM	Watford Junction	British Rail
		NFM	Waterloo City	British Rail
		NFM	Willesden Yard	British Rail
453.9250	460.4250	NFM	London	Docklands Light Rlwy Ch 2
453.9375	460.4375	NFM	Scotland	ScotRail Data Links
453.9625	460.5625	NFM	London	London Taxis
453.9750	460.4750	NFM	Cowley	Rover Security Channel 7
		NFM	Luton	Vauxhall Motors

454.0250 - 454.8250 MHz Private Radio Paging Systems 25 kHz

Base	Mobile	Mode	Location	User & Notes
454.0500		NFM	London	US Embassy
454.0750	447.5750	NFM	Nationwide	Aircall Voice Paging
454.2000		NFM	Nationwide	Medical Pagers

Base	Mobile	Mode	Location	User & Notes
454.3125	454.3125	NFM	Nationwide	Ligier Formula One Team
454.3250		NFM	Oxfordshire	Medical Paging
454.5000		NFM	Nationwide	UK Atomic Energy Autt.
454.6750		NFM	Nationwide	Millicomm Paging
454.7750		NFM	Nationwide	AirCall Paging
454.8250		NFM	Nationwide	Page Boy Paging

454.8500 - 454.9750 MHz Limited MoD & PMR Alloactions

454.9925		WFM	Scotland	Grampian TV O/B Talkback

455.0000 - 455.4500 MHz BBC TV O/B Talkback & Formula One Racing Team Links

Base	Mobile	Mode	Location	User & Notes
455.0000		WFM	Nationwide	Central TV O/B
455.0250		WFM	Nationwide	Central TV O/B
455.0625		WFM		Radio Signal O/B Link
455.0635		WFM		BBC South East
455.0750		WFM	London	Capitol Radio Flying Eye
455.0940		WFM		CTV Outside Broadcast
455.1320		WFM	Scotland	Grampian TV O/B Talkback
455.1375		WFM		CTV Outside Broadcast
455.1625		WFM	London	BBC South East
		WFM		Radio Piccadilly Eye In Sky
		WFM		Radio Signal O/B Link
455.2000		WFM	Nationwide	Central TV O/B
		NFM	Nationwide	Radio Investigations service
455.2250		WFM	Nationwide	Central TV O/B
		NFM	Nationwide	Radio Investigations service
455.2350		NFM	Formula One Racing	Ferrari Team Voice link
455.2500		NFM	Nationwide	Radio Investigations service
455.25625	468.19375	NFM	Nationwide	BBC TV O/B Talkback
455.2850		WFM	Scotland	BBC Scotland O/B
455.3125		WFM	Jersey	BBC Jersey O/B
		WFM	Nationwide	BBC Radio 1 O/B
455.3625		WFM	Jersey	BBC Jersey O/B
		WFM	Nationwide	BBC Radio 1 O/B
455.39375		WFM	Nationwide	BBC TV O/B Talkback

455.4750 - 455.9750 MHz PMR Airport Security & Ground Repeaters

Base	Mobile	Mode	Location	User & Notes
455.4750	460.7750	NFM	Birmingham Airport	Ground Channel 1
455.4875	461.7875	NFM	Bournemouth Airport	Tower
		NFM	East Midlands	Ground
		NFM	Heathrow	Airport Police c/s Hunter D
455.5000		NFM	Guernsey	119.95 MHz Tower Link
455.5125		NFM	Nationwide	Transport Police Bravo Xray
455.5250	460.8250	NFM	Heathrow Ground	Tower Rebroadcast
		NFM	Nationwide	Airport Customs Use
		NFM	Bristol Airport	Customs
		NFM	Manchester Airport	Customs
455.5375	460.5375	NFM	Birmingham Airport	Ground Control
455.5500	461.8500	NFM	Heathrow	Ground Control Inbound

Base	Mobile	Mode	Location	User & Notes
		NFM	Manchester	Ground
		NFM	Stansted Airport	Ground Control
455.5750	461.2250	NFM	Birmingham Airport	Fire Channel 2
		NFM	Heathrow	Security
		NFM	Various Airports	Fire Services
455.6000		NFM	BAe Filton	Security
455.6000	460.9000	NFM	East Midlands	Terminal Security
455.6125	461.9125	NFM	Heathrow	
455.6250	460.9250	NFM	Nationwide	Red Devil Parachute Team
455.6375	460.9375	NFM	Liverpool Airport	Ground Control
		NFM	Luton Airport	
455.6500		NFM	MoD Airfields	Fire Control
455.6500	460.8500	NFM	Birmingham Airport	Airway Crossing/Security
		NFM	Heathrow	Ground Control
		NFM	Manchester	Ground
455.6750	455.6750	NFM	Norwich	Customs & Excise
455.7000	461.0000	NFM	Birmingham Airport	Apron Channel 4
455.7125	461.0125	NFM	Docklands Airport	Ground Control
		NFM	MoD Bedford	Channel 1
455.7250	461.0250	NFM	Heathrow	Armed Police Hunter W
		NFM	Nationwide	Airport Tower Rebroadcasts
455.7625	461.0625	NFM	Birmingham Airport	
455.7750	461.0750	NFM	Gatwick	Ground Control
		NFM	Heathrow	Armed Police Hunter W
455.7750	455.7750	NFM	Nationwide	Airport Customs
455.7875		NFM	MoD Bedford	Channel 2
455.8125	461.1125	NFM	Aberdeen/Dyce	Ground
		NFM	Heathrow	
455.8250	461.1250	NFM	Birmingham Airport	Maglev
		NFM	Luton Airport	MacAlpine Aviation
		NFM	Midlands Airport	Maintenance
455.8250	455.8250	NFM	Nationwide	Airport Customs
455.8375	461.1375	NFM	Glasgow Airport	
455.8500	455.8500	NFM	Nationwide	Airport Customs
455.8750	461.1750	NFM	Heathrow	Approach
455.9875	455.9875	NFM	Scotland	Fire Brigade Channel 7

456.0000 - 457.0000 MHz Transport, Security 12.5 kHz (Split + 5.5 MHz)
Formula One Racing Team Links

Base	Mobile	Mode	Location	User & Notes
456.0250	461.5250	NFM	Aberdeen	
		NFM	Gatwick Airport	Security
		NFM	Nottingham	Streamline Taxis
		NFM	Stansted Airport	Aviation Traders
456.0500	461.5500	NFM	UKAEA Bristol	Power Station Security
456.1000	461.6000	NFM	England	CEGB
456.1250	461.6250	NFM	Birmingham	City Council
456.1750	461.6750	NFM	Dungeness	Power Station
		NFM	Calverton	NCB
456.2000	461.7000	NFM	Berkley	BNFL Command & Control
456.2750	461.7750	NFM	Dungeness	Power Station

Base	Mobile	Mode	Location	User & Notes
456.3250	461.8250	NFM	Dungeness	Power Station
456.3375	461.8375	NFM	Nationwide	BR Transport Police
456.3500	461.8500	NFM	Aberdeen	Buses
		NFM	Eastborne	City Buses
		NFM	Gatwick Airport	Cleaners & Cargo Loaders
		NFM	Moray Firth	Beatrice Bravo Platform
		NFM	Nottingham	City Engineers Channel 1
456.3750	461.8375	NFM	Nationwide	British Trans. Police Ch 2
456.4000	461.9000	NFM	BR Stations	Post Offices
		NFM	Martlesham Heath	British Telecom
		NFM	Nationwide	Airport Customs
		NFM	Manchester	Airport Customs
		NFM	Luton	Airport Customs
		NFM	Sheffield	City Buses
456.4250	461.9250	NFM	Gatwick	
		NFM	London	BR Transport Police Ch 1
		NFM	Birmingham	BR Transport Police
		NFM	Doncaster	BR Transport Police
		NFM	Edinburgh	BR Transport Police
		NFM	Glasgow	BR Transport Police
		NFM	Liverpool	BR Transport Police
456.4500	461.9500	NFM	Aberdeen	Buses
		NFM	IWM Duxford	Security Channel 1
		NFM	Thamesmead	Caretakers & Lift Ops
456.4750	461.9750	NFM	Nationwide	Independent Coach Ops
		NFM	West Mildands	Safari Park Control
456.5000	462.0000	NFM	Dungeness	Power Station
		NFM	Gatwick Airport	Aircraft Tugs
		NFM	Nationwide	Federal Express
		NFM	Nationwide	Customs
456.5250	462.0250	NFM	Gatwick Airport	Aircraft Tugs
		NFM	Ipswich	Port Authority Channel 2
		NFM	Nationwide	Customs
456.5500	462.0500	NFM	BAe Filton	Security
		NFM	Gatwick Airport	
		NFM	Nottingham	Cap Count Victoria Centre
456.5750	462.0750	NFM	Nottingham	Boots Beeston
456.6000	462.1000	NFM	Aberdeen/Dyce	Loading
456.6150		NFM	Formula One	Ferrari Team Voice Link
456.6250	462.1250	NFM	Birmingham Univ.	Security
		NFM	Gatwick Airport	
		NFM	IWM Duxford	Fire Channel 2
456.6250	462.1250	NFM	Oxford	Clarendon Centre
456.6500	462.1500	NFM	Manchester Airport	Ground Control
456.6750	462.1750	NFM	Aberdeen/Dyce	Staff
		NFM	North Thames	Electricity Board
		NFM	Oldbury	BNFL Backup Channel
456.7000	456.2000	NFM	Nationwide	Marconi Communications
		NFM	Nationwide	Motorola
456.7000	462.2000	NFM	Grays	Lakeside Centre Car Park

Base	Mobile	Mode	Location	User & Notes
456.7250	462.2250	NFM	Cowley	Rover Plant Channel 5
		NFM	Gatwick Airport	Avionics Maintenance
		NFM	Nationwide	UHF Demonstration Chan
456.7500	462.2500	NFM	Aberdeen/Dyce	Loading
		NFM	Barrow	VSEL Fire & Nuclear Incid.
		NFM	Bexleyheath	Shopping Centre
		NFM	Glasgow Airport	Cargo Handlers
456.7750	462.2750	NFM	Blackpool	Blackpool Pleasure Beach
		NFM	Glasgow	Strathclyde Police
		NFM	Harwich	Trinity House
		NFM	Redditch	King Fisher Shopping Cntr
456.8000	462.3000	NFM	Nottingham	Boots Beeston
456.8250	462.3250	NFM	Dover	Coastguard Cliff Rescue
		NFM	Nationwide	NCB & Docks
456.8500	462.3500	NFM	Aberdeen	
		NFM	Cowley	Rover Plant
		NFM	Nottingham	Mapperly Hospital
		NFM	Tilbury	Docks Security
456.8750	462.3750	NFM	Gatwick Airport	Handling
		NFM	Grays	Lakeside Centre Security
		NFM	Nottingham	Esso Colwick
456.9000	462.4000	NFM	Barrow	VSEL Works Security Cont
456.9250	462.4250	NFM	Cowley	Rover Plant Channel 6
		NFM	IWM Duxford	General Channel 3
		NFM	Nationwide	UHF Demonstration Chan
		NFM	Nationwide	Airport Security
		NFM	Nationwide	Power Station Security
456.9750	462.4750	NFM	Nationwide	Bus & Coach Operators
		NFM	Aberdeen	
		NFM	Longbridge	British Leyland

457.0000 - 457.5000 MHz UHF Fire Mobile Links 12.5 kHz
Simplex & Duplex Base (Split + 5.5 MHz)
Scottish HydroElectric
Formula One Racing Team Links

Base	Mobile	Mode	Location	User & Notes
457.0125	462.5125	NFM	London	Fire Brigade Channel 3
457.0125	457.0125	NFM	Scotland	Fire Brigade Channel 3
457.0250	462.5250	NFM	Gairloch	Scottish Hydro Electric
457.0250		NFM	Nationwide	Minardi Formula One Voice
457.0375	462.5375	NFM	London	Fire Brigade Channel 1
457.0375	457.0375	NFM	Scotland	Fire Brigade Channel 1
457.0500		NFM	Nationwide	Minardi Formula One Voice
457.0750	462.5750	NFM	Achanshellach	Scottish Hydro Electric
457.0875	462.5875	NFM	London	Fire Brigade Channel 2
		NFM	Scotland	Fire Brigade Channel 2
457.1000	462.6000	NFM	Gairloch	Scottish Hydro Electric
		NFM	Loch A'Burra	Scottish Hydro Electric
457.1375	462.6375	NFM	London	Fire Brigade Channel 5
457.1375	457.1375	NFM	Scotland	Fire Brigade Channel 5
457.1875	462.6875	NFM	London	Fire Brigade Channel 4

Base	Mobile	Mode	Location	User & Notes
457.1875	457.1875	NFM	Scotland	Fire Brigade Channel 4
457.2000	462.7000	NFM	Nationwide	NCB Mine Rescue
457.2375	462.7375	NFM	London	Fire Brigade Channel 6
457.2375	457.2375	NFM	Scotland	Fire Brigade Channel 6
457.3125		NFM	Formula One Racing	Ligier Team Voice Link
457.3500	462.8500	NFM	Gairloch	Scottish Hydro Electric
		NFM	Nationwide	AA UHF Mobiles
457.3750	462.8750	NFM	Gairloch	Scottish Hydro Electric
		NFM	Inverness	Automobile Association
457.4500	462.9500	NFM	Gairloch	Scottish Hydro Electric
457.4750	462.9750	NFM	Garve	Scottish Roads

457.50625 - 458.49375 MHz Fixed Scan Links & Ship Board Handheld Transceivers 6.25 kHz (Split + 5.5 MHz)

Base	Mobile	Mode	Location	User & Notes
457.5250		NFM	Maritime	Ship Communications
457.5500		NFM	Maritime	Ship Communications
457.5750		NFM	Maritime	Ship Communications
457.6000	463.1000	NFM	Garve	Scottish Roads
457.6000		NFM	Maritime	Ship Communications
457.6250	463.1250	NFM	Gatwick Airport	Airport Police - GatPol
457.7000	463.2000	NFM	Ullapool	Scottish Roads
457.7250	463.2250	NFM	Gairloch	Scottish Roads
457.9250	463.4250	NFM	Nationwide	Short Term Hire Equipment
458.1750	463.6750	NFM	Lee Valley	Water Board
458.2750	463.7750	NFM	Oxford	Ambulance Service
458.3500	463.8500	NFM	Lee Valley	Water Board

458.5000 - 459.0000 MHz UHF Low Power Devices

458.5000 - 459.5000 MHz UHF Remote Controlled Model Band

Base	Mobile	Mode	Location	User & Notes
458.5000		NFM	Nationwide	UHF Demonstration Chan
		NFM	Nationwide	Telemetry
458.5125		NFM	Nationwide	Telemetry
458.6500		NFM	Nationwide	Telemetry
458.7000		NFM	Nationwide	Telemetry

458.5125 - 459.4875 MHz Low Power Paging

Base	Mobile	Mode	Location	User & Notes
458.8250		NFM	Fixed Alarm Paging	
458.8375		NFM	Transportable & Mobile Alarm Paging	
458.9000		NFM	Car Theft Alarm Paging	
459.0000		NFM	Medical & Biological Telemetry	
459.1250	161.0000	NFM	Paging	
459.1500	161.0250	NFM	Marina Paging	
459.2500	161.0500	NFM	Paging	
459.3250	161.0125	NFM	Paging	
459.3500	161.0250	NFM	Paging	
459.3750	161.0375	NFM	Paging	
459.4000	161.0500	NFM	Paging	
459.4250	161.0625	NFM	BBC Bush House Pager	
459.4500	161.1000	NFM	Paging	

Base	Mobile	Mode	Location	User & Notes
459.4750	161.1125	NFM	Paging	

459.5000 - 460.5000 MHz Ambulance Handsets and Various PMR Users
Mobile   12.5 kHz (Split-5.5 MHz)
Formula One Racing Team Links

Base	Mobile	Mode	Location	User & Notes
459.6500	453.1500	NFM	Nottingham	Boots Broadmarch Shop Ctr
459.7250	453.2250	NFM	Nationwide	Police and Fire Channel
459.7750	453.2750	NFM	Nationwide	Ambulance UHF to VHF
		NFM	Nationwide	Philips Security
459.8500	453.3500	NFM	Nationwide	Ambulance UHF to VHF
459.9500	453.4500	NFM	Gatwick Airport	Message for Captains
		NFM	Channel Islands	Customs & Excise
460.1000	453.6000	NFM	Bedford	Debenhams Security
		NFM	Birmingham	Debenhams Security
		NFM	Cambridge	Debenhams Security
		NFM	Derby	Debenhams Security
		NFM	Folkestone	Debenhams Security
460.2250	453.7250	NFM	Isle Of Grain	Oil
460.2500	453.7500	NFM	Nottingham	Formans
460.3250		NFM	Formula One Racing	Lotus Team Voice Link
460.4000	453.9000	NFM	Bristol	British Rail
		NFM	Doncater Yard	British Rail
		NFM	Euston	British Rail
		NFM	Gatwick Airport	British Rail
		NFM	Hornsey Depot	British Rail
		NFM	Marylebone	British Rail
		NFM	Victoria	British Rail

460.5000 - 461.5000 MHz UHF Point to Point Links 25 kHz (Split+6.5MHz)

Base	Mobile	Mode	Location	User & Notes
460.50625	467.00625	NFM	Nationwide	Sky TV O/B Talkback
460.5500	467.0500	NFM	Jersey	Air Traffic Control Link
460.6750	467.1750	NFM	Silverstone	French TV Talkback
460.7875	467.2875	NFM	Nationwide	National Air Traffic Service
461.0000	467.5000	NFM	RAE Filton	Ground Crews
461.0250	467.5250	NFM	Nationwide	National Air Traffic Service

461.3000 - 461.4875 MHz PMR Simplex

Base	Mobile	Mode	Location	User & Notes
461.4250	467.9250	NFM	Gatwick Airport	Cellular Link

461.5000 - 462.5000 MHz PMR UHF Band Base 12.5 kHz (Split-5.5 MHz)
Formula One Racing Team Links

Base	Mobile	Mode	Location	User & Notes
461.5750	456.0750	NFM	Nationwide	NCB Security
461.7750	456.2500	NFM	Nottingham	Esso Colwick
461.8500	456.3500	NFM	Nottingham	Technical Services
461.9000	456.4000	NFM	Customs & Excise	Covert Repeater
461.9000	461.9000	NFM	Nationwide	Customs Surveillance
461.9500	456.4500	NFM	Duxford Imp War Museum	Security
462.1250	456.6250	NFM	Duxford Imp War Museum	Security
462.3250	456.8250	NFM	Ipswick Port Authority	Channel 4
462.4250	456.9250	NFM	Formula One Racing	Lotus Team Voice Link

Base	Mobile	Mode	Location	User & Notes		
462.4750	456.9750	NFM	Nationwide	Short Term Hire Equipment		
		NFM	Shireoaks Colliery	NCB Security		
462.5750	457.0750	NFM	Aberdeen	Hydro Electric		

463.0000 - 464.0000 MHz UHF Telemetry Links NFM

464.0000 - 465.0000 MHz British Telecom Links NFM (Split-14.0 MHz)

Base	Mobile	Mode	Location	User & Notes		
464.3500	450.3500	NFM	Edinburgh	EEC Summit		
464.4000	450.4000	NFM	Edinburgh	EEC Summit		
464.4500	450.4500	NFM	Edinburgh	EEC Summit		
464.5000	450.5000	NFM	Edinburgh	EEC Summit		
464.6750	450.6750	NFM	Edinburgh	EEC Summit		
464.7750	450.7750	NFM	Edinburgh	EEC Summit		

465.0000 - 467.0000 MHz Police PR Base & Repeater System (Scotland)
Formula One Racing Team Links

Base	Mobile	Mode	Location	User & Notes		
465.1000	451.1000	NFM	Broxburn	F	
465.1500	451.1500	NFM	South Queensferry	F	
465.2350		NFM	Ferrari Team Voice Link		
465.2750	451.2750	NFM	Edinburgh Westerhailes	CH	Channel 5
465.3000	451.3000	NFM	Fife		
465.4750	451.4750	NFM	Brechin		
		NFM	Montrose		
465.6250	451.6250	NFM	Scottish Special Events		Channel 33
465.6750	451.6750	NFM	Scottish Special Events		Channel 34
465.7250	451.7250	NFM	Scottish Special Events		Channel 35
		NFM	Perth	W	Channel 13
465.7750	451.7750	NFM	Scottish Special Events		Channel 36
466.0750		NFM	Scottish Police Data Link		
		NFM	Cupar		
		NFM	Dundee		
		NFM	Edinburgh		
		NFM	Kirkaldy		
		NFM	Perth		
		NFM	St Andrews		
466.2500	452.2500	NFM	Aberdeen		Channel 18
		NFM	Ayrshire		
		NFM	Edinburgh, St Leonards, Div HQ	B	Channel 2
		NFM	Glasgow		
		NFM	Lanark		
		NFM	Scottish Firearms Support Group		
466.2750	452.2750	NFM	Scottish Traffic Wardens	TW	Channel 15
		NFM	Aberdeen		
		NFM	Ayrshire	TW	
		NFM	Edinburgh Tattoo & Tfc Wardens	C	
		NFM	Glasgow	TW	
		NFM	Lanark	TW	
		NFM	Perth	TW	Channel 80
		NFM	Scottish Special Events		
466.3250	452.3250	NFM	Dumbarton	L	Channel 9

Base	Mobile	Mode	Location	User & Notes	
		NFM	Edinburgh Oxgangs	CO	
466.3500	451.3500	NFM	Glasgow	E	Channel 5
466.3750	452.3750	NFM	Edinburgh		
		NFM	Glasgow Barrhead	K	Channel 8
466.4000	452.4000	NFM	East Lothian	E	
		NFM	HMP Aberdeen		
		NFM	HMP Noranside, Forfar		
		NFM	HMP Perth		
466.4250	452.4250	NFM	Various Areas		
466.4500	452.4500	NFM	Dundee	ZS	Channel 3
		NFM	Glasgow North	C	
466.5000	452.5000	NFM	Arbroath		Channel 1
		NFM	Argyll	A	
		NFM	Edinburgh		
		NFM	Forfar		
		NFM	Glasgow Central	A	
		NFM	Perth	W	
466.5250	452.5250	NFM	Edinburgh, Mayfield	BY	Channel 1
		NFM	St Johnstone FC Security	Y	
466.5500	452.5500	NFM	Argyll	B	Channel 2
		NFM	Glasgow West	B	
		NFM	Kilmarnock	U	
		NFM	Inverness		
466.5750	452.5750	NFM	Edinburgh West End Div HQ	C	Channel 4
466.6000	452.6000	NFM	Aberdeen		Channel 4
		NFM	Cupar	C	
		NFM	Glasgow Easterhill	D	
466.6150		NFM	Ferrari Team Voice Link		
466.6250	452.6250	NFM	Aberdeen		
		NFM	Edinburgh Gayfield	BG	Channel 3
466.6500	452.6500	NFM	Aberdeen		Channel 6
		NFM	Ayr	R	
		NFM	Edinburgh Drylaw	DR	Channel 10
		NFM	Glasgow Killpatrick	M	
		NFM	Glasgow South/East	F	
466.6750	452.6750	NFM	Portobello	DJ	
466.7000	452.7000	NFM	Ayrshire	X	Channel 14
		NFM	Musselburgh		
466.7250	452.7250	NFM	Livingstone	F	
466.7500	452.7500	NFM	Glasgow Ibrox	G	Channel 7
466.7750	452.7750	NFM	Glasgow Hamilton	Q	
466.8000	452.8000	NFM	Scottish Firearms Support Group		Channel 16
466.8250	452.8250	NFM	Edinburgh Leith Div HQ	D	Channel 9
		NFM	Lanark	N	Channel 11
466.9000	452.9000	NFM	Scottish Firearms Support Group		Channel 17

467.0000 - 467.8250 MHz UHF Point to Point Simplex Links Ship Communications, Repeaters & BBC Local Radio

467.3000		WFM	Nationwide	BBC Radio 5 OB	
467.5250	467.5250	NFM	Ship Communications		

Base	Mobile	Mode	Location	User & Notes
467.5500	467.5500	NFM	Ship Communications	
467.5750	467.5750	NFM	Ship Communications	
467.6125	467.6625	WFM	London	Capital Radio O/B Link
467.6625	467.6125	WFM	London	Capital Radio Flying Eye
467.6750	467.6750	NFM	Ship Communications	
467.7500	467.7500	NFM	Ship Communications	
467.7750	467.7750	NFM	Ship Communications	
467.8000	467.8000	NFM	Ship Communications	
467.8250	467.8250	NFM	Ship Communications	

467.8250 - 468.0000 MHz UHF Point to Point Links (Split - 6.0 MHz)

Base	Mobile	Mode	Location	User & Notes
467.8750	461.8750	NFM	Nationwide	Pye Telecom Mobiles
467.9250	461.9250	NFM	Nationwide	Pye Telecom Mobiles
467.9500	461.9500	NFM	Derbyshire	County Council Mobiles

468.0000 - 468.3500 MHz PMR Reserve 25 kHz O/B Talkback

Base	Mobile	Mode	Location	User & Notes
468.0000	141.0000	NFM	London	LWT
468.0250		NFM	Dartford Tunnel	Police
468.0625		NFM	Sussex	Fire Brigade
468.0750		NFM	Invicta Radio	
468.1500		NFM	Nationwide	BBC Radio 1 OB
468.2750		NFM	Nationwide	BBC Radio 1 OB
468.3125		NFM	Gatwick Airport	Police Gatpol

468.3500 - 469.4750 MHz Outside Broadcast Talkback

Base	Mobile	Mode	Location	User & Notes
468.3750	468.3750	NFM	Nationwide	ITV Cameras
468.4250	468.4250	NFM	England	Central TV Base Link Input
468.4625	468.4625	NFM	Inverness	Moray Firth Radio OB
468.4750	468.4750	NFM	Nationwide	IBA Riggers
468.5000	468.5000	NFM	Nationwide	ITV OB
469.0125	469.0125	NFM	Shropshire	BBC Radio Shropshire OB
469.0625	469.0625	NFM	Nationwide	BBC Radio 5 OB
469.2400	469.2400	NFM	London	Executive Buses
469.2125	469.2125	NFM	Nationwide	BBC Radio 5 OB
469.3125	469.3125	NFM	Sussex	Radio Mercury
469.4125	469.4125	NFM	Essex	Saxon Radio

469.5000 - 469.8500 MHz Limited PMR Allocations

470.0000 - 582.0000 MHz UK TV Channels (Video/Sound)
Local Radio Talkback and Theatre
Radiomicrophones (10mW Max)

Base	Mobile	Mode	Location	User & Notes
470.5065		NFM	Nationwide	British Telecom
471.2500	477.2500	WFM	Nationwide	Channel 21
473.0750		NFM	Canvey	Police Car Link
473.5500		NFM	Grays	Police Car Link
474.4750		NFM	Basildon	Police Car Link
474.8750		NFM	Dagenham	Police Car Link
478.0000		NFM	London	Radio London
478.7000		NFM	Nationwide	Theatre Radiomicrophone

Base	Mobile	Mode	Location	User & Notes
479.2500	485.2500	WFM	Nationwide	Channel 22
479.6500		NFM	Nationwide	Theatre Radiomicrophone
480.2000		NFM	Nationwide	Theatre Radiomicrophone
480.4000		NFM	Nationwide	Theatre Radiomicrophone
487.2500	493.2500	WFM	Nationwide	Channel 23
495.2500	501.2500	WFM	Nationwide	Channel 24
497.5000		NFM	Nationwide	Theatre Radiomicrophone
497.7000		NFM	Nationwide	Theatre Radiomicrophone
498.4800		NFM	Nationwide	Theatre Radiomicrophone
498.7800		NFM	Nationwide	Theatre Radiomicrophone
499.6100		NFM	Nationwide	Theatre Radiomicrophone
500.0000		AM	Nationwide	NATO Mayday Discreet
500.2800		NFM	Nationwide	Theatre Radiomicrophone
502.4400		NFM	Nationwide	Theatre Radiomicrophone
502.6900		NFM	Nationwide	Theatre Radiomicrophone
503.2500	509.2500	WFM	Nationwide	Channel 25
511.2500	517.2500	WFM	Nationwide	Channel 26
519.2500	525.2500	WFM	Nationwide	Channel 27
527.2500	533.2500	WFM	Nationwide	Channel 28
535.2500	541.2500	WFM	Nationwide	Channel 29
543.2500	549.2590	WFM	Nationwide	Channel 30
551.2500	557.2500	WFM	Nationwide	Channel 31
559.2500	565.2500	WFM	Nationwide	Channel 32
567.2500	573.2500	WFM	Nationwide	Channel 33
575.2500	581.2500	WFM	Nationwide	Channel 34
583.6900		NFM	Nationwide	Theatre Radiomicrophone
584.1500		NFM	Nationwide	Theatre Radiomicrophone

590.0000 - 598.0000 MHz Civil & Defence Radar

598.0000 - 606.0000 MHz Future TV Channel 5 Allocation

606.0000 - 614.0000 MHz Radio Astronomy

614.0000 - 854.0000 MHz UK WFM TV Channels (Video/Sound)

Base	Mobile	Mode	Location	User & Notes
615.2500	621.2500	NFM	Nationwide	Channel 39
623.2500	629.2500	NFM	Nationwide	Channel 40
631.2500	637.2500	NFM	Nationwide	Channel 41
639.2500	645.2500	NFM	Nationwide	Channel 42
647.2500	653.2500	NFM	Nationwide	Channel 43
655.2500	661.2500	NFM	Nationwide	Channel 44
663.2500	669.2500	NFM	Nationwide	Channel 45
671.2500	677.2500	NFM	Nationwide	Channel 46
679.2500	685.2500	NFM	Nationwide	Channel 47
687.2500	693.2500	NFM	Nationwide	Channel 48
695.2500	701.2500	NFM	Nationwide	Channel 49
703.2500	709.2500	NFM	Nationwide	Channel 50
711.2500	717.2500	NFM	Nationwide	Channel 51
719.2500	725.2500	NFM	Nationwide	Channel 52
727.2500	733.2500	NFM	Nationwide	Channel 53

Base	Mobile	Mode	Location	User & Notes
735.2500	741.2500	NFM	Nationwide	Channel 54
743.2500	749.2500	NFM	Nationwide	Channel 55
751.2500	757.2500	NFM	Nationwide	Channel 56
759.2500	765.2500	NFM	Nationwide	Channel 57
767.2500	773.2500	NFM	Nationwide	Channel 58
775.2500	781.2500	NFM	Nationwide	Channel 59
783.2500	789.2500	NFM	Nationwide	Channel 60
791.2500	797.2500	NFM	Nationwide	Channel 61
799.2500	805.2500	NFM	Nationwide	Channel 62
807.2500	813.2500	NFM	Nationwide	Channel 63
815.2500	821.2500	NFM	Nationwide	Channel 64
823.2500	829.2500	NFM	Nationwide	Channel 65
831.2500	837.2500	NFM	Nationwide	Channel 66
839.2500	845.2500	NFM	Nationwide	Channel 67
846.0000			RAF Boulmer	R80 Defence Radar
847.2500	853.2500	NFM	Nationwide	Channel 68
854.5000		WFM	Sutton Coldfield	BBC O/B Link

854.7500 - 855.2500 MHz Radio Microphones

Base	Mobile	Mode	Location	User & Notes
854.7500		WFM	Nationwide	Channel 1
854.7750		WFM	Nationwide	Channel 2
854.8000		WFM	Nationwide	Channel 3
854.8250		WFM	Nationwide	Channel 4
854.8500		WFM	Nationwide	Channel 5
854.8750		WFM	Nationwide	Channel 6
854.9000		WFM	Nationwide	Channel 7
854.9250		WFM	Nationwide	Channel 8
854.9500		WFM	Nationwide	Channel 9
854.9750		WFM	Nationwide	Channel 10
855.0000		WFM	Nationwide	Channel 11
855.0250		WFM	Nationwide	Channel 12
855.0500		WFM	Nationwide	Channel 13
855.0750		WFM	Nationwide	Channel 14
855.1000		WFM	Nationwide	Channel 15
855.1250		WFM	Nationwide	Channel 16
855.1500		WFM	Nationwide	Channel 17
855.1750		WFM	Nationwide	Channel 18
855.2000		WFM	Nationwide	Channel 19
855.2250		WFM	Nationwide	Channel 20
859.8000		WFM	Sutton Coldfield	BBC O/B Link

860.2500 - 860.7500 MHz Radio Microphones

Base	Mobile	Mode	Location	User & Notes
860.2500		WFM	Nationwide	Channel 1
860.2750		WFM	Nationwide	Channel 2
860.3000		WFM	Nationwide	Channel 3
860.3250		WFM	Nationwide	Channel 4
860.3500		WFM	Nationwide	Channel 5
860.3750		WFM	Nationwide	Channel 6
860.4000		WFM	Nationwide	Channel 7
860.4250		WFM	Nationwide	Channel 8

Base	Mobile	Mode	Location	User & Notes
860.4500		WFM	Nationwide	Channel 9
860.4750		WFM	Nationwide	Channel 10
860.5000		WFM	Nationwide	Channel 11
860.5250		WFM	Nationwide	Channel 12
860.5500		WFM	Nationwide	Channel 13
860.5750		WFM	Nationwide	Channel 14
860.6000		WFM	Nationwide	Channel 15
860.6250		WFM	Nationwide	Channel 16
860.6500		WFM	Nationwide	Channel 17
860.6750		WFM	Nationwide	Channel 18
860.7000		WFM	Nationwide	Channel 19
860.7250		WFM	Nationwide	Channel 20

863.0000 - 864.0000 MHz Land Mobile Experimental Use

865.0500 - 867.9500 MHz New 'Zonephone' Cordless Telephone Band

Base		Mode	Location	User & Notes
865.0500		WFM	Nationwide	Channel 01
865.1500		WFM	Nationwide	Channel 02
865.2500		WFM	Nationwide	Channel 03
865.3500		WFM	Nationwide	Channel 04
865.4500		WFM	Nationwide	Channel 05
865.5500		WFM	Nationwide	Channel 06
865.6500		WFM	Nationwide	Channel 07
865.7500		WFM	Nationwide	Channel 08
865.8500		WFM	Nationwide	Channel 09
865.9500		WFM	Nationwide	Channel 09
866.0500		WFM	Nationwide	Channel 11
866.1500		WFM	Nationwide	Channel 12
866.2500		WFM	Nationwide	Channel 13
866.3500		WFM	Nationwide	Channel 14
866.4500		WFM	Nationwide	Channel 15
866.5500		WFM	Nationwide	Channel 16
866.6500		WFM	Nationwide	Channel 17
866.7500		WFM	Nationwide	Channel 18
866.8500		WFM	Nationwide	Channel 19
866.9500		WFM	Nationwide	Channel 20
867.0500		WFM	Nationwide	Channel 21
867.1500		WFM	Nationwide	Channel 22
867.2500		WFM	Nationwide	Channel 23
867.3500		WFM	Nationwide	Channel 24
867.4500		WFM	Nationwide	Channel 25
867.5500		WFM	Nationwide	Channel 26
867.6500		WFM	Nationwide	Channel 27
867.7500		WFM	Nationwide	Channel 28
867.8500		WFM	Nationwide	Channel 29
867.9500		WFM	Nationwide	Channel 30

868.0000 - 869.9750 MHz Deregulated Band For Low Power Devices

Base	Mobile	Mode	Location	User & Notes	
872.0000 - 904.9875 MHz UHF ETACS Cellular Telephone Mobiles					
875.0000		WFM	Nationwide	WFM BBC Music Link	
880.0000		WFM	Nationwide	WFM BBC Music Link	

905.0000 - 915.0000 MHz Pan European Digital Cellular Service Mobiles

914.0125 - 914.9875 MHz New Cybernet/Uniden Cordless Telephones Handset

Base	Mobile	Mode	Location	User & Notes	
917.0125 - 949.9875 MHz UHF Cellular ETACS (Extended Total Access Communications System) Telephone Nodes					
917.0125	872.0125	NFM	Nationwide	Vodafone	Channel 1329
917.0375	872.0375	NFM	Nationwide	Vodafone	Channel 1330
917.0625	872.0625	NFM	Nationwide	Vodafone	Channel 1331
917.0875	872.0875	NFM	Nationwide	Vodafone	Channel 1332
917.1125	872.1125	NFM	Nationwide	Vodafone	Channel 1333
917.1375	872.1375	NFM	Nationwide	Vodafone	Channel 1334
917.1625	872.1625	NFM	Nationwide	Vodafone	Channel 1335
917.1875	872.1875	NFM	Nationwide	Vodafone	Channel 1336
917.2125	872.2125	NFM	Nationwide	Vodafone	Channel 1337
917.2375	872.2375	NFM	Nationwide	Vodafone	Channel 1338
917.2625	872.2625	NFM	Nationwide	Vodafone	Channel 1339
917.2875	872.2875	NFM	Nationwide	Vodafone	Channel 1340
917.3125	872.3125	NFM	Nationwide	Vodafone	Channel 1341
917.3375	872.3375	NFM	Nationwide	Vodafone	Channel 1342
917.3625	872.3625	NFM	Nationwide	Vodafone	Channel 1343
917.3875	872.3875	NFM	Nationwide	Vodafone	Channel 1344
917.4125	872.4125	NFM	Nationwide	Vodafone	Channel 1345
917.4375	872.4375	NFM	Nationwide	Vodafone	Channel 1346
917.4625	872.4625	NFM	Nationwide	Vodafone	Channel 1347
917.4875	872.4875	NFM	Nationwide	Vodafone	Channel 1348
917.5125	872.5125	NFM	Nationwide	Vodafone	Channel 1349
917.5375	872.5375	NFM	Nationwide	Vodafone	Channel 1350
917.5625	872.5625	NFM	Nationwide	Vodafone	Channel 1351
917.5875	872.5875	NFM	Nationwide	Vodafone	Channel 1352
917.6125	872.6125	NFM	Nationwide	Vodafone	Channel 1353
917.6375	872.6375	NFM	Nationwide	Vodafone	Channel 1354
917.6625	872.6625	NFM	Nationwide	Vodafone	Channel 1355
917.6875	872.6875	NFM	Nationwide	Vodafone	Channel 1356
917.7125	872.7125	NFM	Nationwide	Vodafone	Channel 1357
917.7375	872.7375	NFM	Nationwide	Vodafone	Channel 1358
917.7625	872.7625	NFM	Nationwide	Vodafone	Channel 1359
917.7875	872.7875	NFM	Nationwide	Vodafone	Channel 1360
917.8125	872.8125	NFM	Nationwide	Vodafone	Channel 1361
917.8375	872.8375	NFM	Nationwide	Vodafone	Channel 1362
917.8625	872.8625	NFM	Nationwide	Vodafone	Channel 1363
917.8875	872.8875	NFM	Nationwide	Vodafone	Channel 1364
917.9125	872.9125	NFM	Nationwide	Vodafone	Channel 1365
917.9375	872.9375	NFM	Nationwide	Vodafone	Channel 1366
917.9625	872.9625	NFM	Nationwide	Vodafone	Channel 1367

Base	Mobile	Mode	Location	User & Notes	
917.9875	872.9875	NFM	Nationwide	Vodafone	Channel 1368
918.0125	873.0125	NFM	Nationwide	Vodafone	Channel 1369
918.0375	873.0375	NFM	Nationwide	Vodafone	Channel 1370
918.0625	873.0625	NFM	Nationwide	Vodafone	Channel 1371
918.0875	873.0875	NFM	Nationwide	Vodafone	Channel 1372
918.1125	873.1125	NFM	Nationwide	Vodafone	Channel 1373
918.1375	873.1375	NFM	Nationwide	Vodafone	Channel 1374
918.1625	873.1625	NFM	Nationwide	Vodafone	Channel 1375
918.1875	873.1875	NFM	Nationwide	Vodafone	Channel 1376
918.2125	873.2125	NFM	Nationwide	Vodafone	Channel 1377
918.2375	873.2375	NFM	Nationwide	Vodafone	Channel 1378
918.2625	873.2625	NFM	Nationwide	Vodafone	Channel 1379
918.2875	873.2875	NFM	Nationwide	Vodafone	Channel 1380
918.3125	873.3125	NFM	Nationwide	Vodafone	Channel 1381
918.3375	873.3375	NFM	Nationwide	Vodafone	Channel 1382
918.3625	873.3625	NFM	Nationwide	Vodafone	Channel 1383
918.3875	873.3875	NFM	Nationwide	Vodafone	Channel 1384
918.4125	873.4125	NFM	Nationwide	Vodafone	Channel 1385
918.4375	873.4375	NFM	Nationwide	Vodafone	Channel 1386
918.4625	873.4625	NFM	Nationwide	Vodafone	Channel 1387
918.4875	873.4875	NFM	Nationwide	Vodafone	Channel 1388
918.5125	873.5125	NFM	Nationwide	Vodafone	Channel 1389
918.5375	873.5375	NFM	Nationwide	Vodafone	Channel 1390
918.5625	873.5625	NFM	Nationwide	Vodafone	Channel 1391
918.5875	873.5875	NFM	Nationwide	Vodafone	Channel 1392
918.6125	873.6125	NFM	Nationwide	Vodafone	Channel 1393
918.6375	873.6375	NFM	Nationwide	Vodafone	Channel 1394
918.6625	873.6625	NFM	Nationwide	Vodafone	Channel 1395
918.6875	873.6875	NFM	Nationwide	Vodafone	Channel 1396
918.7125	873.7125	NFM	Nationwide	Vodafone	Channel 1397
918.7375	873.7375	NFM	Nationwide	Vodafone	Channel 1398
918.7625	873.7625	NFM	Nationwide	Vodafone	Channel 1399
918.7875	873.7875	NFM	Nationwide	Vodafone	Channel 1400
918.8125	873.8125	NFM	Nationwide	Vodafone	Channel 1401
918.8375	873.8375	NFM	Nationwide	Vodafone	Channel 1402
918.8625	873.8625	NFM	Nationwide	Vodafone	Channel 1403
918.8875	873.8875	NFM	Nationwide	Vodafone	Channel 1404
918.9125	873.9125	NFM	Nationwide	Vodafone	Channel 1405
918.9375	873.9375	NFM	Nationwide	Vodafone	Channel 1406
918.9625	873.9625	NFM	Nationwide	Vodafone	Channel 1407
918.9875	873.9875	NFM	Nationwide	Vodafone	Channel 1408
919.0125	874.0125	NFM	Nationwide	Vodafone	Channel 1409
919.0375	874.0375	NFM	Nationwide	Vodafone	Channel 1410
919.0625	874.0625	NFM	Nationwide	Vodafone	Channel 1411
919.0875	874.0875	NFM	Nationwide	Vodafone	Channel 1412
919.1125	874.1125	NFM	Nationwide	Vodafone	Channel 1413
919.1375	874.1375	NFM	Nationwide	Vodafone	Channel 1414
919.1625	874.1625	NFM	Nationwide	Vodafone	Channel 1415
919.1875	874.1875	NFM	Nationwide	Vodafone	Channel 1416
919.2125	874.2125	NFM	Nationwide	Vodafone	Channel 1417

Base	Mobile	Mode	Location	User & Notes	
919.2375	874.2375	NFM	Nationwide	Vodafone	Channel 1418
919.2625	874.2625	NFM	Nationwide	Vodafone	Channel 1419
919.2875	874.2875	NFM	Nationwide	Vodafone	Channel 1420
919.3125	874.3125	NFM	Nationwide	Vodafone	Channel 1421
919.3375	874.3375	NFM	Nationwide	Vodafone	Channel 1422
919.3625	874.3625	NFM	Nationwide	Vodafone	Channel 1423
919.3875	874.3875	NFM	Nationwide	Vodafone	Channel 1424
919.4125	874.4125	NFM	Nationwide	Vodafone	Channel 1425
919.4375	874.4375	NFM	Nationwide	Vodafone	Channel 1426
919.4625	874.4625	NFM	Nationwide	Vodafone	Channel 1427
919.4875	874.4875	NFM	Nationwide	Vodafone	Channel 1428
919.5125	874.5125	NFM	Nationwide	Vodafone	Channel 1429
919.5375	874.5375	NFM	Nationwide	Vodafone	Channel 1430
919.5625	874.5625	NFM	Nationwide	Vodafone	Channel 1431
919.5875	874.5875	NFM	Nationwide	Vodafone	Channel 1432
919.6125	874.6125	NFM	Nationwide	Vodafone	Channel 1433
919.6375	874.6375	NFM	Nationwide	Vodafone	Channel 1434
919.6625	874.6625	NFM	Nationwide	Vodafone	Channel 1435
919.6875	874.6875	NFM	Nationwide	Vodafone	Channel 1436
919.7125	874.7125	NFM	Nationwide	Vodafone	Channel 1437
919.7375	874.7375	NFM	Nationwide	Vodafone	Channel 1438
919.7625	874.7625	NFM	Nationwide	Vodafone	Channel 1439
919.7875	874.7875	NFM	Nationwide	Vodafone	Channel 1440
919.8125	874.8125	NFM	Nationwide	Vodafone	Channel 1441
919.8375	874.8375	NFM	Nationwide	Vodafone	Channel 1442
919.8625	874.8625	NFM	Nationwide	Vodafone	Channel 1443
919.8875	874.8875	NFM	Nationwide	Vodafone	Channel 1444
919.9125	874.9125	NFM	Nationwide	Vodafone	Channel 1445
919.9375	874.9375	NFM	Nationwide	Vodafone	Channel 1446
919.9625	874.9625	NFM	Nationwide	Vodafone	Channel 1447
919.9875	874.9875	NFM	Nationwide	Vodafone	Channel 1448
920.0125	875.0125	NFM	Nationwide	Vodafone	Channel 1449
920.0375	875.0375	NFM	Nationwide	Vodafone	Channel 1450
920.0625	875.0625	NFM	Nationwide	Vodafone	Channel 1451
920.0875	875.0875	NFM	Nationwide	Vodafone	Channel 1452
920.1125	875.1125	NFM	Nationwide	Vodafone	Channel 1453
920.1375	875.1375	NFM	Nationwide	Vodafone	Channel 1454
920.1625	875.1625	NFM	Nationwide	Vodafone	Channel 1455
920.1875	875.1875	NFM	Nationwide	Vodafone	Channel 1456
920.2125	875.2125	NFM	Nationwide	Vodafone	Channel 1457
920.2375	875.2375	NFM	Nationwide	Vodafone	Channel 1458
920.2625	875.2625	NFM	Nationwide	Vodafone	Channel 1459
920.2875	875.2875	NFM	Nationwide	Vodafone	Channel 1460
920.3125	875.3125	NFM	Nationwide	Vodafone	Channel 1461
920.3375	875.3375	NFM	Nationwide	Vodafone	Channel 1462
920.3625	875.3625	NFM	Nationwide	Vodafone	Channel 1463
920.3875	875.3875	NFM	Nationwide	Vodafone	Channel 1464
920.4125	875.4125	NFM	Nationwide	Vodafone	Channel 1465
920.4375	875.4375	NFM	Nationwide	Vodafone	Channel 1466
920.4625	875.4625	NFM	Nationwide	Vodafone	Channel 1467

Base	Mobile	Mode	Location	User & Notes	
920.4875	875.4875	NFM	Nationwide	Vodafone	Channel 1468
920.5125	875.5125	NFM	Nationwide	Vodafone	Channel 1469
920.5375	875.5375	NFM	Nationwide	Vodafone	Channel 1470
920.5625	875.5625	NFM	Nationwide	Vodafone	Channel 1471
920.5875	875.5875	NFM	Nationwide	Vodafone	Channel 1472
920.6125	875.6125	NFM	Nationwide	Vodafone	Channel 1473
920.6375	875.6375	NFM	Nationwide	Vodafone	Channel 1474
920.6625	875.6625	NFM	Nationwide	Vodafone	Channel 1475
920.6875	875.6875	NFM	Nationwide	Vodafone	Channel 1476
920.7125	875.7125	NFM	Nationwide	Vodafone	Channel 1477
920.7375	875.7375	NFM	Nationwide	Vodafone	Channel 1478
920.7625	875.7625	NFM	Nationwide	Vodafone	Channel 1479
920.7875	875.7875	NFM	Nationwide	Vodafone	Channel 1480
920.8125	875.8125	NFM	Nationwide	Vodafone	Channel 1481
920.8375	875.8375	NFM	Nationwide	Vodafone	Channel 1482
920.8625	875.8625	NFM	Nationwide	Vodafone	Channel 1483
920.8875	875.8875	NFM	Nationwide	Vodafone	Channel 1484
920.9125	875.9125	NFM	Nationwide	Vodafone	Channel 1485
920.9375	875.9375	NFM	Nationwide	Vodafone	Channel 1486
920.9625	875.9625	NFM	Nationwide	Vodafone	Channel 1487
920.9875	875.9875	NFM	Nationwide	Vodafone	Channel 1488
921.0125	876.0125	NFM	Nationwide	Vodafone	Channel 1489
921.0375	876.0375	NFM	Nationwide	Vodafone	Channel 1490
921.0625	876.0625	NFM	Nationwide	Vodafone	Channel 1491
921.0875	876.0875	NFM	Nationwide	Vodafone	Channel 1492
921.1125	876.1125	NFM	Nationwide	Vodafone	Channel 1493
921.1375	876.1375	NFM	Nationwide	Vodafone	Channel 1494
921.1625	876.1625	NFM	Nationwide	Vodafone	Channel 1495
921.1875	876.1875	NFM	Nationwide	Vodafone	Channel 1496
921.2125	876.2125	NFM	Nationwide	Vodafone	Channel 1497
921.2375	876.2375	NFM	Nationwide	Vodafone	Channel 1498
921.2625	876.2625	NFM	Nationwide	Vodafone	Channel 1499
921.2875	876.2875	NFM	Nationwide	Vodafone	Channel 1500
921.3125	876.3125	NFM	Nationwide	Vodafone	Channel 1501
921.3375	876.3375	NFM	Nationwide	Vodafone	Channel 1502
921.3625	876.3625	NFM	Nationwide	Vodafone	Channel 1503
921.3875	876.3875	NFM	Nationwide	Vodafone	Channel 1504
921.4125	876.4125	NFM	Nationwide	Vodafone	Channel 1505
921.4375	876.4375	NFM	Nationwide	Vodafone	Channel 1506
921.4625	876.4625	NFM	Nationwide	Vodafone	Channel 1507
921.4875	876.4875	NFM	Nationwide	Vodafone	Channel 1508
921.5125	876.5125	NFM	Nationwide	Vodafone	Channel 1509
921.5375	876.5375	NFM	Nationwide	Vodafone	Channel 1510
921.5625	876.5625	NFM	Nationwide	Vodafone	Channel 1511
921.5875	876.5875	NFM	Nationwide	Vodafone	Channel 1512
921.6125	876.6125	NFM	Nationwide	Vodafone	Channel 1513
921.6375	876.6375	NFM	Nationwide	Vodafone	Channel 1514
921.6625	876.6625	NFM	Nationwide	Vodafone	Channel 1515
921.6875	876.6875	NFM	Nationwide	Vodafone	Channel 1516
921.7125	876.7125	NFM	Nationwide	Vodafone	Channel 1517

Base	Mobile	Mode	Location	User & Notes	
921.7375	876.7375	NFM	Nationwide	Vodafone	Channel 1518
921.7625	876.7625	NFM	Nationwide	Vodafone	Channel 1519
921.7875	876.7875	NFM	Nationwide	Vodafone	Channel 1520
921.8125	876.8125	NFM	Nationwide	Vodafone	Channel 1521
921.8375	876.8375	NFM	Nationwide	Vodafone	Channel 1522
921.8625	876.8625	NFM	Nationwide	Vodafone	Channel 1523
921.8875	876.8875	NFM	Nationwide	Vodafone	Channel 1524
921.9125	876.9125	NFM	Nationwide	Vodafone	Channel 1525
921.9375	876.9375	NFM	Nationwide	Vodafone	Channel 1526
921.9625	876.9625	NFM	Nationwide	Vodafone	Channel 1527
921.9875	876.9875	NFM	Nationwide	Vodafone	Channel 1528
922.0125	877.0125	NFM	Nationwide	Vodafone	Channel 1529
922.0375	877.0375	NFM	Nationwide	Vodafone	Channel 1530
922.0625	877.0625	NFM	Nationwide	Vodafone	Channel 1531
922.0875	877.0875	NFM	Nationwide	Vodafone	Channel 1532
922.1125	877.1125	NFM	Nationwide	Vodafone	Channel 1533
922.1375	877.1375	NFM	Nationwide	Vodafone	Channel 1534
922.1625	877.1625	NFM	Nationwide	Vodafone	Channel 1535
922.1875	877.1875	NFM	Nationwide	Vodafone	Channel 1536
922.2125	877.2125	NFM	Nationwide	Vodafone	Channel 1537
922.2375	877.2375	NFM	Nationwide	Vodafone	Channel 1538
922.2625	877.2625	NFM	Nationwide	Vodafone	Channel 1539
922.2875	877.2875	NFM	Nationwide	Vodafone	Channel 1540
922.3125	877.3125	NFM	Nationwide	Vodafone	Channel 1541
922.3375	877.3375	NFM	Nationwide	Vodafone	Channel 1542
922.3625	877.3625	NFM	Nationwide	Vodafone	Channel 1543
922.3875	877.3875	NFM	Nationwide	Vodafone	Channel 1544
922.4125	877.4125	NFM	Nationwide	Vodafone	Channel 1545
922.4375	877.4375	NFM	Nationwide	Vodafone	Channel 1546
922.4625	877.4625	NFM	Nationwide	Vodafone	Channel 1547
922.4875	877.4875	NFM	Nationwide	Vodafone	Channel 1548
922.5125	877.5125	NFM	Nationwide	Vodafone	Channel 1549
922.5375	877.5375	NFM	Nationwide	Vodafone	Channel 1550
922.5625	877.5625	NFM	Nationwide	Vodafone	Channel 1551
922.5875	877.5875	NFM	Nationwide	Vodafone	Channel 1552
922.6125	877.6125	NFM	Nationwide	Vodafone	Channel 1553
922.6375	877.6375	NFM	Nationwide	Vodafone	Channel 1554
922.6625	877.6625	NFM	Nationwide	Vodafone	Channel 1555
922.6875	877.6875	NFM	Nationwide	Vodafone	Channel 1556
922.7125	877.7125	NFM	Nationwide	Vodafone	Channel 1557
922.7375	877.7375	NFM	Nationwide	Vodafone	Channel 1558
922.7625	877.7625	NFM	Nationwide	Vodafone	Channel 1559
922.7875	877.7875	NFM	Nationwide	Vodafone	Channel 1560
922.8125	877.8125	NFM	Nationwide	Vodafone	Channel 1561
922.8375	877.8375	NFM	Nationwide	Vodafone	Channel 1562
922.8625	877.8625	NFM	Nationwide	Vodafone	Channel 1563
922.8875	877.8875	NFM	Nationwide	Vodafone	Channel 1564
922.9125	877.9125	NFM	Nationwide	Vodafone	Channel 1565
922.9375	877.9375	NFM	Nationwide	Vodafone	Channel 1566
922.9625	877.9625	NFM	Nationwide	Vodafone	Channel 1567

Base	Mobile	Mode	Location	User & Notes	
922.9875	877.9875	NFM	Nationwide	Vodafone	Channel 1568
923.0125	878.0125	NFM	Nationwide	Vodafone	Channel 1569
923.0375	878.0375	NFM	Nationwide	Vodafone	Channel 1570
923.0625	878.0625	NFM	Nationwide	Vodafone	Channel 1571
923.0875	878.0875	NFM	Nationwide	Vodafone	Channel 1572
923.1125	878.1125	NFM	Nationwide	Vodafone	Channel 1573
923.1375	878.1375	NFM	Nationwide	Vodafone	Channel 1574
923.1625	878.1625	NFM	Nationwide	Vodafone	Channel 1575
923.1875	878.1875	NFM	Nationwide	Vodafone	Channel 1576
923.2125	878.2125	NFM	Nationwide	Vodafone	Channel 1577
923.2375	878.2375	NFM	Nationwide	Vodafone	Channel 1578
923.2625	878.2625	NFM	Nationwide	Vodafone	Channel 1579
923.2875	878.2875	NFM	Nationwide	Vodafone	Channel 1580
923.3125	878.3125	NFM	Nationwide	Vodafone	Channel 1581
923.3375	878.3375	NFM	Nationwide	Vodafone	Channel 1582
923.3625	878.3625	NFM	Nationwide	Vodafone	Channel 1583
923.3875	878.3875	NFM	Nationwide	Vodafone	Channel 1584
923.4125	878.4125	NFM	Nationwide	Vodafone	Channel 1585
923.4375	878.4375	NFM	Nationwide	Vodafone	Channel 1586
923.4625	878.4625	NFM	Nationwide	Vodafone	Channel 1587
923.4875	878.4875	NFM	Nationwide	Vodafone	Channel 1588
923.5125	878.5125	NFM	Nationwide	Vodafone	Channel 1589
923.5375	878.5375	NFM	Nationwide	Vodafone	Channel 1590
923.5625	878.5625	NFM	Nationwide	Vodafone	Channel 1591
923.5875	878.5875	NFM	Nationwide	Vodafone	Channel 1592
923.6125	878.6125	NFM	Nationwide	Vodafone	Channel 1593
923.6375	878.6375	NFM	Nationwide	Vodafone	Channel 1594
923.6625	878.6625	NFM	Nationwide	Vodafone	Channel 1595
923.6875	878.6875	NFM	Nationwide	Vodafone	Channel 1596
923.7125	878.7125	NFM	Nationwide	Vodafone	Channel 1597
923.7375	878.7375	NFM	Nationwide	Vodafone	Channel 1598
923.7625	878.7625	NFM	Nationwide	Vodafone	Channel 1599
923.7875	878.7875	NFM	Nationwide	Vodafone	Channel 1600
923.8125	878.8125	NFM	Nationwide	Vodafone	Channel 1601
923.8375	878.8375	NFM	Nationwide	Vodafone	Channel 1602
923.8625	878.8625	NFM	Nationwide	Vodafone	Channel 1603
923.8875	878.8875	NFM	Nationwide	Vodafone	Channel 1604
923.9125	878.9125	NFM	Nationwide	Vodafone	Channel 1605
923.9375	878.9375	NFM	Nationwide	Vodafone	Channel 1606
923.9625	878.9625	NFM	Nationwide	Vodafone	Channel 1607
923.9875	878.9875	NFM	Nationwide	Vodafone	Channel 1608
924.0125	879.0125	NFM	Nationwide	Vodafone	Channel 1609
924.0375	879.0375	NFM	Nationwide	Vodafone	Channel 1610
924.0625	879.0625	NFM	Nationwide	Vodafone	Channel 1611
924.0875	879.0875	NFM	Nationwide	Vodafone	Channel 1612
924.1125	879.1125	NFM	Nationwide	Vodafone	Channel 1613
924.1375	879.1375	NFM	Nationwide	Vodafone	Channel 1614
924.1625	879.1625	NFM	Nationwide	Vodafone	Channel 1615
924.1875	879.1875	NFM	Nationwide	Vodafone	Channel 1616
924.2125	879.2125	NFM	Nationwide	Vodafone	Channel 1617

Base	Mobile	Mode	Location	User & Notes	
924.2375	879.2375	NFM	Nationwide	Vodafone	Channel 1618
924.2625	879.2625	NFM	Nationwide	Vodafone	Channel 1619
924.2875	879.2875	NFM	Nationwide	Vodafone	Channel 1620
924.3125	879.3125	NFM	Nationwide	Vodafone	Channel 1621
924.3375	879.3375	NFM	Nationwide	Vodafone	Channel 1622
924.3625	879.3625	NFM	Nationwide	Vodafone	Channel 1623
924.3875	879.3875	NFM	Nationwide	Vodafone	Channel 1624
924.4125	879.4125	NFM	Nationwide	Vodafone	Channel 1625
924.4375	879.4375	NFM	Nationwide	Vodafone	Channel 1626
924.4625	879.4625	NFM	Nationwide	Vodafone	Channel 1627
924.4875	879.4875	NFM	Nationwide	Vodafone	Channel 1628
924.5125	879.5125	NFM	Nationwide	Vodafone	Channel 1629
924.5375	879.5375	NFM	Nationwide	Vodafone	Channel 1630
924.5625	879.5625	NFM	Nationwide	Vodafone	Channel 1631
924.5875	879.5875	NFM	Nationwide	Vodafone	Channel 1632
924.6125	879.6125	NFM	Nationwide	Vodafone	Channel 1633
924.6375	879.6375	NFM	Nationwide	Vodafone	Channel 1634
924.6625	879.6625	NFM	Nationwide	Vodafone	Channel 1635
924.6875	879.6875	NFM	Nationwide	Vodafone	Channel 1636
924.7125	879.7125	NFM	Nationwide	Vodafone	Channel 1637
924.7375	879.7375	NFM	Nationwide	Vodafone	Channel 1638
924.7625	879.7625	NFM	Nationwide	Vodafone	Channel 1639
924.7875	879.7875	NFM	Nationwide	Vodafone	Channel 1640
924.8125	879.8125	NFM	Nationwide	Vodafone	Channel 1641
924.8375	879.8375	NFM	Nationwide	Vodafone	Channel 1642
924.8625	879.8625	NFM	Nationwide	Vodafone	Channel 1643
924.8875	879.8875	NFM	Nationwide	Vodafone	Channel 1644
924.9125	879.9125	NFM	Nationwide	Vodafone	Channel 1645
924.9375	879.9375	NFM	Nationwide	Vodafone	Channel 1646
924.9625	879.9625	NFM	Nationwide	Vodafone	Channel 1647
924.9875	879.9875	NFM	Nationwide	Vodafone	Channel 1648
925.0125	880.0125	NFM	Nationwide	Vodafone	Channel 1649
925.0375	880.0375	NFM	Nationwide	Vodafone	Channel 1650
925.0625	880.0625	NFM	Nationwide	Vodafone	Channel 1651
925.0875	880.0875	NFM	Nationwide	Vodafone	Channel 1652
925.1125	880.1125	NFM	Nationwide	Vodafone	Channel 1653
925.1375	880.1375	NFM	Nationwide	Vodafone	Channel 1654
925.1625	880.1625	NFM	Nationwide	Vodafone	Channel 1655
925.1875	880.1875	NFM	Nationwide	Vodafone	Channel 1656
925.2125	880.2125	NFM	Nationwide	Vodafone	Channel 1657
925.2375	880.2375	NFM	Nationwide	Vodafone	Channel 1658
925.2625	880.2625	NFM	Nationwide	Vodafone	Channel 1659
925.2875	880.2875	NFM	Nationwide	Vodafone	Channel 1660
925.3125	880.3125	NFM	Nationwide	Vodafone	Channel 1661
925.3375	880.3375	NFM	Nationwide	Vodafone	Channel 1662
925.3625	880.3625	NFM	Nationwide	Vodafone	Channel 1663
925.3875	880.3875	NFM	Nationwide	Vodafone	Channel 1664
925.4125	880.4125	NFM	Nationwide	Vodafone	Channel 1665
925.4375	880.4375	NFM	Nationwide	Vodafone	Channel 1666
925.4625	880.4625	NFM	Nationwide	Vodafone	Channel 1667

Base	Mobile	Mode	Location	User & Notes	
925.4875	880.4875	NFM	Nationwide	Vodafone	Channel 1668
925.5125	880.5125	NFM	Nationwide	Vodafone	Channel 1669
925.5375	880.5375	NFM	Nationwide	Vodafone	Channel 1670
925.5625	880.5625	NFM	Nationwide	Vodafone	Channel 1671
925.5875	880.5875	NFM	Nationwide	Vodafone	Channel 1672
925.6125	880.6125	NFM	Nationwide	Vodafone	Channel 1673
925.6375	880.6375	NFM	Nationwide	Vodafone	Channel 1674
925.6625	880.6625	NFM	Nationwide	Vodafone	Channel 1675
925.6875	880.6875	NFM	Nationwide	Vodafone	Channel 1676
925.7125	880.7125	NFM	Nationwide	Vodafone	Channel 1677
925.7375	880.7375	NFM	Nationwide	Vodafone	Channel 1678
925.7625	880.7625	NFM	Nationwide	Vodafone	Channel 1679
925.7875	880.7875	NFM	Nationwide	Vodafone	Channel 1680
925.8125	880.8125	NFM	Nationwide	Vodafone	Channel 1681
925.8375	880.8375	NFM	Nationwide	Vodafone	Channel 1682
925.8625	880.8625	NFM	Nationwide	Vodafone	Channel 1683
925.8875	880.8875	NFM	Nationwide	Vodafone	Channel 1684
925.9125	880.9125	NFM	Nationwide	Vodafone	Channel 1685
925.9375	880.9375	NFM	Nationwide	Vodafone	Channel 1686
925.9625	880.9625	NFM	Nationwide	Vodafone	Channel 1687
925.9875	880.9875	NFM	Nationwide	Cellnet	Channel 1688
926.0125	881.0125	NFM	Nationwide	Cellnet	Channel 1689
926.0375	881.0375	NFM	Nationwide	Cellnet	Channel 1690
926.0625	881.0625	NFM	Nationwide	Cellnet	Channel 1691
926.0875	881.0875	NFM	Nationwide	Cellnet	Channel 1692
926.1125	881.1125	NFM	Nationwide	Cellnet	Channel 1693
926.1375	881.1375	NFM	Nationwide	Cellnet	Channel 1694
926.1625	881.1625	NFM	Nationwide	Cellnet	Channel 1695
926.1875	881.1875	NFM	Nationwide	Cellnet	Channel 1696
926.2125	881.2125	NFM	Nationwide	Cellnet	Channel 1697
926.2375	881.2375	NFM	Nationwide	Cellnet	Channel 1698
926.2625	881.2625	NFM	Nationwide	Cellnet	Channel 1699
926.2875	881.2875	NFM	Nationwide	Cellnet	Channel 1700
926.3125	881.3125	NFM	Nationwide	Cellnet	Channel 1701
926.3375	881.3375	NFM	Nationwide	Cellnet	Channel 1702
926.3625	881.3625	NFM	Nationwide	Cellnet	Channel 1703
926.3875	881.3875	NFM	Nationwide	Cellnet	Channel 1704
926.4125	881.4125	NFM	Nationwide	Cellnet	Channel 1705
926.4375	881.4375	NFM	Nationwide	Cellnet	Channel 1706
926.4625	881.4625	NFM	Nationwide	Cellnet	Channel 1707
926.4875	881.4875	NFM	Nationwide	Cellnet	Channel 1708
926.5125	881.5125	NFM	Nationwide	Cellnet	Channel 1709
926.5375	881.5375	NFM	Nationwide	Cellnet	Channel 1710
926.5625	881.5625	NFM	Nationwide	Cellnet	Channel 1711
926.5875	881.5875	NFM	Nationwide	Cellnet	Channel 1712
926.6125	881.6125	NFM	Nationwide	Cellnet	Channel 1713
926.6375	881.6375	NFM	Nationwide	Cellnet	Channel 1714
926.6625	881.6625	NFM	Nationwide	Cellnet	Channel 1715
926.6875	881.6875	NFM	Nationwide	Cellnet	Channel 1716
926.7125	881.7125	NFM	Nationwide	Cellnet	Channel 1717

Base	Mobile	Mode	Location	User & Notes	
926.7375	881.7375	NFM	Nationwide	Cellnet	Channel 1718
926.7625	881.7625	NFM	Nationwide	Cellnet	Channel 1719
926.7875	881.7875	NFM	Nationwide	Cellnet	Channel 1720
926.8125	881.8125	NFM	Nationwide	Cellnet	Channel 1721
926.8375	881.8375	NFM	Nationwide	Cellnet	Channel 1722
926.8625	881.8625	NFM	Nationwide	Cellnet	Channel 1723
926.8875	881.8875	NFM	Nationwide	Cellnet	Channel 1724
926.9125	881.9125	NFM	Nationwide	Cellnet	Channel 1725
926.9375	881.9375	NFM	Nationwide	Cellnet	Channel 1726
926.9625	881.9625	NFM	Nationwide	Cellnet	Channel 1727
926.9875	881.9875	NFM	Nationwide	Cellnet	Channel 1728
927.0125	882.0125	NFM	Nationwide	Cellnet	Channel 1729
927.0375	882.0375	NFM	Nationwide	Cellnet	Channel 1730
927.0625	882.0625	NFM	Nationwide	Cellnet	Channel 1731
927.0875	882.0875	NFM	Nationwide	Cellnet	Channel 1732
927.1125	882.1125	NFM	Nationwide	Cellnet	Channel 1733
927.1375	882.1375	NFM	Nationwide	Cellnet	Channel 1734
927.1625	882.1625	NFM	Nationwide	Cellnet	Channel 1735
927.1875	882.1875	NFM	Nationwide	Cellnet	Channel 1736
927.2125	882.2125	NFM	Nationwide	Cellnet	Channel 1737
927.2375	882.2375	NFM	Nationwide	Cellnet	Channel 1738
927.2625	882.2625	NFM	Nationwide	Cellnet	Channel 1739
927.2875	882.2875	NFM	Nationwide	Cellnet	Channel 1740
927.3125	882.3125	NFM	Nationwide	Cellnet	Channel 1741
927.3375	882.3375	NFM	Nationwide	Cellnet	Channel 1742
927.3625	882.3625	NFM	Nationwide	Cellnet	Channel 1743
927.3875	882.3875	NFM	Nationwide	Cellnet	Channel 1744
927.4125	882.4125	NFM	Nationwide	Cellnet	Channel 1745
927.4375	882.4375	NFM	Nationwide	Cellnet	Channel 1746
927.4625	882.4625	NFM	Nationwide	Cellnet	Channel 1747
927.4875	882.4875	NFM	Nationwide	Cellnet	Channel 1748
927.5125	882.5125	NFM	Nationwide	Cellnet	Channel 1749
927.5375	882.5375	NFM	Nationwide	Cellnet	Channel 1750
927.5625	882.5625	NFM	Nationwide	Cellnet	Channel 1751
927.5875	882.5875	NFM	Nationwide	Cellnet	Channel 1752
927.6125	882.6125	NFM	Nationwide	Cellnet	Channel 1753
927.6375	882.6375	NFM	Nationwide	Cellnet	Channel 1754
927.6625	882.6625	NFM	Nationwide	Cellnet	Channel 1755
927.6875	882.6875	NFM	Nationwide	Cellnet	Channel 1756
927.7125	882.7125	NFM	Nationwide	Cellnet	Channel 1757
927.7375	882.7375	NFM	Nationwide	Cellnet	Channel 1758
927.7625	882.7625	NFM	Nationwide	Cellnet	Channel 1759
927.7875	882.7875	NFM	Nationwide	Cellnet	Channel 1760
927.8125	882.8125	NFM	Nationwide	Cellnet	Channel 1761
927.8375	882.8375	NFM	Nationwide	Cellnet	Channel 1762
927.8625	882.8625	NFM	Nationwide	Cellnet	Channel 1763
927.8875	882.8875	NFM	Nationwide	Cellnet	Channel 1764
927.9125	882.9125	NFM	Nationwide	Cellnet	Channel 1765
927.9375	882.9375	NFM	Nationwide	Cellnet	Channel 1766
927.9625	882.9625	NFM	Nationwide	Cellnet	Channel 1767

Base	Mobile	Mode	Location	User & Notes	
927.9875	882.9875	NFM	Nationwide	Cellnet	Channel 1768
928.0125	883.0125	NFM	Nationwide	Cellnet	Channel 1769
928.0375	883.0375	NFM	Nationwide	Cellnet	Channel 1770
928.0625	883.0625	NFM	Nationwide	Cellnet	Channel 1771
928.0875	883.0875	NFM	Nationwide	Cellnet	Channel 1772
928.1125	883.1125	NFM	Nationwide	Cellnet	Channel 1773
928.1375	883.1375	NFM	Nationwide	Cellnet	Channel 1774
928.1625	883.1625	NFM	Nationwide	Cellnet	Channel 1775
928.1875	883.1875	NFM	Nationwide	Cellnet	Channel 1776
928.2125	883.2125	NFM	Nationwide	Cellnet	Channel 1777
928.2375	883.2375	NFM	Nationwide	Cellnet	Channel 1778
928.2625	883.2625	NFM	Nationwide	Cellnet	Channel 1779
928.2875	883.2875	NFM	Nationwide	Cellnet	Channel 1780
928.3125	883.3125	NFM	Nationwide	Cellnet	Channel 1781
928.3375	883.3375	NFM	Nationwide	Cellnet	Channel 1782
928.3625	883.3625	NFM	Nationwide	Cellnet	Channel 1783
928.3875	883.3875	NFM	Nationwide	Cellnet	Channel 1784
928.4125	883.4125	NFM	Nationwide	Cellnet	Channel 1785
928.4375	883.4375	NFM	Nationwide	Cellnet	Channel 1786
928.4625	883.4625	NFM	Nationwide	Cellnet	Channel 1787
928.4875	883.4875	NFM	Nationwide	Cellnet	Channel 1788
928.5125	883.5125	NFM	Nationwide	Cellnet	Channel 1789
928.5375	883.5375	NFM	Nationwide	Cellnet	Channel 1790
928.5625	883.5625	NFM	Nationwide	Cellnet	Channel 1791
928.5875	883.5875	NFM	Nationwide	Cellnet	Channel 1792
928.6125	883.6125	NFM	Nationwide	Cellnet	Channel 1793
928.6375	883.6375	NFM	Nationwide	Cellnet	Channel 1794
928.6625	883.6625	NFM	Nationwide	Cellnet	Channel 1795
928.6875	883.6875	NFM	Nationwide	Cellnet	Channel 1796
928.7125	883.7125	NFM	Nationwide	Cellnet	Channel 1797
928.7375	883.7375	NFM	Nationwide	Cellnet	Channel 1798
928.7625	883.7625	NFM	Nationwide	Cellnet	Channel 1799
928.7875	883.7875	NFM	Nationwide	Cellnet	Channel 1800
928.8125	883.8125	NFM	Nationwide	Cellnet	Channel 1801
928.8375	883.8375	NFM	Nationwide	Cellnet	Channel 1802
928.8625	883.8625	NFM	Nationwide	Cellnet	Channel 1803
928.8875	883.8875	NFM	Nationwide	Cellnet	Channel 1804
928.9125	883.9125	NFM	Nationwide	Cellnet	Channel 1805
928.9375	883.9375	NFM	Nationwide	Cellnet	Channel 1806
928.9625	883.9625	NFM	Nationwide	Cellnet	Channel 1807
928.9875	883.9875	NFM	Nationwide	Cellnet	Channel 1808
929.0125	884.0125	NFM	Nationwide	Cellnet	Channel 1809
929.0375	884.0375	NFM	Nationwide	Cellnet	Channel 1810
929.0625	884.0625	NFM	Nationwide	Cellnet	Channel 1811
929.0875	884.0875	NFM	Nationwide	Cellnet	Channel 1812
929.1125	884.1125	NFM	Nationwide	Cellnet	Channel 1813
929.1375	884.1375	NFM	Nationwide	Cellnet	Channel 1814
929.1625	884.1625	NFM	Nationwide	Cellnet	Channel 1815
929.1875	884.1875	NFM	Nationwide	Cellnet	Channel 1816
929.2125	884.2125	NFM	Nationwide	Cellnet	Channel 1817

Base	Mobile	Mode	Location	User & Notes	
929.2375	884.2375	NFM	Nationwide	Cellnet	Channel 1818
929.2625	884.2625	NFM	Nationwide	Cellnet	Channel 1819
929.2875	884.2875	NFM	Nationwide	Cellnet	Channel 1820
929.3125	884.3125	NFM	Nationwide	Cellnet	Channel 1821
929.3375	884.3375	NFM	Nationwide	Cellnet	Channel 1822
929.3625	884.3625	NFM	Nationwide	Cellnet	Channel 1823
929.3875	884.3875	NFM	Nationwide	Cellnet	Channel 1824
929.4125	884.4125	NFM	Nationwide	Cellnet	Channel 1825
929.4375	884.4375	NFM	Nationwide	Cellnet	Channel 1826
929.4625	884.4625	NFM	Nationwide	Cellnet	Channel 1827
929.4875	884.4875	NFM	Nationwide	Cellnet	Channel 1828
929.5125	884.5125	NFM	Nationwide	Cellnet	Channel 1829
929.5375	884.5375	NFM	Nationwide	Cellnet	Channel 1830
929.5625	884.5625	NFM	Nationwide	Cellnet	Channel 1831
929.5875	884.5875	NFM	Nationwide	Cellnet	Channel 1832
929.6125	884.6125	NFM	Nationwide	Cellnet	Channel 1833
929.6375	884.6375	NFM	Nationwide	Cellnet	Channel 1834
929.6625	884.6625	NFM	Nationwide	Cellnet	Channel 1835
929.6875	884.6875	NFM	Nationwide	Cellnet	Channel 1836
929.7125	884.7125	NFM	Nationwide	Cellnet	Channel 1837
929.7375	884.7375	NFM	Nationwide	Cellnet	Channel 1838
929.7625	884.7625	NFM	Nationwide	Cellnet	Channel 1839
929.7875	884.7875	NFM	Nationwide	Cellnet	Channel 1840
929.8125	884.8125	NFM	Nationwide	Cellnet	Channel 1841
929.8375	884.8375	NFM	Nationwide	Cellnet	Channel 1842
929.8625	884.8625	NFM	Nationwide	Cellnet	Channel 1843
929.8875	884.8875	NFM	Nationwide	Cellnet	Channel 1844
929.9125	884.9125	NFM	Nationwide	Cellnet	Channel 1845
929.9375	884.9375	NFM	Nationwide	Cellnet	Channel 1846
929.9625	884.9625	NFM	Nationwide	Cellnet	Channel 1847
929.9875	884.9875	NFM	Nationwide	Cellnet	Channel 1848
930.0125	885.0125	NFM	Nationwide	Cellnet	Channel 1849
930.0375	885.0375	NFM	Nationwide	Cellnet	Channel 1850
930.0625	885.0625	NFM	Nationwide	Cellnet	Channel 1851
930.0875	885.0875	NFM	Nationwide	Cellnet	Channel 1852
930.1125	885.1125	NFM	Nationwide	Cellnet	Channel 1853
930.1375	885.1375	NFM	Nationwide	Cellnet	Channel 1854
930.1625	885.1625	NFM	Nationwide	Cellnet	Channel 1855
930.1875	885.1875	NFM	Nationwide	Cellnet	Channel 1856
930.2125	885.2125	NFM	Nationwide	Cellnet	Channel 1857
930.2375	885.2375	NFM	Nationwide	Cellnet	Channel 1858
930.2625	885.2625	NFM	Nationwide	Cellnet	Channel 1859
930.2875	885.2875	NFM	Nationwide	Cellnet	Channel 1860
930.3125	885.3125	NFM	Nationwide	Cellnet	Channel 1861
930.3375	885.3375	NFM	Nationwide	Cellnet	Channel 1862
930.3625	885.3625	NFM	Nationwide	Cellnet	Channel 1863
930.3875	885.3875	NFM	Nationwide	Cellnet	Channel 1864
930.4125	885.4125	NFM	Nationwide	Cellnet	Channel 1865
930.4375	885.4375	NFM	Nationwide	Cellnet	Channel 1866
930.4625	885.4625	NFM	Nationwide	Cellnet	Channel 1867

The UK Scanning Directory

Base	Mobile	Mode	Location	User & Notes	
930.4875	885.4875	NFM	Nationwide	Cellnet	Channel 1868
930.5125	885.5125	NFM	Nationwide	Cellnet	Channel 1869
930.5375	885.5375	NFM	Nationwide	Cellnet	Channel 1870
930.5625	885.5625	NFM	Nationwide	Cellnet	Channel 1871
930.5875	885.5875	NFM	Nationwide	Cellnet	Channel 1872
930.6125	885.6125	NFM	Nationwide	Cellnet	Channel 1873
930.6375	885.6375	NFM	Nationwide	Cellnet	Channel 1874
930.6625	885.6625	NFM	Nationwide	Cellnet	Channel 1875
930.6875	885.6875	NFM	Nationwide	Cellnet	Channel 1876
930.7125	885.7125	NFM	Nationwide	Cellnet	Channel 1877
930.7375	885.7375	NFM	Nationwide	Cellnet	Channel 1878
930.7625	885.7625	NFM	Nationwide	Cellnet	Channel 1879
930.7875	885.7875	NFM	Nationwide	Cellnet	Channel 1880
930.8125	885.8125	NFM	Nationwide	Cellnet	Channel 1881
930.8375	885.8375	NFM	Nationwide	Cellnet	Channel 1882
930.8625	885.8625	NFM	Nationwide	Cellnet	Channel 1883
930.8875	885.8875	NFM	Nationwide	Cellnet	Channel 1884
930.9125	885.9125	NFM	Nationwide	Cellnet	Channel 1885
930.9375	885.9375	NFM	Nationwide	Cellnet	Channel 1886
930.9625	885.9625	NFM	Nationwide	Cellnet	Channel 1887
930.9875	885.9875	NFM	Nationwide	Cellnet	Channel 1888
931.0125	886.0125	NFM	Nationwide	Cellnet	Channel 1889
931.0375	886.0375	NFM	Nationwide	Cellnet	Channel 1890
931.0625	886.0625	NFM	Nationwide	Cellnet	Channel 1891
931.0875	886.0875	NFM	Nationwide	Cellnet	Channel 1892
931.1125	886.1125	NFM	Nationwide	Cellnet	Channel 1893
931.1375	886.1375	NFM	Nationwide	Cellnet	Channel 1894
931.1625	886.1625	NFM	Nationwide	Cellnet	Channel 1895
931.1875	886.1875	NFM	Nationwide	Cellnet	Channel 1896
931.2125	886.2125	NFM	Nationwide	Cellnet	Channel 1897
931.2375	886.2375	NFM	Nationwide	Cellnet	Channel 1898
931.2625	886.2625	NFM	Nationwide	Cellnet	Channel 1899
931.2875	886.2875	NFM	Nationwide	Cellnet	Channel 1900
931.3125	886.3125	NFM	Nationwide	Cellnet	Channel 1901
931.3375	886.3375	NFM	Nationwide	Cellnet	Channel 1902
931.3625	886.3625	NFM	Nationwide	Cellnet	Channel 1903
931.3875	886.3875	NFM	Nationwide	Cellnet	Channel 1904
931.4125	886.4125	NFM	Nationwide	Cellnet	Channel 1905
931.4375	886.4375	NFM	Nationwide	Cellnet	Channel 1906
931.4625	886.4625	NFM	Nationwide	Cellnet	Channel 1907
931.4875	886.4875	NFM	Nationwide	Cellnet	Channel 1908
931.5125	886.5125	NFM	Nationwide	Cellnet	Channel 1909
931.5375	886.5375	NFM	Nationwide	Cellnet	Channel 1910
931.5625	886.5625	NFM	Nationwide	Cellnet	Channel 1911
931.5875	886.5875	NFM	Nationwide	Cellnet	Channel 1912
931.6125	886.6125	NFM	Nationwide	Cellnet	Channel 1913
931.6375	886.6375	NFM	Nationwide	Cellnet	Channel 1914
931.6625	886.6625	NFM	Nationwide	Cellnet	Channel 1915
931.6875	886.6875	NFM	Nationwide	Cellnet	Channel 1916
931.7125	886.7125	NFM	Nationwide	Cellnet	Channel 1917

Base	Mobile	Mode	Location	User & Notes	
931.7375	886.7375	NFM	Nationwide	Cellnet	Channel 1918
931.7625	886.7625	NFM	Nationwide	Cellnet	Channel 1919
931.7875	886.7875	NFM	Nationwide	Cellnet	Channel 1920
931.8125	886.8125	NFM	Nationwide	Cellnet	Channel 1921
931.8375	886.8375	NFM	Nationwide	Cellnet	Channel 1922
931.8625	886.8625	NFM	Nationwide	Cellnet	Channel 1923
931.8875	886.8875	NFM	Nationwide	Cellnet	Channel 1924
931.9125	886.9125	NFM	Nationwide	Cellnet	Channel 1925
931.9375	886.9375	NFM	Nationwide	Cellnet	Channel 1926
931.9625	886.9625	NFM	Nationwide	Cellnet	Channel 1927
931.9875	886.9875	NFM	Nationwide	Cellnet	Channel 1928
932.0125	887.0125	NFM	Nationwide	Cellnet	Channel 1929
932.0375	887.0375	NFM	Nationwide	Cellnet	Channel 1930
932.0625	887.0625	NFM	Nationwide	Cellnet	Channel 1931
932.0875	887.0875	NFM	Nationwide	Cellnet	Channel 1932
932.1125	887.1125	NFM	Nationwide	Cellnet	Channel 1933
932.1375	887.1375	NFM	Nationwide	Cellnet	Channel 1934
932.1625	887.1625	NFM	Nationwide	Cellnet	Channel 1935
932.1875	887.1875	NFM	Nationwide	Cellnet	Channel 1936
932.2125	887.2125	NFM	Nationwide	Cellnet	Channel 1937
932.2375	887.2375	NFM	Nationwide	Cellnet	Channel 1938
932.2625	887.2625	NFM	Nationwide	Cellnet	Channel 1939
932.2875	887.2875	NFM	Nationwide	Cellnet	Channel 1940
932.3125	887.3125	NFM	Nationwide	Cellnet	Channel 1941
932.3375	887.3375	NFM	Nationwide	Cellnet	Channel 1942
932.3625	887.3625	NFM	Nationwide	Cellnet	Channel 1943
932.3875	887.3875	NFM	Nationwide	Cellnet	Channel 1944
932.4125	887.4125	NFM	Nationwide	Cellnet	Channel 1945
932.4375	887.4375	NFM	Nationwide	Cellnet	Channel 1946
932.4625	887.4625	NFM	Nationwide	Cellnet	Channel 1947
932.4875	887.4875	NFM	Nationwide	Cellnet	Channel 1948
932.5125	887.5125	NFM	Nationwide	Cellnet	Channel 1949
932.5375	887.5375	NFM	Nationwide	Cellnet	Channel 1950
932.5625	887.5625	NFM	Nationwide	Cellnet	Channel 1951
932.5875	887.5875	NFM	Nationwide	Cellnet	Channel 1952
932.6125	887.6125	NFM	Nationwide	Cellnet	Channel 1953
932.6375	887.6375	NFM	Nationwide	Cellnet	Channel 1954
932.6625	887.6625	NFM	Nationwide	Cellnet	Channel 1955
932.6875	887.6875	NFM	Nationwide	Cellnet	Channel 1956
932.7125	887.7125	NFM	Nationwide	Cellnet	Channel 1957
932.7375	887.7375	NFM	Nationwide	Cellnet	Channel 1958
932.7625	887.7625	NFM	Nationwide	Cellnet	Channel 1959
932.7875	887.7875	NFM	Nationwide	Cellnet	Channel 1960
932.8125	887.8125	NFM	Nationwide	Cellnet	Channel 1961
932.8375	887.8375	NFM	Nationwide	Cellnet	Channel 1962
932.8625	887.8625	NFM	Nationwide	Cellnet	Channel 1963
932.8875	887.8875	NFM	Nationwide	Cellnet	Channel 1964
932.9125	887.9125	NFM	Nationwide	Cellnet	Channel 1965
932.9375	887.9375	NFM	Nationwide	Cellnet	Channel 1966
932.9625	887.9625	NFM	Nationwide	Cellnet	Channel 1967

Base	Mobile	Mode	Location	User & Notes	
932.9875	887.9875	NFM	Nationwide	Cellnet	Channel 1968
933.0125	888.0125	NFM	Nationwide	Cellnet	Channel 1969
933.0375	888.0375	NFM	Nationwide	Cellnet	Channel 1970
933.0625	888.0625	NFM	Nationwide	Cellnet	Channel 1971
933.0875	888.0875	NFM	Nationwide	Cellnet	Channel 1972
933.1125	888.1125	NFM	Nationwide	Cellnet	Channel 1973
933.1375	888.1375	NFM	Nationwide	Cellnet	Channel 1974
933.1625	888.1625	NFM	Nationwide	Cellnet	Channel 1975
933.1875	888.1875	NFM	Nationwide	Cellnet	Channel 1976
933.2125	888.2125	NFM	Nationwide	Cellnet	Channel 1977
933.2375	888.2375	NFM	Nationwide	Cellnet	Channel 1978
933.2625	888.2625	NFM	Nationwide	Cellnet	Channel 1979
933.2875	888.2875	NFM	Nationwide	Cellnet	Channel 1980
933.3125	888.3125	NFM	Nationwide	Cellnet	Channel 1981
933.3375	888.3375	NFM	Nationwide	Cellnet	Channel 1982
933.3625	888.3625	NFM	Nationwide	Cellnet	Channel 1983
933.3875	888.3875	NFM	Nationwide	Cellnet	Channel 1984
933.4125	888.4125	NFM	Nationwide	Cellnet	Channel 1985
933.4375	888.4375	NFM	Nationwide	Cellnet	Channel 1986
933.4625	888.4625	NFM	Nationwide	Cellnet	Channel 1987
933.4875	888.4875	NFM	Nationwide	Cellnet	Channel 1988
933.5125	888.5125	NFM	Nationwide	Cellnet	Channel 1989
933.5375	888.5375	NFM	Nationwide	Cellnet	Channel 1990
933.5625	888.5625	NFM	Nationwide	Cellnet	Channel 1991
933.5875	888.5875	NFM	Nationwide	Cellnet	Channel 1992
933.6125	888.6125	NFM	Nationwide	Cellnet	Channel 1993
933.6375	888.6375	NFM	Nationwide	Cellnet	Channel 1994
933.6625	888.6625	NFM	Nationwide	Cellnet	Channel 1995
933.6875	888.6875	NFM	Nationwide	Cellnet	Channel 1996
933.7125	888.7125	NFM	Nationwide	Cellnet	Channel 1997
933.7375	888.7375	NFM	Nationwide	Cellnet	Channel 1998
933.7625	888.7625	NFM	Nationwide	Cellnet	Channel 1999
933.7875	888.7875	NFM	Nationwide	Cellnet	Channel 2000
933.8125	888.8125	NFM	Nationwide	Cellnet	Channel 2001
933.8375	888.8375	NFM	Nationwide	Cellnet	Channel 2002
933.8625	888.8625	NFM	Nationwide	Cellnet	Channel 2003
933.8875	888.8875	NFM	Nationwide	Cellnet	Channel 2004
933.9125	888.9125	NFM	Nationwide	Cellnet	Channel 2005
933.9375	888.9375	NFM	Nationwide	Cellnet	Channel 2006
933.9625	888.9625	NFM	Nationwide	Cellnet	Channel 2007
933.9875	888.9875	NFM	Nationwide	Cellnet	Channel 2008
934.0125	889.0125	NFM	Nationwide	Cellnet	Channel 2009
934.0375	889.0375	NFM	Nationwide	Cellnet	Channel 2010
934.0625	889.0625	NFM	Nationwide	Cellnet	Channel 2011
934.0875	889.0875	NFM	Nationwide	Cellnet	Channel 2012
934.1125	889.1125	NFM	Nationwide	Cellnet	Channel 2013
934.1375	889.1375	NFM	Nationwide	Cellnet	Channel 2014
934.1625	889.1625	NFM	Nationwide	Cellnet	Channel 2015
934.1875	889.1875	NFM	Nationwide	Cellnet	Channel 2016
934.2125	889.2125	NFM	Nationwide	Cellnet	Channel 2017

Base	Mobile	Mode	Location	User & Notes	
934.2375	889.2375	NFM	Nationwide	Cellnet	Channel 2018
934.2625	889.2625	NFM	Nationwide	Cellnet	Channel 2019
934.2875	889.2875	NFM	Nationwide	Cellnet	Channel 2020
934.3125	889.3125	NFM	Nationwide	Cellnet	Channel 2021
934.3375	889.3375	NFM	Nationwide	Cellnet	Channel 2022
934.3625	889.3625	NFM	Nationwide	Cellnet	Channel 2023
934.3875	889.3875	NFM	Nationwide	Cellnet	Channel 2024
934.4125	889.4125	NFM	Nationwide	Cellnet	Channel 2025
934.4375	889.4375	NFM	Nationwide	Cellnet	Channel 2026
934.4625	889.4625	NFM	Nationwide	Cellnet	Channel 2027
934.4875	889.4875	NFM	Nationwide	Cellnet	Channel 2028
934.5125	889.5125	NFM	Nationwide	Cellnet	Channel 2029
934.5375	889.5375	NFM	Nationwide	Cellnet	Channel 2030
934.5625	889.5625	NFM	Nationwide	Cellnet	Channel 2031
934.5875	889.5875	NFM	Nationwide	Cellnet	Channel 2032
934.6125	889.6125	NFM	Nationwide	Cellnet	Channel 2033
934.6375	889.6375	NFM	Nationwide	Cellnet	Channel 2034
934.6625	889.6625	NFM	Nationwide	Cellnet	Channel 2035
934.6875	889.6875	NFM	Nationwide	Cellnet	Channel 2036
934.7125	889.7125	NFM	Nationwide	Cellnet	Channel 2037
934.7375	889.7375	NFM	Nationwide	Cellnet	Channel 2038
934.7625	889.7625	NFM	Nationwide	Cellnet	Channel 2039
934.7875	889.7875	NFM	Nationwide	Cellnet	Channel 2040
934.8125	889.8125	NFM	Nationwide	Cellnet	Channel 2041
934.8375	889.8375	NFM	Nationwide	Cellnet	Channel 2042
934.8625	889.8625	NFM	Nationwide	Cellnet	Channel 2043
934.8875	889.8875	NFM	Nationwide	Cellnet	Channel 2044
934.9125	889.9125	NFM	Nationwide	Cellnet	Channel 2045
934.9375	889.9375	NFM	Nationwide	Cellnet	Channel 2046
934.9625	889.9625	NFM	Nationwide	Cellnet	Channel 2047
934.9875	889.9875	NFM	Nationwide	(Not Used)	Channel 0
935.0125	890.0125	NFM	Nationwide	Vodafone	Channel 1
935.0375	890.0375	NFM	Nationwide	Vodafone	Channel 2
935.0625	890.0625	NFM	Nationwide	Vodafone	Channel 3
935.0875	890.0875	NFM	Nationwide	Vodafone	Channel 4
935.1125	890.1125	NFM	Nationwide	Vodafone	Channel 5
935.1375	890.1375	NFM	Nationwide	Vodafone	Channel 6
935.1625	890.1625	NFM	Nationwide	Vodafone	Channel 7
935.1875	890.1875	NFM	Nationwide	Vodafone	Channel 8
935.2125	890.2125	NFM	Nationwide	Vodafone	Channel 9
935.2375	890.2375	NFM	Nationwide	Vodafone	Channel 10
935.2625	890.2625	NFM	Nationwide	Vodafone	Channel 11
935.2875	890.2875	NFM	Nationwide	Vodafone	Channel 12
935.3125	890.3125	NFM	Nationwide	Vodafone	Channel 13
935.3375	890.3375	NFM	Nationwide	Vodafone	Channel 14
935.3625	890.3625	NFM	Nationwide	Vodafone	Channel 15
935.3875	890.3875	NFM	Nationwide	Vodafone	Channel 16
935.4125	890.4125	NFM	Nationwide	Vodafone	Channel 17
935.4375	890.4375	NFM	Nationwide	Vodafone	Channel 18
935.4625	890.4625	NFM	Nationwide	Vodafone	Channel 19

Base	Mobile	Mode	Location	User & Notes	
935.4875	890.4875	NFM	Nationwide	Vodafone	Channel 20
935.5125	890.5125	NFM	Nationwide	Vodafone	Channel 21
935.5375	890.5375	NFM	Nationwide	Vodafone	Channel 22
935.5625	890.5625	NFM	Nationwide	Vodafone	Channel 23
		NFM	Nationwide	Vodafone	Data Control
935.5875	890.5875	NFM	Nationwide	Vodafone	Channel 24
		NFM	Nationwide	Vodafone	Data Control
935.6125	890.6125	NFM	Nationwide	Vodafone	Channel 25
		NFM	Nationwide	Vodafone	Data Control
935.6375	890.6375	NFM	Nationwide	Vodafone	Channel 26
		NFM	Nationwide	Vodafone	Data Control
935.6625	890.6625	NFM	Nationwide	Vodafone	Channel 27
		NFM	Nationwide	Vodafone	Data Control
935.6875	890.6875	NFM	Nationwide	Vodafone	Channel 28
		NFM	Nationwide	Vodafone	Data Control
935.7125	890.7125	NFM	Nationwide	Vodafone	Channel 29
		NFM	Nationwide	Vodafone	Data Control
935.7375	890.7375	NFM	Nationwide	Vodafone	Channel 30
		NFM	Nationwide	Vodafone	Data Control
935.7625	890.7625	NFM	Nationwide	Vodafone	Channel 31
		NFM	Perth	Vodafone	Data Control
935.7875	890.7875	NFM	Nationwide	Vodafone	Channel 32
		NFM	Nationwide	Vodafone	Data Control
935.8125	890.8125	NFM	Nationwide	Vodafone	Channel 33
		NFM	Nationwide	Vodafone	Data Control
935.8375	890.8375	NFM	Nationwide	Vodafone	Channel 34
		NFM	Nationwide	Vodafone	Data Control
935.8625	890.8625	NFM	Nationwide	Vodafone	Channel 35
		NFM	Nationwide	Vodafone	Data Control
935.8875	890.8875	NFM	Nationwide	Vodafone	Channel 36
		NFM	Nationwide	Vodafone	Data Control
935.9125	890.9125	NFM	Nationwide	Vodafone	Channel 37
		NFM	Nationwide	Vodafone	Data Control
935.9375	890.9375	NFM	Nationwide	Vodafone	Channel 38
		NFM	Nationwide	Vodafone	Data Control
935.9625	890.9625	NFM	Nationwide	Vodafone	Channel 39
		NFM	Nationwide	Vodafone	Data Control
935.9875	890.9875	NFM	Nationwide	Vodafone	Channel 40
		NFM	Nationwide	Vodafone	Data Control
936.0125	891.0125	NFM	Nationwide	Vodafone	Channel 41
		NFM	Nationwide	Vodafone	Data Control
936.0375	891.0375	NFM	Nationwide	Vodafone	Channel 42
		NFM	Nationwide	Vodafone	Data Control
936.0625	891.0625	NFM	Nationwide	Vodafone	Channel 43
		NFM	Edinburgh	Vodafone	Data Control
936.0875	891.0875	NFM	Nationwide	Vodafone	Channel 44
936.1125	891.1125	NFM	Nationwide	Vodafone	Channel 45
936.1375	891.1375	NFM	Nationwide	Vodafone	Channel 46
936.1625	891.1625	NFM	Nationwide	Vodafone	Channel 47
936.1875	891.1875	NFM	Nationwide	Vodafone	Channel 48

Base	Mobile	Mode	Location	User & Notes	
936.2125	891.2125	NFM	Nationwide	Vodafone	Channel 49
936.2375	891.2375	NFM	Nationwide	Vodafone	Channel 50
936.2625	891.2625	NFM	Nationwide	Vodafone	Channel 51
936.2875	891.2875	NFM	Nationwide	Vodafone	Channel 52
936.3125	891.3125	NFM	Nationwide	Vodafone	Channel 53
936.3375	891.3375	NFM	Nationwide	Vodafone	Channel 54
936.3625	891.3625	NFM	Nationwide	Vodafone	Channel 55
936.3875	891.3875	NFM	Nationwide	Vodafone	Channel 56
936.4125	891.4125	NFM	Nationwide	Vodafone	Channel 57
936.4375	891.4375	NFM	Nationwide	Vodafone	Channel 58
936.4625	891.4625	NFM	Nationwide	Vodafone	Channel 59
936.4875	891.4875	NFM	Nationwide	Vodafone	Channel 60
936.5125	891.5125	NFM	Nationwide	Vodafone	Channel 61
936.5375	891.5375	NFM	Nationwide	Vodafone	Channel 62
936.5625	891.5625	NFM	Nationwide	Vodafone	Channel 63
936.5875	891.5875	NFM	Nationwide	Vodafone	Channel 64
936.6125	891.6125	NFM	Nationwide	Vodafone	Channel 65
936.6375	891.6375	NFM	Nationwide	Vodafone	Channel 66
936.6625	891.6625	NFM	Nationwide	Vodafone	Channel 67
936.6875	891.6875	NFM	Nationwide	Vodafone	Channel 68
936.7125	891.7125	NFM	Nationwide	Vodafone	Channel 69
936.7375	891.7375	NFM	Nationwide	Vodafone	Channel 70
936.7625	891.7625	NFM	Nationwide	Vodafone	Channel 71
936.7875	891.7875	NFM	Nationwide	Vodafone	Channel 72
936.8125	891.8125	NFM	Nationwide	Vodafone	Channel 73
936.8375	891.8375	NFM	Nationwide	Vodafone	Channel 74
936.8625	891.8625	NFM	Nationwide	Vodafone	Channel 75
936.8875	891.8875	NFM	Nationwide	Vodafone	Channel 76
936.9125	891.9125	NFM	Nationwide	Vodafone	Channel 77
936.9375	891.9375	NFM	Nationwide	Vodafone	Channel 78
936.9625	891.9625	NFM	Nationwide	Vodafone	Channel 79
936.9875	891.9875	NFM	Nationwide	Vodafone	Channel 80
937.0125	892.0125	NFM	Nationwide	Vodafone	Channel 81
937.0375	892.0375	NFM	Nationwide	Vodafone	Channel 82
937.0625	892.0625	NFM	Nationwide	Vodafone	Channel 83
937.0875	892.0875	NFM	Nationwide	Vodafone	Channel 84
937.1125	892.1125	NFM	Nationwide	Vodafone	Channel 85
937.1375	892.1375	NFM	Nationwide	Vodafone	Channel 86
937.1625	892.1625	NFM	Nationwide	Vodafone	Channel 87
937.1875	892.1875	NFM	Nationwide	Vodafone	Channel 88
937.2125	892.2125	NFM	Nationwide	Vodafone	Channel 89
937.2375	892.2375	NFM	Nationwide	Vodafone	Channel 90
937.2625	892.2625	NFM	Nationwide	Vodafone	Channel 91
937.2875	892.2875	NFM	Nationwide	Vodafone	Channel 92
937.3125	892.3125	NFM	Nationwide	Vodafone	Channel 93
937.3375	892.3375	NFM	Nationwide	Vodafone	Channel 94
937.3625	892.3625	NFM	Nationwide	Vodafone	Channel 95
937.3875	892.3875	NFM	Nationwide	Vodafone	Channel 96
937.4125	892.4125	NFM	Nationwide	Vodafone	Channel 97
937.4375	892.4375	NFM	Nationwide	Vodafone	Channel 98

This picture illustrates the construction of a typical cellular telephone relay tower. Note the microwave dish half way up the structure. This dish provides the radio link with a network of similar towers. The smaller mast along side shows a wealth of VHF PMR antennas as well as a dedicated microwave link.

Base	Mobile	Mode	Location	User & Notes	
937.4625	892.4625	NFM	Nationwide	Vodafone	Channel 99
937.4875	892.4875	NFM	Nationwide	Vodafone	Channel 100
937.5125	892.5125	NFM	Nationwide	Vodafone	Channel 101
937.5375	892.5375	NFM	Nationwide	Vodafone	Channel 102
937.5625	892.5625	NFM	Nationwide	Vodafone	Channel 103
937.5875	892.5875	NFM	Nationwide	Vodafone	Channel 104
937.6125	892.6125	NFM	Nationwide	Vodafone	Channel 105
937.6375	892.6375	NFM	Nationwide	Vodafone	Channel 106
937.6625	892.6625	NFM	Nationwide	Vodafone	Channel 107
937.6875	892.6875	NFM	Nationwide	Vodafone	Channel 108
937.7125	892.7125	NFM	Nationwide	Vodafone	Channel 109
937.7375	892.7375	NFM	Nationwide	Vodafone	Channel 110
937.7625	892.7625	NFM	Nationwide	Vodafone	Channel 111
937.7875	892.7875	NFM	Nationwide	Vodafone	Channel 112
937.8125	892.8125	NFM	Nationwide	Vodafone	Channel 113
937.8375	892.8375	NFM	Nationwide	Vodafone	Channel 114
937.8625	892.8625	NFM	Nationwide	Vodafone	Channel 115
937.8875	892.8875	NFM	Nationwide	Vodafone	Channel 116
937.9125	892.9125	NFM	Nationwide	Vodafone	Channel 117
937.9375	892.9375	NFM	Nationwide	Vodafone	Channel 118
937.9625	892.9625	NFM	Nationwide	Vodafone	Channel 119
937.9875	892.9875	NFM	Nationwide	Vodafone	Channel 120
938.0125	893.0125	NFM	Nationwide	Vodafone	Channel 121
938.0375	893.0375	NFM	Nationwide	Vodafone	Channel 122
938.0625	893.0625	NFM	Nationwide	Vodafone	Channel 123
938.0875	893.0875	NFM	Nationwide	Vodafone	Channel 124
938.1125	893.1125	NFM	Nationwide	Vodafone	Channel 125
938.1375	893.1375	NFM	Nationwide	Vodafone	Channel 126
938.1625	893.1625	NFM	Nationwide	Vodafone	Channel 127
938.1875	893.1875	NFM	Nationwide	Vodafone	Channel 128
938.2125	893.2125	NFM	Nationwide	Vodafone	Channel 129
938.2375	893.2375	NFM	Nationwide	Vodafone	Channel 130
938.2625	893.2625	NFM	Nationwide	Vodafone	Channel 131
938.2875	893.2875	NFM	Nationwide	Vodafone	Channel 132
938.3125	893.3125	NFM	Nationwide	Vodafone	Channel 133
938.3375	893.3375	NFM	Nationwide	Vodafone	Channel 134
938.3625	893.3625	NFM	Nationwide	Vodafone	Channel 135
938.3875	893.3875	NFM	Nationwide	Vodafone	Channel 136
938.4125	893.4125	NFM	Nationwide	Vodafone	Channel 137
938.4375	893.4375	NFM	Nationwide	Vodafone	Channel 138
938.4625	893.4625	NFM	Nationwide	Vodafone	Channel 139
938.4875	893.4875	NFM	Nationwide	Vodafone	Channel 140
938.5125	893.5125	NFM	Nationwide	Vodafone	Channel 141
938.5375	893.5375	NFM	Nationwide	Vodafone	Channel 142
938.5625	893.5625	NFM	Nationwide	Vodafone	Channel 143
938.5875	893.5875	NFM	Nationwide	Vodafone	Channel 144
938.6125	893.6125	NFM	Nationwide	Vodafone	Channel 145
938.6375	893.6375	NFM	Nationwide	Vodafone	Channel 146
938.6625	893.6625	NFM	Nationwide	Vodafone	Channel 147
938.6875	893.6875	NFM	Nationwide	Vodafone	Channel 148

Base	Mobile	Mode	Location	User & Notes	
938.7125	893.7125	NFM	Nationwide	Vodafone	Channel 149
938.7375	893.7375	NFM	Nationwide	Vodafone	Channel 150
938.7625	893.7625	NFM	Nationwide	Vodafone	Channel 151
938.7875	893.7875	NFM	Nationwide	Vodafone	Channel 152
938.8125	893.8125	NFM	Nationwide	Vodafone	Channel 153
938.8375	893.8375	NFM	Nationwide	Vodafone	Channel 154
938.8625	893.8625	NFM	Nationwide	Vodafone	Channel 155
938.8875	893.8875	NFM	Nationwide	Vodafone	Channel 156
938.9125	893.9125	NFM	Nationwide	Vodafone	Channel 157
938.9375	893.9375	NFM	Nationwide	Vodafone	Channel 158
938.9625	893.9625	NFM	Nationwide	Vodafone	Channel 159
938.9875	893.9875	NFM	Nationwide	Vodafone	Channel 160
939.0125	894.0125	NFM	Nationwide	Vodafone	Channel 161
939.0375	894.0375	NFM	Nationwide	Vodafone	Channel 162
939.0625	894.0625	NFM	Nationwide	Vodafone	Channel 163
939.0875	894.0875	NFM	Nationwide	Vodafone	Channel 164
939.1125	894.1125	NFM	Nationwide	Vodafone	Channel 165
939.1375	894.1375	NFM	Nationwide	Vodafone	Channel 166
939.1625	894.1625	NFM	Nationwide	Vodafone	Channel 167
939.1875	894.1875	NFM	Nationwide	Vodafone	Channel 168
939.2125	894.2125	NFM	Nationwide	Vodafone	Channel 169
939.2375	894.2375	NFM	Nationwide	Vodafone	Channel 170
939.2625	894.2625	NFM	Nationwide	Vodafone	Channel 171
939.2875	894.2875	NFM	Nationwide	Vodafone	Channel 172
939.3125	894.3125	NFM	Nationwide	Vodafone	Channel 173
939.3375	894.3375	NFM	Nationwide	Vodafone	Channel 174
939.3625	894.3625	NFM	Nationwide	Vodafone	Channel 175
939.3875	894.3875	NFM	Nationwide	Vodafone	Channel 176
939.4125	894.4125	NFM	Nationwide	Vodafone	Channel 177
939.4375	894.4375	NFM	Nationwide	Vodafone	Channel 178
939.4625	894.4625	NFM	Nationwide	Vodafone	Channel 179
939.4875	894.4875	NFM	Nationwide	Vodafone	Channel 180
939.5125	894.5125	NFM	Nationwide	Vodafone	Channel 181
939.5375	894.5375	NFM	Nationwide	Vodafone	Channel 182
939.5625	894.5625	NFM	Nationwide	Vodafone	Channel 183
939.5875	894.5875	NFM	Nationwide	Vodafone	Channel 184
939.6125	894.6125	NFM	Nationwide	Vodafone	Channel 185
939.6375	894.6375	NFM	Nationwide	Vodafone	Channel 186
939.6625	894.6625	NFM	Nationwide	Vodafone	Channel 187
939.6875	894.6875	NFM	Nationwide	Vodafone	Channel 188
939.7125	894.7125	NFM	Nationwide	Vodafone	Channel 189
939.7375	894.7375	NFM	Nationwide	Vodafone	Channel 190
939.7625	894.7625	NFM	Nationwide	Vodafone	Channel 191
939.7875	894.7875	NFM	Nationwide	Vodafone	Channel 192
939.8125	894.8125	NFM	Nationwide	Vodafone	Channel 193
939.8375	894.8375	NFM	Nationwide	Vodafone	Channel 194
939.8625	894.8625	NFM	Nationwide	Vodafone	Channel 195
939.8875	894.8875	NFM	Nationwide	Vodafone	Channel 196
939.9125	894.9125	NFM	Nationwide	Vodafone	Channel 197
939.9375	894.9375	NFM	Nationwide	Vodafone	Channel 198

Base	Mobile	Mode	Location	User & Notes	
939.9625	894.9625	NFM	Nationwide	Vodafone	Channel 199
939.9875	894.9875	NFM	Nationwide	Vodafone	Channel 200
940.0125	895.0125	NFM	Nationwide	Vodafone	Channel 201
940.0375	895.0375	NFM	Nationwide	Vodafone	Channel 202
940.0625	895.0625	NFM	Nationwide	Vodafone	Channel 203
940.0875	895.0875	NFM	Nationwide	Vodafone	Channel 204
940.1125	895.1125	NFM	Nationwide	Vodafone	Channel 205
940.1375	895.1375	NFM	Nationwide	Vodafone	Channel 206
940.1625	895.1625	NFM	Nationwide	Vodafone	Channel 207
940.1875	895.1875	NFM	Nationwide	Vodafone	Channel 208
940.2125	895.2125	NFM	Nationwide	Vodafone	Channel 209
940.2375	895.2375	NFM	Nationwide	Vodafone	Channel 210
940.2625	895.2625	NFM	Nationwide	Vodafone	Channel 211
940.2875	895.2875	NFM	Nationwide	Vodafone	Channel 212
940.3125	895.3125	NFM	Nationwide	Vodafone	Channel 213
940.3375	895.3375	NFM	Nationwide	Vodafone	Channel 214
940.3625	895.3625	NFM	Nationwide	Vodafone	Channel 215
940.3875	895.3875	NFM	Nationwide	Vodafone	Channel 216
940.4125	895.4125	NFM	Nationwide	Vodafone	Channel 217
940.4375	895.4375	NFM	Nationwide	Vodafone	Channel 218
940.4625	895.4625	NFM	Nationwide	Vodafone	Channel 219
940.4875	895.4875	NFM	Nationwide	Vodafone	Channel 220
940.5125	895.5125	NFM	Nationwide	Vodafone	Channel 221
940.5375	895.5375	NFM	Nationwide	Vodafone	Channel 222
940.5625	895.5625	NFM	Nationwide	Vodafone	Channel 223
940.5875	895.5875	NFM	Nationwide	Vodafone	Channel 224
940.6125	895.6125	NFM	Nationwide	Vodafone	Channel 225
940.6375	895.6375	NFM	Nationwide	Vodafone	Channel 226
940.6625	895.6625	NFM	Nationwide	Vodafone	Channel 227
940.6875	895.6875	NFM	Nationwide	Vodafone	Channel 228
940.7125	895.7125	NFM	Nationwide	Vodafone	Channel 229
940.7375	895.7375	NFM	Nationwide	Vodafone	Channel 230
940.7625	895.7625	NFM	Nationwide	Vodafone	Channel 231
940.7875	895.7875	NFM	Nationwide	Vodafone	Channel 232
940.8125	895.8125	NFM	Nationwide	Vodafone	Channel 233
940.8375	895.8375	NFM	Nationwide	Vodafone	Channel 234
940.8625	895.8625	NFM	Nationwide	Vodafone	Channel 235
940.8875	895.8875	NFM	Nationwide	Vodafone	Channel 236
940.9125	895.9125	NFM	Nationwide	Vodafone	Channel 237
940.9375	895.9375	NFM	Nationwide	Vodafone	Channel 238
940.9625	895.9625	NFM	Nationwide	Vodafone	Channel 239
940.9875	895.9875	NFM	Nationwide	Vodafone	Channel 240
941.0125	896.0125	NFM	Nationwide	Vodafone	Channel 241
941.0375	896.0375	NFM	Nationwide	Vodafone	Channel 242
941.0625	896.0625	NFM	Nationwide	Vodafone	Channel 243
941.0875	896.0875	NFM	Nationwide	Vodafone	Channel 244
941.1125	896.1125	NFM	Nationwide	Vodafone	Channel 245
941.1375	896.1375	NFM	Nationwide	Vodafone	Channel 246
941.1625	896.1625	NFM	Nationwide	Vodafone	Channel 247
941.1875	896.1875	NFM	Nationwide	Vodafone	Channel 248

Base	Mobile	Mode	Location	User & Notes	
941.2125	896.2125	NFM	Nationwide	Vodafone	Channel 249
941.2375	896.2375	NFM	Nationwide	Vodafone	Channel 250
941.2625	896.2625	NFM	Nationwide	Vodafone	Channel 251
941.2875	896.2875	NFM	Nationwide	Vodafone	Channel 252
941.3125	896.3125	NFM	Nationwide	Vodafone	Channel 253
941.3375	896.3375	NFM	Nationwide	Vodafone	Channel 254
941.3625	896.3625	NFM	Nationwide	Vodafone	Channel 255
941.3875	896.3875	NFM	Nationwide	Vodafone	Channel 256
941.4125	896.4125	NFM	Nationwide	Vodafone	Channel 257
941.4375	896.4375	NFM	Nationwide	Vodafone	Channel 258
941.4625	896.4625	NFM	Nationwide	Vodafone	Channel 259
941.4875	896.4875	NFM	Nationwide	Vodafone	Channel 260
941.5125	896.5125	NFM	Nationwide	Vodafone	Channel 261
941.5375	896.5375	NFM	Nationwide	Vodafone	Channel 262
941.5625	896.5625	NFM	Nationwide	Vodafone	Channel 263
941.5875	896.5875	NFM	Nationwide	Vodafone	Channel 264
941.6125	896.6125	NFM	Nationwide	Vodafone	Channel 265
941.6375	896.6375	NFM	Nationwide	Vodafone	Channel 266
941.6625	896.6625	NFM	Nationwide	Vodafone	Channel 267
941.6875	896.6875	NFM	Nationwide	Vodafone	Channel 268
941.7125	896.7125	NFM	Nationwide	Vodafone	Channel 269
941.7375	896.7375	NFM	Nationwide	Vodafone	Channel 270
941.7625	896.7625	NFM	Nationwide	Vodafone	Channel 271
941.7875	896.7875	NFM	Nationwide	Vodafone	Channel 272
941.8125	896.8125	NFM	Nationwide	Vodafone	Channel 273
941.8375	896.8375	NFM	Nationwide	Vodafone	Channel 274
941.8625	896.8625	NFM	Nationwide	Vodafone	Channel 275
941.8875	896.8875	NFM	Nationwide	Vodafone	Channel 276
941.9125	896.9125	NFM	Nationwide	Vodafone	Channel 277
941.9375	896.9375	NFM	Nationwide	Vodafone	Channel 278
941.9625	896.9625	NFM	Nationwide	Vodafone	Channel 279
941.9875	896.9875	NFM	Nationwide	Vodafone	Channel 280
942.0125	897.0125	NFM	Nationwide	Vodafone	Channel 281
942.0375	897.0375	NFM	Nationwide	Vodafone	Channel 282
942.0625	897.0625	NFM	Nationwide	Vodafone	Channel 283
942.0875	897.0875	NFM	Nationwide	Vodafone	Channel 284
942.1125	897.1125	NFM	Nationwide	Vodafone	Channel 285
942.1375	897.1375	NFM	Nationwide	Vodafone	Channel 286
942.1625	897.1625	NFM	Nationwide	Vodafone	Channel 287
942.1875	897.1875	NFM	Nationwide	Vodafone	Channel 288
942.2125	897.2125	NFM	Nationwide	Vodafone	Channel 289
942.2375	897.2375	NFM	Nationwide	Vodafone	Channel 290
942.2625	897.2625	NFM	Nationwide	Vodafone	Channel 291
942.2875	897.2875	NFM	Nationwide	Vodafone	Channel 292
942.3125	897.3125	NFM	Nationwide	Vodafone	Channel 293
942.3375	897.3375	NFM	Nationwide	Vodafone	Channel 294
942.3625	897.3625	NFM	Nationwide	Vodafone	Channel 295
942.3875	897.3875	NFM	Nationwide	Vodafone	Channel 296
942.4125	897.4125	NFM	Nationwide	Vodafone	Channel 297
942.4375	897.4375	NFM	Nationwide	Vodafone	Channel 298

Base	Mobile	Mode	Location	User & Notes	
942.4625	897.4625	NFM	Nationwide	Vodafone	Channel 299
942.4875	897.4875	NFM	Nationwide	Vodafone	Channel 300
942.5125	897.5125	NFM	Nationwide	Cellnet	Channel 301
942.5375	897.5375	NFM	Nationwide	Cellnet	Channel 302
942.5625	897.5625	NFM	Nationwide	Cellnet	Channel 303
942.5875	897.5875	NFM	Nationwide	Cellnet	Channel 304
942.6125	897.6125	NFM	Nationwide	Cellnet	Channel 305
942.6375	897.6375	NFM	Nationwide	Cellnet	Channel 306
942.6625	897.6625	NFM	Nationwide	Cellnet	Channel 307
942.6875	897.6875	NFM	Nationwide	Cellnet	Channel 308
942.7125	897.7125	NFM	Nationwide	Cellnet	Channel 309
942.7375	897.7375	NFM	Nationwide	Cellnet	Channel 310
942.7625	897.7625	NFM	Nationwide	Cellnet	Channel 311
942.7875	897.7875	NFM	Nationwide	Cellnet	Channel 312
942.8125	897.8125	NFM	Nationwide	Cellnet	Channel 313
942.8375	897.8375	NFM	Nationwide	Cellnet	Channel 314
942.8625	897.8625	NFM	Nationwide	Cellnet	Channel 315
942.8875	897.8875	NFM	Nationwide	Cellnet	Channel 316
942.9125	897.9125	NFM	Nationwide	Cellnet	Channel 317
942.9375	897.9375	NFM	Nationwide	Cellnet	Channel 318
942.9625	897.9625	NFM	Nationwide	Cellnet	Channel 319
942.9875	897.9875	NFM	Nationwide	Cellnet	Channel 320
943.0125	898.0125	NFM	Nationwide	Cellnet	Channel 321
943.0375	898.0375	NFM	Nationwide	Cellnet	Channel 322
943.0625	898.0625	NFM	Nationwide	Cellnet	Channel 323
		NFM	Edinburgh	Cellnet	Data Control
943.0875	898.0875	NFM	Nationwide	Cellnet	Channel 324
		NFM	La Chasse, Jersey	Cellnet	Data Control
943.1125	898.1125	NFM	Nationwide	Cellnet	Channel 325
		NFM	Nationwide	Cellnet	Data Control
943.1375	898.1375	NFM	Nationwide	Cellnet	Channel 326
		NFM	Guernsey	Cellnet	Data Control
943.1625	898.1625	NFM	Nationwide	Cellnet	Channel 327
		NFM	Nationwide	Cellnet	Data Control
943.1875	898.1875	NFM	Nationwide	Cellnet	Channel 328
		NFM	Five Oaks, Jersey	Cellnet	Data Control
943.2125	898.2125	NFM	Nationwide	Cellnet	Channel 329
		NFM	Nationwide	Cellnet	Data Control
943.2375	898.2375	NFM	Nationwide	Cellnet	Channel 330
		NFM	Nationwide	Cellnet	Data Control
943.2625	898.2625	NFM	Nationwide	Cellnet	Channel 331
		NFM	Nationwide	Cellnet	Data Control
943.2875	898.2875	NFM	Nationwide	Cellnet	Channel 332
		NFM	Guernsey	Cellnet	Data Control
943.3125	898.3125	NFM	Nationwide	Cellnet	Channel 333
		NFM	Nationwide	Cellnet	Data Control
943.3375	898.3375	NFM	Nationwide	Cellnet	Channel 334
		NFM	Edinburgh	Cellnet	Data Control
		NFM	Gorey Hill, Jersey	Cellnet	Data Control
943.3625	898.3625	NFM	Nationwide	Cellnet	Channel 335

Base	Mobile	Mode	Location	User & Notes	
		NFM	Perth	Cellnet	Data Control
943.3875	898.3875	NFM	Nationwide	Cellnet	Channel 336
		NFM	Fort Regent, Jersey	Cellnet	Data Control
943.4125	898.4125	NFM	Nationwide	Cellnet	Channel 337
		NFM	Nationwide	Cellnet	Data Control
943.4375	898.4375	NFM	Nationwide	Cellnet	Channel 338
		NFM	Nationwide	Cellnet	Data Control
943.4625	898.4625	NFM	Nationwide	Cellnet	Channel 339
		NFM	Nationwide	Cellnet	Data Control
943.4875	898.4875	NFM	Nationwide	Cellnet	Channel 340
		NFM	Nationwide	Cellnet	Data Control
943.5125	898.5125	NFM	Nationwide	Cellnet	Channel 341
		NFM	Nationwide	Cellnet	Data Control
943.5375	898.5375	NFM	Nationwide	Cellnet	Channel 342
		NFM	Nationwide	Cellnet	Data Control
943.5625	898.5625	NFM	Nationwide	Cellnet	Channel 343
		NFM	Nationwide	Cellnet	Data Control
943.5875	898.5875	NFM	Nationwide	Cellnet	Channel 344
943.6125	898.6125	NFM	Nationwide	Cellnet	Channel 345
943.6375	898.6375	NFM	Nationwide	Cellnet	Channel 346
943.6625	898.6625	NFM	Nationwide	Cellnet	Channel 347
943.6875	898.6875	NFM	Nationwide	Cellnet	Channel 348
943.7125	898.7125	NFM	Nationwide	Cellnet	Channel 349
943.7375	898.7375	NFM	Nationwide	Cellnet	Channel 350
943.7625	898.7625	NFM	Nationwide	Cellnet	Channel 351
943.7875	898.7875	NFM	Nationwide	Cellnet	Channel 352
943.8125	898.8125	NFM	Nationwide	Cellnet	Channel 353
943.8375	898.8375	NFM	Nationwide	Cellnet	Channel 354
943.8625	898.8625	NFM	Nationwide	Cellnet	Channel 355
943.8875	898.8875	NFM	Nationwide	Cellnet	Channel 356
943.9125	898.9125	NFM	Nationwide	Cellnet	Channel 357
943.9375	898.9375	NFM	Nationwide	Cellnet	Channel 358
943.9625	898.9625	NFM	Nationwide	Cellnet	Channel 359
943.9875	898.9875	NFM	Nationwide	Cellnet	Channel 360
944.0125	899.0125	NFM	Nationwide	Cellnet	Channel 361
944.0375	899.0375	NFM	Nationwide	Cellnet	Channel 362
944.0625	899.0625	NFM	Nationwide	Cellnet	Channel 363
944.0875	899.0875	NFM	Nationwide	Cellnet	Channel 364
944.1125	899.1125	NFM	Nationwide	Cellnet	Channel 365
944.1375	899.1375	NFM	Nationwide	Cellnet	Channel 366
944.1625	899.1625	NFM	Nationwide	Cellnet	Channel 367
944.1875	899.1875	NFM	Nationwide	Cellnet	Channel 368
944.2125	899.2125	NFM	Nationwide	Cellnet	Channel 369
944.2375	899.2375	NFM	Nationwide	Cellnet	Channel 370
944.2625	899.2625	NFM	Nationwide	Cellnet	Channel 371
944.2875	899.2875	NFM	Nationwide	Cellnet	Channel 372
944.3125	899.3125	NFM	Nationwide	Cellnet	Channel 373
944.3375	899.3375	NFM	Nationwide	Cellnet	Channel 374
944.3625	899.3625	NFM	Nationwide	Cellnet	Channel 375
944.3875	899.3875	NFM	Nationwide	Cellnet	Channel 376

Base	Mobile	Mode	Location	User & Notes	
944.4125	899.4125	NFM	Nationwide	Cellnet	Channel 377
944.4375	899.4375	NFM	Nationwide	Cellnet	Channel 378
944.4625	899.4625	NFM	Nationwide	Cellnet	Channel 379
944.4875	899.4875	NFM	Nationwide	Cellnet	Channel 380
944.5125	899.5125	NFM	Nationwide	Cellnet	Channel 381
944.5375	899.5375	NFM	Nationwide	Cellnet	Channel 382
944.5625	899.5625	NFM	Nationwide	Cellnet	Channel 383
944.5875	899.5875	NFM	Nationwide	Cellnet	Channel 384
944.6125	899.6125	NFM	Nationwide	Cellnet	Channel 385
944.6375	899.6375	NFM	Nationwide	Cellnet	Channel 386
944.6625	899.6625	NFM	Nationwide	Cellnet	Channel 387
944.6875	899.6875	NFM	Nationwide	Cellnet	Channel 388
944.7125	899.7125	NFM	Nationwide	Cellnet	Channel 389
944.7375	899.7375	NFM	Nationwide	Cellnet	Channel 390
944.7625	899.7625	NFM	Nationwide	Cellnet	Channel 391
944.7875	899.7875	NFM	Nationwide	Cellnet	Channel 392
944.8125	899.8125	NFM	Nationwide	Cellnet	Channel 393
944.8375	899.8375	NFM	Nationwide	Cellnet	Channel 394
944.8625	899.8625	NFM	Nationwide	Cellnet	Channel 395
944.8875	899.8875	NFM	Nationwide	Cellnet	Channel 396
944.9125	899.9125	NFM	Nationwide	Cellnet	Channel 397
944.9375	899.9375	NFM	Nationwide	Cellnet	Channel 398
944.9625	899.9625	NFM	Nationwide	Cellnet	Channel 399
944.9875	899.9875	NFM	Nationwide	Cellnet	Channel 400
945.0125	900.0125	NFM	Nationwide	Cellnet	Channel 401
945.0375	900.0375	NFM	Nationwide	Cellnet	Channel 402
945.0625	900.0625	NFM	Nationwide	Cellnet	Channel 403
945.0875	900.0875	NFM	Nationwide	Cellnet	Channel 404
945.1125	900.1125	NFM	Nationwide	Cellnet	Channel 405
945.1375	900.1375	NFM	Nationwide	Cellnet	Channel 406
945.1625	900.1625	NFM	Nationwide	Cellnet	Channel 407
945.1875	900.1875	NFM	Nationwide	Cellnet	Channel 408
945.2125	900.2125	NFM	Nationwide	Cellnet	Channel 409
945.2375	900.2375	NFM	Nationwide	Cellnet	Channel 410
945.2625	900.2625	NFM	Nationwide	Cellnet	Channel 411
945.2875	900.2875	NFM	Nationwide	Cellnet	Channel 412
945.3125	900.3125	NFM	Nationwide	Cellnet	Channel 413
945.3375	900.3375	NFM	Nationwide	Cellnet	Channel 414
945.3625	900.3625	NFM	Nationwide	Cellnet	Channel 415
945.3875	900.3875	NFM	Nationwide	Cellnet	Channel 416
945.4125	900.4125	NFM	Nationwide	Cellnet	Channel 417
945.4375	900.4375	NFM	Nationwide	Cellnet	Channel 418
945.4625	900.4625	NFM	Nationwide	Cellnet	Channel 419
945.4875	900.4875	NFM	Nationwide	Cellnet	Channel 420
945.5125	900.5125	NFM	Nationwide	Cellnet	Channel 421
945.5375	900.5375	NFM	Nationwide	Cellnet	Channel 422
945.5625	900.5625	NFM	Nationwide	Cellnet	Channel 423
945.5875	900.5875	NFM	Nationwide	Cellnet	Channel 424
945.6125	900.6125	NFM	Nationwide	Cellnet	Channel 425
945.6375	900.6375	NFM	Nationwide	Cellnet	Channel 426

The UK Scanning Directory

Base	Mobile	Mode	Location	User & Notes	
945.6625	900.6625	NFM	Nationwide	Cellnet	Channel 427
945.6875	900.6875	NFM	Nationwide	Cellnet	Channel 428
945.7125	900.7125	NFM	Nationwide	Cellnet	Channel 429
945.7375	900.7375	NFM	Nationwide	Cellnet	Channel 430
945.7625	900.7625	NFM	Nationwide	Cellnet	Channel 431
945.7875	900.7875	NFM	Nationwide	Cellnet	Channel 432
945.8125	900.8125	NFM	Nationwide	Cellnet	Channel 433
945.8375	900.8375	NFM	Nationwide	Cellnet	Channel 434
945.8625	900.8625	NFM	Nationwide	Cellnet	Channel 435
945.8875	900.8875	NFM	Nationwide	Cellnet	Channel 436
945.9125	900.9125	NFM	Nationwide	Cellnet	Channel 437
945.9375	900.9375	NFM	Nationwide	Cellnet	Channel 438
945.9625	900.9625	NFM	Nationwide	Cellnet	Channel 439
945.9875	900.9875	NFM	Nationwide	Cellnet	Channel 440
946.0125	901.0125	NFM	Nationwide	Cellnet	Channel 441
946.0375	901.0375	NFM	Nationwide	Cellnet	Channel 442
946.0625	901.0625	NFM	Nationwide	Cellnet	Channel 443
946.0875	901.0875	NFM	Nationwide	Cellnet	Channel 444
946.1125	901.1125	NFM	Nationwide	Cellnet	Channel 445
946.1375	901.1375	NFM	Nationwide	Cellnet	Channel 446
946.1625	901.1625	NFM	Nationwide	Cellnet	Channel 447
946.1875	901.1875	NFM	Nationwide	Cellnet	Channel 448
946.2125	901.2125	NFM	Nationwide	Cellnet	Channel 449
946.2375	901.2375	NFM	Nationwide	Cellnet	Channel 450
946.2625	901.2625	NFM	Nationwide	Cellnet	Channel 451
946.2875	901.2875	NFM	Nationwide	Cellnet	Channel 452
946.3125	901.3125	NFM	Nationwide	Cellnet	Channel 453
946.3375	901.3375	NFM	Nationwide	Cellnet	Channel 454
946.3625	901.3625	NFM	Nationwide	Cellnet	Channel 455
946.3875	901.3875	NFM	Nationwide	Cellnet	Channel 456
946.4125	901.4125	NFM	Nationwide	Cellnet	Channel 457
946.4375	901.4375	NFM	Nationwide	Cellnet	Channel 458
946.4625	901.4625	NFM	Nationwide	Cellnet	Channel 459
946.4875	901.4875	NFM	Nationwide	Cellnet	Channel 460
946.5125	901.5125	NFM	Nationwide	Cellnet	Channel 461
946.5375	901.5375	NFM	Nationwide	Cellnet	Channel 462
946.5625	901.5625	NFM	Nationwide	Cellnet	Channel 463
946.5875	901.5875	NFM	Nationwide	Cellnet	Channel 464
946.6125	901.6125	NFM	Nationwide	Cellnet	Channel 465
946.6375	901.6375	NFM	Nationwide	Cellnet	Channel 466
946.6625	901.6625	NFM	Nationwide	Cellnet	Channel 467
946.6875	901.6875	NFM	Nationwide	Cellnet	Channel 468
946.7125	901.7125	NFM	Nationwide	Cellnet	Channel 469
946.7375	901.7375	NFM	Nationwide	Cellnet	Channel 470
946.7625	901.7625	NFM	Nationwide	Cellnet	Channel 471
946.7875	901.7875	NFM	Nationwide	Cellnet	Channel 472
946.8125	901.8125	NFM	Nationwide	Cellnet	Channel 473
946.8375	901.8375	NFM	Nationwide	Cellnet	Channel 474
946.8625	901.8625	NFM	Nationwide	Cellnet	Channel 475
946.8875	901.8875	NFM	Nationwide	Cellnet	Channel 476

Base	Mobile	Mode	Location	User & Notes	
946.9125	901.9125	NFM	Nationwide	Cellnet	Channel 477
946.9375	901.9375	NFM	Nationwide	Cellnet	Channel 478
946.9625	901.9625	NFM	Nationwide	Cellnet	Channel 479
946.9875	901.9875	NFM	Nationwide	Cellnet	Channel 480
947.0125	902.0125	NFM	Nationwide	Cellnet	Channel 481
947.0375	902.0375	NFM	Nationwide	Cellnet	Channel 482
947.0625	902.0625	NFM	Nationwide	Cellnet	Channel 483
947.0875	902.0875	NFM	Nationwide	Cellnet	Channel 484
947.1125	902.1125	NFM	Nationwide	Cellnet	Channel 485
947.1375	902.1375	NFM	Nationwide	Cellnet	Channel 486
947.1625	902.1625	NFM	Nationwide	Cellnet	Channel 487
947.1875	902.1875	NFM	Nationwide	Cellnet	Channel 488
947.2125	902.2125	NFM	Nationwide	Cellnet	Channel 489
947.2375	902.2375	NFM	Nationwide	Cellnet	Channel 490
947.2625	902.2625	NFM	Nationwide	Cellnet	Channel 491
947.2875	902.2875	NFM	Nationwide	Cellnet	Channel 492
947.3125	902.3125	NFM	Nationwide	Cellnet	Channel 493
947.3375	902.3375	NFM	Nationwide	Cellnet	Channel 494
947.3625	902.3625	NFM	Nationwide	Cellnet	Channel 495
947.3875	902.3875	NFM	Nationwide	Cellnet	Channel 496
947.4125	902.4125	NFM	Nationwide	Cellnet	Channel 497
947.4375	902.4375	NFM	Nationwide	Cellnet	Channel 498
947.4625	902.4625	NFM	Nationwide	Cellnet	Channel 499
947.4875	902.4875	NFM	Nationwide	Cellnet	Channel 500
947.5125	902.5125	NFM	Nationwide	Cellnet	Channel 501
947.5375	902.5375	NFM	Nationwide	Cellnet	Channel 502
947.5625	902.5625	NFM	Nationwide	Cellnet	Channel 503
947.5875	902.5875	NFM	Nationwide	Cellnet	Channel 504
947.6125	902.6125	NFM	Nationwide	Cellnet	Channel 505
947.6375	902.6375	NFM	Nationwide	Cellnet	Channel 506
947.6625	902.6625	NFM	Nationwide	Cellnet	Channel 507
947.6875	902.6875	NFM	Nationwide	Cellnet	Channel 508
947.7125	902.7125	NFM	Nationwide	Cellnet	Channel 509
947.7375	902.7375	NFM	Nationwide	Cellnet	Channel 510
947.7625	902.7625	NFM	Nationwide	Cellnet	Channel 511
947.7875	902.7875	NFM	Nationwide	Cellnet	Channel 512
947.8125	902.8125	NFM	Nationwide	Cellnet	Channel 513
947.8375	902.8375	NFM	Nationwide	Cellnet	Channel 514
947.8625	902.8625	NFM	Nationwide	Cellnet	Channel 515
947.8875	902.8875	NFM	Nationwide	Cellnet	Channel 516
947.9125	902.9125	NFM	Nationwide	Cellnet	Channel 517
947.9375	902.9375	NFM	Nationwide	Cellnet	Channel 518
947.9625	902.9625	NFM	Nationwide	Cellnet	Channel 519
947.9875	902.9875	NFM	Nationwide	Cellnet	Channel 520
948.0125	903.0125	NFM	Nationwide	Cellnet	Channel 521
948.0375	903.0375	NFM	Nationwide	Cellnet	Channel 522
948.0625	903.0625	NFM	Nationwide	Cellnet	Channel 523
948.0875	903.0875	NFM	Nationwide	Cellnet	Channel 524
948.1125	903.1125	NFM	Nationwide	Cellnet	Channel 525
948.1375	903.1375	NFM	Nationwide	Cellnet	Channel 526

Base	Mobile	Mode	Location	User & Notes	
948.1625	903.1625	NFM	Nationwide	Cellnet	Channel 527
948.1875	903.1875	NFM	Nationwide	Cellnet	Channel 528
948.2125	903.2125	NFM	Nationwide	Cellnet	Channel 529
948.2375	903.2375	NFM	Nationwide	Cellnet	Channel 530
948.2625	903.2625	NFM	Nationwide	Cellnet	Channel 531
948.2875	903.2875	NFM	Nationwide	Cellnet	Channel 532
948.3125	903.3125	NFM	Nationwide	Cellnet	Channel 533
948.3375	903.3375	NFM	Nationwide	Cellnet	Channel 534
948.3625	903.3625	NFM	Nationwide	Cellnet	Channel 535
948.3875	903.3875	NFM	Nationwide	Cellnet	Channel 536
948.4125	903.4125	NFM	Nationwide	Cellnet	Channel 537
948.4375	903.4375	NFM	Nationwide	Cellnet	Channel 538
948.4625	903.4625	NFM	Nationwide	Cellnet	Channel 539
948.4875	903.4875	NFM	Nationwide	Cellnet	Channel 540
948.5125	903.5125	NFM	Nationwide	Cellnet	Channel 541
948.5375	903.5375	NFM	Nationwide	Cellnet	Channel 542
948.5625	903.5625	NFM	Nationwide	Cellnet	Channel 543
948.5875	903.5875	NFM	Nationwide	Cellnet	Channel 544
948.6125	903.6125	NFM	Nationwide	Cellnet	Channel 545
948.6375	903.6375	NFM	Nationwide	Cellnet	Channel 546
948.6625	903.6625	NFM	Nationwide	Cellnet	Channel 547
948.6875	903.6875	NFM	Nationwide	Cellnet	Channel 548
948.7125	903.7125	NFM	Nationwide	Cellnet	Channel 549
948.7375	903.7375	NFM	Nationwide	Cellnet	Channel 550
948.7625	903.7625	NFM	Nationwide	Cellnet	Channel 551
948.7875	903.7875	NFM	Nationwide	Cellnet	Channel 552
948.8125	903.8125	NFM	Nationwide	Cellnet	Channel 553
948.8375	903.8375	NFM	Nationwide	Cellnet	Channel 554
948.8625	903.8625	NFM	Nationwide	Cellnet	Channel 555
948.8875	903.8875	NFM	Nationwide	Cellnet	Channel 556
948.9125	903.9125	NFM	Nationwide	Cellnet	Channel 557
948.9375	903.9375	NFM	Nationwide	Cellnet	Channel 558
948.9625	903.9625	NFM	Nationwide	Cellnet	Channel 559
948.9875	903.9875	NFM	Nationwide	Cellnet	Channel 560
949.0125	904.0125	NFM	Nationwide	Cellnet	Channel 561
949.0375	904.0375	NFM	Nationwide	Cellnet	Channel 562
949.0625	904.0625	NFM	Nationwide	Cellnet	Channel 563
949.0875	904.0875	NFM	Nationwide	Cellnet	Channel 564
949.1125	904.1125	NFM	Nationwide	Cellnet	Channel 565
949.1375	904.1375	NFM	Nationwide	Cellnet	Channel 566
949.1625	904.1625	NFM	Nationwide	Cellnet	Channel 567
949.1875	904.1875	NFM	Nationwide	Cellnet	Channel 568
949.2125	904.2125	NFM	Nationwide	Cellnet	Channel 569
949.2375	904.2375	NFM	Nationwide	Cellnet	Channel 570
949.2625	904.2625	NFM	Nationwide	Cellnet	Channel 571
949.2875	904.2875	NFM	Nationwide	Cellnet	Channel 572
949.3125	904.3125	NFM	Nationwide	Cellnet	Channel 573
949.3375	904.3375	NFM	Nationwide	Cellnet	Channel 574
949.3625	904.3625	NFM	Nationwide	Cellnet	Channel 575
949.3875	904.3875	NFM	Nationwide	Cellnet	Channel 576

Base	Mobile	Mode	Location	User & Notes	
949.4125	904.4125	NFM	Nationwide	Cellnet	Channel 577
949.4375	904.4375	NFM	Nationwide	Cellnet	Channel 578
949.4625	904.4625	NFM	Nationwide	Cellnet	Channel 579
949.4875	904.4875	NFM	Nationwide	Cellnet	Channel 580
949.5125	904.5125	NFM	Nationwide	Cellnet	Channel 581
949.5375	904.5375	NFM	Nationwide	Cellnet	Channel 582
949.5625	904.5625	NFM	Nationwide	Cellnet	Channel 583
949.5875	904.5875	NFM	Nationwide	Cellnet	Channel 584
949.6125	904.6125	NFM	Nationwide	Cellnet	Channel 585
949.6375	904.6375	NFM	Nationwide	Cellnet	Channel 586
949.6625	904.6625	NFM	Nationwide	Cellnet	Channel 587
949.6875	904.6875	NFM	Nationwide	Cellnet	Channel 588
949.7125	904.7125	NFM	Nationwide	Cellnet	Channel 589
949.7375	904.7375	NFM	Nationwide	Cellnet	Channel 590
949.7625	904.7625	NFM	Nationwide	Cellnet	Channel 591
949.7875	904.7875	NFM	Nationwide	Cellnet	Channel 592
949.8125	904.8125	NFM	Nationwide	Cellnet	Channel 593
949.8375	904.8375	NFM	Nationwide	Cellnet	Channel 594
949.8625	904.8625	NFM	Nationwide	Cellnet	Channel 595
949.8875	904.8875	NFM	Nationwide	Cellnet	Channel 596
949.9125	904.9125	NFM	Nationwide	Cellnet	Channel 597
949.9375	904.9375	NFM	Nationwide	Cellnet	Channel 598
949.9625	904.9625	NFM	Nationwide	Cellnet	Channel 599
949.9875	904.9875	NFM	Nationwide	Cellnet	Channel 600

950.0000 - 960.0000 MHz Pan European Digital Cellular Service

959.0125 - 959.9875 MHz New Cybernet/Uniden Cordless Telephone Base

Base	Mobile	Mode	Location	User & Notes
959.0125	914.0125	NFM	Nationwide	Channel 1
959.0250	914.0250	NFM	Nationwide	Channel 2
959.0375	914.0375	NFM	Nationwide	Channel 3
959.0500	914.0500	NFM	Nationwide	Channel 4
959.0625	914.0625	NFM	Nationwide	Channel 5
959.0750	914.0750	NFM	Nationwide	Channel 6
959.0875	914.0875	NFM	Nationwide	Channel 7
959.1000	914.1000	NFM	Nationwide	Channel 8
959.1125	914.1125	NFM	Nationwide	Channel 9
959.1250	914.1250	NFM	Nationwide	Channel 10
959.1375	914.1375	NFM	Nationwide	Channel 11
959.1500	914.1500	NFM	Nationwide	Channel 12
959.1625	914.1625	NFM	Nationwide	Channel 13
959.1750	914.1750	NFM	Nationwide	Channel 14
959.1875	914.1875	NFM	Nationwide	Channel 15
959.2000	914.2000	NFM	Nationwide	Channel 16
959.2125	914.2125	NFM	Nationwide	Channel 17
959.2250	914.2250	NFM	Nationwide	Channel 18
959.2375	914.2375	NFM	Nationwide	Channel 19
959.2500	914.2500	NFM	Nationwide	Channel 20
959.2625	914.2625	NFM	Nationwide	Channel 21
959.2750	914.2750	NFM	Nationwide	Channel 22

Base	Mobile	Mode	Location	User & Notes
959.2875	914.2875	NFM	Nationwide	Channel 23
959.3000	914.3000	NFM	Nationwide	Channel 24
959.3125	914.3125	NFM	Nationwide	Channel 25
959.3250	914.3250	NFM	Nationwide	Channel 26
959.3375	914.3375	NFM	Nationwide	Channel 27
959.3500	914.3500	NFM	Nationwide	Channel 28
959.3625	914.3625	NFM	Nationwide	Channel 29
959.3750	914.3750	NFM	Nationwide	Channel 30
959.3875	914.3875	NFM	Nationwide	Channel 31
959.4000	914.4000	NFM	Nationwide	Channel 32
959.4125	914.4125	NFM	Nationwide	Channel 33
959.4250	914.4250	NFM	Nationwide	Channel 34
959.4375	914.4375	NFM	Nationwide	Channel 35
959.4500	914.4500	NFM	Nationwide	Channel 36
959.4625	914.4625	NFM	Nationwide	Channel 37
959.4750	914.4750	NFM	Nationwide	Channel 38
959.4875	914.4875	NFM	Nationwide	Channel 39
959.5000	914.5000	NFM	Nationwide	Channel 40
959.5125	914.5125	NFM	Nationwide	Channel 41
959.5250	914.5250	NFM	Nationwide	Channel 42
959.5375	914.5375	NFM	Nationwide	Channel 43
959.5500	914.5500	NFM	Nationwide	Channel 44
959.5625	914.5625	NFM	Nationwide	Channel 45
959.5750	914.5750	NFM	Nationwide	Channel 46
959.5875	914.5875	NFM	Nationwide	Channel 47
959.6000	914.6000	NFM	Nationwide	Channel 48
959.6125	914.6125	NFM	Nationwide	Channel 49
959.6250	914.6250	NFM	Nationwide	Channel 50
959.6375	914.6375	NFM	Nationwide	Channel 51
959.6500	914.6500	NFM	Nationwide	Channel 52
959.6625	914.6625	NFM	Nationwide	Channel 53
959.6750	914.6750	NFM	Nationwide	Channel 54
959.6875	914.6875	NFM	Nationwide	Channel 55
959.7000	914.7000	NFM	Nationwide	Channel 56
959.7125	914.7125	NFM	Nationwide	Channel 57
959.7250	914.7250	NFM	Nationwide	Channel 58
959.7375	914.7375	NFM	Nationwide	Channel 59
959.7500	914.7500	NFM	Nationwide	Channel 60
959.7625	914.7625	NFM	Nationwide	Channel 61
959.7750	914.7750	NFM	Nationwide	Channel 62
959.7875	914.7875	NFM	Nationwide	Channel 63
959.8000	914.8000	NFM	Nationwide	Channel 64
959.8125	914.8125	NFM	Nationwide	Channel 65
959.8250	914.8250	NFM	Nationwide	Channel 66
959.8375	914.8375	NFM	Nationwide	Channel 67
959.8500	914.8500	NFM	Nationwide	Channel 68
959.8625	914.8625	NFM	Nationwide	Channel 69
959.8750	914.8750	NFM	Nationwide	Channel 70
959.8875	914.8875	NFM	Nationwide	Channel 71
959.9000	914.9000	NFM	Nationwide	Channel 72

Base	Mobile	Mode	Location	User & Notes
959.9125	914.9125	NFM	Nationwide	Channel 73
959.9250	914.9250	NFM	Nationwide	Channel 74
959.9375	914.9375	NFM	Nationwide	Channel 75
959.9500	914.9500	NFM	Nationwide	Channel 76
959.9625	914.9625	NFM	Nationwide	Channel 77
959.9750	914.9750	NFM	Nationwide	Channel 78
959.9875	914.9875	NFM	Nationwide	Channel 79

960.0000 - 1215.0000 MHz DME Aeronautical Radio Navigation and Transponder Equipment

Base	Mobile	Mode	Location	User & Notes
962.0000	1025.0000	AM	Nationwide	DME Channel 1X Not Used
963.0000	1026.0000	AM	Nationwide	DME Channel 2X Not Used
964.0000	1027.0000	AM	Nationwide	DME Channel 3X Not Used
965.0000	1028.0000	AM	Nationwide	DME Channel 4X Not Used
966.0000	1029.0000	AM	Nationwide	DME Channel 5X Not Used
967.0000	1030.0000	AM	Nationwide	DME Channel 6X Not Used
968.0000	1031.0000	AM	Nationwide	DME Channel 7X Not Used
969.0000	1032.0000	AM	Nationwide	DME Channel 8X Not Used
970.0000	1033.0000	AM	Nationwide	DME Channel 9X Not Used
971.0000	1034.0000	AM	Nationwide	DME Channel 10X Not Used
972.0000	1035.0000	AM	Nationwide	DME Channel 11X Not Used
973.0000	1036.0000	AM	Nationwide	DME Channel 12X Not Used
974.0000	1037.0000	AM	Nationwide	DME Channel 13X Not Used
975.0000	1038.0000	AM	Nationwide	DME Channel 14X Not Used
976.0000	1039.0000	AM	Nationwide	DME Channel 15X Not Used
977.0000	1040.0000	AM	Nationwide	DME Channel 16X Not Used
978.0000	1041.0000	AM	Nationwide	DME Ch 17X (108.00 MHz)
		AM	RAF Greenham Common	TACAN
979.0000	1042.0000	AM	Nationwide	DME Ch 18X (108.10 MHz)
		AM	Belfast	DME
		AM	Dundee	DME
		AM	RAF Cottesmore	TACAN
980.0000	1043.0000	AM	Nationwide	DME Ch 19X (108.20 MHz)
		AM	RAF Boscombe Down	TACAN
981.0000	1044.0000	AM	Nationwide	DME Ch 20X (108.30 MHz)
982.0000	1045.0000	AM	Nationwide	DME Ch 21X (108.40 MHz)
		AM	RAF Valley	TACAN
983.0000	1046.0000	AM	Nationwide	DME Ch 22X (108.50 MHz)
984.0000	1047.0000	AM	Nationwide	DME Ch 23X (108.60 MHz)
		AM	Kirkwall	DME
		AM	RAF Bentwaters	TACAN
985.0000	1048.0000	AM	Nationwide	DME Ch 24X (108.70 MHz)
		AM	Newton Point	TACAN
986.0000	1049.0000	AM	Nationwide	DME Ch 25X (108.80 MHz)
		AM	Weathersfield	TACAN
987.0000	1050.0000	AM	Nationwide	DME Ch 26X (108.90 MHz)
		AM	Edinburgh	DME
		AM	Kerry	DME
		AM	RAF Woodbridge	DME
		AM	Ventnor	TACAN

Base	Mobile	Mode	Location	User & Notes
988.0000	1051.0000	AM	Nationwide	DME Ch 27X (109.00 MHz)
		AM	RAF Alconbury	TACAN
989.0000	1052.0000	AM	Nationwide	DME Ch 28X (109.10 MHz)
990.0000	1053.0000	AM	Nationwide	DME Ch 29X (109.20 MHz)
		AM	Inverness	DME
		AM	Swansea	DME
991.0000	1054.0000	AM	Nationwide	DME Ch 30X (109.30 MHz)
		AM	RAF Wattisham	TACAN
992.0000	1055.0000	AM	Nationwide	DME Ch 31X (109.40 MHz)
		AM	Barrow	DME
		AM	Guernsey	DME
993.0000	1056.0000	AM	Nationwide	DME Ch 32X (109.50 MHz)
		AM	London/Heathrow	DME
		AM	Manchester	DME
		AM	Plymouth	DME
994.0000	1057.0000	AM	Nationwide	DME Ch 33X (109.60 MHz)
		AM	RAF Linton-On-Ouse	TACAN
		AM	RAF Odiham	TACAN
995.0000	1058.0000	AM	Nationwide	DME Ch 34X (109.70 MHz)
996.0000	1059.0000	AM	Nationwide	DME Ch 35X (109.80 MHz)
		AM	RAF Kinloss	TACAN
		AM	RAF Lyneham	TACAN
997.0000	1060.0000	AM	Nationwide	DME Ch 36X (109.90 MHz)
		AM	Warton	DME
998.0000	1061.0000	AM	Nationwide	DME Ch 37X (110.00 MHz)
999.0000	1062.0000	AM	Nationwide	DME Ch 38X (110.10 MHz)
1000.0000	1063.0000	AM	Nationwide	DME Ch 39X (110.20 MHz)
		AM	RAF Lakenheath	TACAN
1001.0000	1064.0000	AM	Nationwide	DME Ch 40X (110.30 MHz)
		AM	London/Heathrow	DME
1002.0000	1065.0000	AM	Nationwide	DME Ch 41X (110.40 MHz)
1003.0000	1066.0000	AM	Nationwide	DME Ch 42X (110.50 MHz)
		AM	London/Stansted	DME
		AM	RAF Leuchars	TACAN
1004.0000	1067.0000	AM	Nationwide	DME Ch 43X (110.60 MHz)
1005.0000	1068.0000	AM	Nationwide	DME Ch 44X (110.70 MHz)
		AM	Carlisle	DME
1006.0000	1069.0000	AM	Nationwide	DME Ch 45X (110.80 MHz)
1007.0000	1070.0000	AM	Nationwide	DME Ch 46X (110.90 MHz)
		AM	London/Gatwick	DME
		AM	Ronaldsway	DME
1008.0000	1071.0000	AM	Nationwide	DME Ch 47X (111.00 MHz)
		AM	RN Yeovilton	TACAN
1009.0000	1072.0000	AM	Nationwide	DME Ch 48X (111.10 MHz)
		AM	RAF Coningsby	TACAN
1010.0000	1073.0000	AM	Nationwide	DME Ch 49X (111.20 MHz)
1011.0000	1074.0000	AM	Nationwide	DME Ch 50X (111.30 MHz)
1012.0000	1075.0000	AM	Nationwide	DME Ch 51X (111.40 MHz)
		AM	RAF Binbrook	TACAN
1013.0000	1076.0000	AM	Nationwide	DME Ch 52X (111.50 MHz)

Base	Mobile	Mode	Location	User & Notes
		AM	Newcastle	DME
		AM	RAF Fairford	TACAN
1014.0000	1077.0000	AM	Nationwide	DME Ch 53X (111.60 MHz)
		AM	RAF Chivenor	TACAN
1015.0000	1078.0000	AM	Nationwide	DME Ch 54X (111.70 MHz)
1016.0000	1079.0000	AM	Nationwide	DME Ch 55X (111.80 MHz)
1017.0000	1080.0000	AM	Nationwide	DME Ch 56X (111.90 MHz)
		AM	RAF Brize Norton	TACAN
1018.0000	1081.0000	AM	Nationwide	DME Ch 57X (112.00 MHz)
1019.0000	1082.0000	AM	Nationwide	DME Ch 58X (112.10 MHz)
		AM	Pole Hill	DME
1020.0000	1083.0000	AM	Nationwide	DME Ch 59X (112.20 MHz)
		AM	Isle Of Man	DME
		AM	Jersey	DME
1021.0000	1084.0000	AM	Nationwide	DME Ch 60X (112.30 MHz)
1022.0000	1085.0000	AM	Nationwide	DME Ch 61X Not Used
1023.0000	1086.0000	AM	Nationwide	DME Ch 62X Not Used
1024.0000	1087.0000	AM	Nationwide	DME Ch 63X Not Used
1025.0000	1088.0000	AM	Nationwide	DME Ch 64X Not Used
1026.0000	1089.0000	AM	Nationwide	DME Ch 65X Not Used
1027.0000	1090.0000	AM	Nationwide	DME Ch 66X Not Used
1028.0000	1091.0000	AM	Nationwide	DME Ch 67X Not Used
1029.0000	1092.0000	AM	Nationwide	DME Ch 68X Not Used
1030.0000	1090.0000	AM	Nationwide	Transponder Interrogation/ Reply
1030.0000	1093.0000	AM	Nationwide	DME Ch 69X Not Used
1031.0000	1094.0000	AM	Nationwide	DME Ch 70X (112.30 MHz)
1032.0000	1095.0000	AM	Nationwide	DME Ch 71X (112.40 MHz)
1033.0000	1096.0000	AM	Nationwide	DME Ch 72X (112.50 MHz)
		AM	St Abbs	DME
1034.0000	1097.0000	AM	Nationwide	DME Ch 73X (112.60 MHz)
		AM	RAF St Mawgan	TACAN
1035.0000	1098.0000	AM	Nationwide	DME Ch 74X (112.70 MHz)
		AM	Berry Head	DME
1036.0000	1099.0000	AM	Nationwide	DME Ch 75X (112.80 MHz)
		AM	Gamston	DME
1037.0000	1100.0000	AM	Nationwide	DME Ch 76X (112.90 MHz)
1038.0000	1101.0000	AM	Nationwide	DME Ch 77X (113.00 MHz)
1039.0000	1102.0000	AM	Nationwide	DME Ch 78X (113.10 MHz)
		AM	Strumble	DME
1040.0000	1103.0000	AM	Nationwide	DME Ch 79X (113.20 MHz)
		AM	St Anthony	DME
		AM	Warton	TACAN
1041.0000	1104.0000	AM	Nationwide	DME Ch 80X (113.30 MHz)
		AM	Shannon	DME
1042.0000	1105.0000	AM	Nationwide	DME Ch 81X (113.40 MHz)
1043.0000	1106.0000	AM	Nationwide	DME Ch 82X (113.50 MHz)
1044.0000	1107.0000	AM	Nationwide	DME Ch 83X (113.60 MHz)
		AM	London	DME
		AM	Wick	TACAN

Base	Mobile	Mode	Location	User & Notes
1045.0000	1108.0000	AM	Nationwide	DME Ch 84X (113.70 MHz)
		AM	RAF Upper Heyford	TACAN
1046.0000	1109.0000	AM	Nationwide	DME Ch 85X (113.80 MHz)
		AM	Talla	DME
1047.0000	1110.0000	AM	Nationwide	DME Ch 86X (113.90 MHz)
		AM	Ottringham	DME
1048.0000	1111.0000	AM	Nationwide	DME Ch 87X (114.00 MHz)
		AM	Midhurst	DME
1049.0000	1112.0000	AM	Nationwide	DME Ch 88X (114.10 MHz)
		AM	Wallasey	DME
1050.0000	1113.0000	AM	Nationwide	DME Ch 89X (114.20 MHz)
		AM	Lands End	DME
1051.0000	1114.0000	AM	Nationwide	DME Ch 90X (114.30 MHz)
		AM	Aberdeen/Dyce	DME
1052.0000	1115.0000	AM	Nationwide	DME Ch 91X (114.40 MHz)
		AM	Benbecula	TACAN
1053.0000	1116.0000	AM	Nationwide	DME Ch 92X (114.50 MHz)
1054.0000	1117.0000	AM	Nationwide	DME Ch 93X (114.60 MHz)
		AM	Cork	DME
1055.0000	1118.0000	AM	Nationwide	DME Ch 94X (114.70 MHz)
1056.0000	1119.0000	AM	Nationwide	DME Ch 95X (114.80 MHz)
		AM	RAF Sculthorpe	TACAN
1057.0000	1120.0000	AM	Nationwide	DME Ch 96X (114.90 MHz)
		AM	Vallafield	TACAN
1058.0000	1121.0000	AM	Nationwide	DME Ch 97X (115.00 MHz)
1059.0000	1122.0000	AM	Nationwide	DME Ch 98X (115.10 MHz)
		AM	Stornoway	TACAN
1060.0000	1123.0000	AM	Nationwide	DME Ch 99X (115.20 MHz)
		AM	Dean Cross	DME
1061.0000	1124.0000	AM	Nationwide	DME Ch 100X (115.30 MHz)
		AM	Ockham	DME
1062.0000	1125.0000	AM	Nationwide	DME Ch 101X (115.40 MHz)
		AM	Glasgow	DME
1063.0000	1126.0000	AM	Nationwide	DME Ch 102X (115.50 MHz)
1064.0000	1127.0000	AM	Nationwide	DME Ch 103X (115.60 MHz)
		AM	Lambourne	DME
1065.0000	1128.0000	AM	Nationwide	DME Ch 104X (115.70 MHz)
		AM	Trent	DME
1066.0000	1129.0000	AM	Nationwide	DME Ch 105X (115.80 MHz)
1067.0000	1130.0000	AM	Nationwide	DME Ch 106X (115.90 MHz)
		AM	RAF Mildenhall	TACAN
1068.0000	1131.0000	AM	Nationwide	DME Ch 107X (116.00 MHz)
		AM	RAF Machrihanish	TACAN
1069.0000	1132.0000	AM	Nationwide	DME Ch 108X (116.10 MHz)
		AM	RAF Church Fenton	TACAN
1070.0000	1133.0000	AM	Nationwide	DME Ch 109X (116.20 MHz)
		AM	Blackbushe	DME
1071.0000	1134.0000	AM	Nationwide	DME Ch 110X (116.30 MHz)
1072.0000	1135.0000	AM	Nationwide	DME Ch 111X (116.40 MHz)
		AM	Daventry	DME

Base	Mobile	Mode	Location	User & Notes
1073.0000	1136.0000	AM	Nationwide	DME Ch 112X (116.50 MHz)
		AM	RAF Coltishall	TACAN
1074.0000	1137.0000	AM	Nationwide	DME Ch 113X (116.60 MHz)
		AM	RAF Brawdy	TACAN
1075.0000	1138.0000	AM	Nationwide	DME Ch 114X (116.70 MHz)
1076.0000	1139.0000	AM	Nationwide	DME Ch 115X (116.80 MHz)
1077.0000	1140.0000	AM	Nationwide	DME Ch 116X (116.90 MHz)
1078.0000	1141.0000	AM	Nationwide	DME Ch 117X (117.00 MHz)
		AM	Seaford	DME
1079.0000	1142.0000	AM	Nationwide	DME Ch 118X (117.10 MHz)
		AM	RAF Woodbridge	TACAN
1080.0000	1143.0000	AM	Nationwide	DME Ch 119X (117.20 MHz)
		AM	Belfast	DME
1081.0000	1144.0000	AM	Nationwide	DME Ch 120X (117.30 MHz)
		AM	Detling	DME
1082.0000	1145.0000	AM	Nationwide	DME Ch 121X (117.40 MHz)
		AM	Connaught	DME
		AM	RAF Cranwell	TACAN
1083.0000	1146.0000	AM	Nationwide	DME Ch 122X (117.50 MHz)
		AM	Brookmans Park	DME
		AM	Turnberry	DME
1084.0000	1147.0000	AM	Nationwide	DME Ch 123X (117.60 MHz)
		AM	RAF Wittering	TACAN
1085.0000	1148.0000	AM	Nationwide	DME Ch 124X (117.70 MHz)
		AM	Oxford	DME
		AM	Tiree	DME
1086.0000	1149.0000	AM	Nationwide	DME Ch 125X (117.80 MHz)
1087.0000	1150.0000	AM	Nationwide	DME Ch 126X (117.90 MHz)
		AM	Mayfield	DME
1088.0000	1025.0000	AM	Nationwide	DME Channel 1Y Not Used
1089.0000	1026.0000	AM	Nationwide	DME Channel 2Y Not Used
1090.0000	1027.0000	AM	Nationwide	DME Channel 3Y Not Used
1091.0000	1028.0000	AM	Nationwide	DME Channel 4Y Not Used
1092.0000	1029.0000	AM	Nationwide	DME Channel 5Y Not Used
1093.0000	1030.0000	AM	Nationwide	DME Channel 6Y Not Used
1094.0000	1031.0000	AM	Nationwide	DME Channel 7Y Not Used
1095.0000	1032.0000	AM	Nationwide	DME Channel 8Y Not Used
1096.0000	1033.0000	AM	Nationwide	DME Channel 9Y Not Used
1097.0000	1034.0000	AM	Nationwide	DME Channel 10Y Not Used
1098.0000	1035.0000	AM	Nationwide	DME Channel 11Y Not Used
1099.0000	1036.0000	AM	Nationwide	DME Channel 12Y Not Used
1100.0000	1037.0000	AM	Nationwide	DME Channel 13Y Not Used
1101.0000	1038.0000	AM	Nationwide	DME Channel 14Y Not Used
1102.0000	1039.0000	AM	Nationwide	DME Channel 15Y Not Used
1103.0000	1040.0000	AM	Nationwide	DME Channel 16Y Not Used
1104.0000	1041.0000	AM	Nationwide	DME Ch 17Y (108.05 MHz)
		AM	Lydd	DME
1105.0000	1042.0000	AM	Nationwide	DME Ch 18Y (108.15 MHz)
1106.0000	1043.0000	AM	Nationwide	DME Ch 19Y (108.25 MHz)
1107.0000	1044.0000	AM	Nationwide	DME Ch 20Y (108.35 MHz)

The UK Scanning Directory

Base	Mobile	Mode	Location	User & Notes
1108.0000	1045.0000	AM	Nationwide	DME Ch 21Y (108.45 MHz)
1109.0000	1046.0000	AM	Nationwide	DME Ch 22Y (108.55 MHz)
1110.0000	1047.0000	AM	Nationwide	DME Ch 23Y (108.65 MHz)
1111.0000	1048.0000	AM	Nationwide	DME Ch 24Y (108.75 MHz)
		AM	Humberside	DME
1112.0000	1049.0000	AM	Nationwide	DME Ch 25Y (108.85 MHz)
1113.0000	1050.0000	AM	Nationwide	DME Ch26Y (108.95 MHz)
		AM	Woodford	DME
1114.0000	1051.0000	AM	Nationwide	DME Ch 27Y (109.05 MHz)
		AM	Yeovil	DME
1115.0000	1052.0000	AM	Nationwide	DME Ch 28Y (109.15 MHz)
		AM	Luton	DME
1116.0000	1053.0000	AM	Nationwide	DME Ch 29Y (109.25 MHz)
1117.0000	1054.0000	AM	Nationwide	DME Ch 30Y (109.35 MHz)
		AM	Biggin Hill	DME
1118.0000	1055.0000	AM	Nationwide	DME Ch 31Y (109.45 MHz)
1119.0000	1056.0000	AM	Nationwide	DME Ch 32Y (109.55 MHz)
1120.0000	1057.0000	AM	Nationwide	DME Ch 33Y (109.65 MHz)
1121.0000	1058.0000	AM	Nationwide	DME Ch 34Y (109.75 MHz)
1122.0000	1059.0000	AM	Nationwide	DME Ch 35Y (109.85 MHz)
		AM	Fair Oaks	DME
1123.0000	1060.0000	AM	Nationwide	DME Ch36Y (109.95 MHz)
1124.0000	1061.0000	AM	Nationwide	DME Ch 37Y (110.05 MHz)
1125.0000	1062.0000	AM	Nationwide	DME Ch 38Y (110.15 MHz)
1126.0000	1063.0000	AM	Nationwide	DME Ch39Y (110.25 MHz)
1127.0000	1064.0000	AM	Nationwide	DME Ch 40Y (110.35 MHz)
1128.0000	1065.0000	AM	Nationwide	DME Ch 41Y (110.45 MHz)
1129.0000	1066.0000	AM	Nationwide	DME Ch 42Y (110.55 MHz)
1130.0000	1067.0000	AM	Nationwide	DME Ch 43Y (110.65 MHz)
1131.0000	1068.0000	AM	Nationwide	DME Ch 44Y (110.75 MHz)
1132.0000	1069.0000	AM	Nationwide	DME Ch 45Y (110.85 MHz)
1133.0000	1070.0000	AM	Nationwide	DME Ch 46Y (110.95 MHz)
1134.0000	1071.0000	AM	Nationwide	DME Ch 47Y (111.05 MHz)
1135.0000	1072.0000	AM	Nationwide	DME Ch 48Y (111.15 MHz)
1136.0000	1073.0000	AM	Nationwide	DME Ch 49Y (111.25 MHz)
1137.0000	1074.0000	AM	Nationwide	DME Ch 50Y (111.35 MHz)
1138.0000	1075.0000	AM	Nationwide	DME Ch 51Y (111.45 MHz)
1139.0000	1076.0000	AM	Nationwide	DME Ch 52Y (111.55 MHz)
1140.0000	1077.0000	AM	Nationwide	DME Ch 53Y (111.65 MHz)
1141.0000	1078.0000	AM	Nationwide	DME Ch 54Y (111.75 MHz)
		AM	Liverpool	DME
1142.0000	1079.0000	AM	Nationwide	DME Ch 55Y (111.85 MHz)
1143.0000	1080.0000	AM	Nationwide	DME Ch 56Y (111.95 MHz)
1144.0000	1081.0000	AM	Nationwide	DME Ch 57Y (112.05 MHz)
1145.0000	1082.0000	AM	Nationwide	DME Ch 58Y (112.15 MHz)
1146.0000	1083.0000	AM	Nationwide	DME Ch 59Y (112.25 MHz)
1147.0000	1084.0000	AM	Nationwide	DME Channel 60Y Not Used
1148.0000	1085.0000	AM	Nationwide	DME Channel 61Y Not Used
1149.0000	1086.0000	AM	Nationwide	DME Channel 62Y Not Used
1150.0000	1087.0000	AM	Nationwide	DME Channel 63Y Not Used

Base	Mobile	Mode	Location	User & Notes
1151.0000	1088.0000	AM	Nationwide	DME Channel 64Y Not Used
1152.0000	1089.0000	AM	Nationwide	DME Channel 65Y Not Used
1153.0000	1090.0000	AM	Nationwide	DME Channel 66Y Not Used
1154.0000	1091.0000	AM	Nationwide	DME Channel 67Y Not Used
1155.0000	1092.0000	AM	Nationwide	DME Channel 68Y Not Used
1156.0000	1093.0000	AM	Nationwide	DME Channel 69Y Not Used
1157.0000	1094.0000	AM	Nationwide	DME Ch 70Y (112.35 MHz)
1158.0000	1095.0000	AM	Nationwide	DME Ch 71Y (112.45 MHz)
1159.0000	1096.0000	AM	Nationwide	DME Ch 72Y (112.55 MHz)
1160.0000	1097.0000	AM	Nationwide	DME Ch 73Y (112.65 MHz)
1161.0000	1098.0000	AM	Nationwide	DME Ch 74Y (112.75 MHz)
1162.0000	1099.0000	AM	Nationwide	DME Ch 75Y (112.85 MHz)
1163.0000	1100.0000	AM	Nationwide	DME Ch 76Y (112.95 MHz)
1164.0000	1101.0000	AM	Nationwide	DME Ch 77Y (113.05 MHz)
1165.0000	1102.0000	AM	Nationwide	DME Ch 78Y (113.15 MHz)
1166.0000	1103.0000	AM	Nationwide	DME Ch 79Y (113.25 MHz)
1167.0000	1104.0000	AM	Nationwide	DME Ch 80Y (113.35 MHz)
		AM	Southampton	DME
1168.0000	1105.0000	AM	Nationwide	DME Ch 81Y (113.45 MHz)
1169.0000	1106.0000	AM	Nationwide	DME Ch 82Y (113.55 MHz)
		AM	Manchester	DME
1170.0000	1107.0000	AM	Nationwide	DME Ch 83Y (113.65 MHz)
		AM	Honiley	DME
1171.0000	1108.0000	AM	Nationwide	DME Ch 84Y (113.75 MHz)
		AM	Bovingdon	DME
1172.0000	1109.0000	AM	Nationwide	DME Ch 85Y (113.85 MHz)
1173.0000	1110.0000	AM	Nationwide	DME Ch 86Y (113.95 MHz)
1174.0000	1111.0000	AM	Nationwide	DME Ch 87Y (114.05 MHz)
1175.0000	1112.0000	AM	Nationwide	DME Ch 88Y (114.15 MHz)
1176.0000	1113.0000	AM	Nationwide	DME Ch 89Y (114.25 MHz)
		AM	Newcastle	DME
1177.0000	1114.0000	AM	Nationwide	DME Ch90Y (114.35 MHz)
		AM	Compton	DME
1178.0000	1115.0000	AM	Nationwide	DME Ch 91Y (114.45 MHz)
1179.0000	1116.0000	AM	Nationwide	DME Ch 92Y (114.55 MHz)
		AM	Clacton	DME
1180.0000	1117.0000	AM	Nationwide	DME Ch 93Y (114.65 MHz)
1181.0000	1118.0000	AM	Nationwide	DME Ch 94Y (114.75 MHz)
1182.0000	1119.0000	AM	Nationwide	DME Ch 95Y (114.85 MHz)
1183.0000	1120.0000	AM	Nationwide	DME Ch 96Y (114.95 MHz)
		AM	Dover	DME
1184.0000	1121.0000	AM	Nationwide	DME Ch 97Y (115.05 MHz)
1185.0000	1122.0000	AM	Nationwide	DME Ch 98Y (115.15 MHz)
1186.0000	1123.0000	AM	Nationwide	DME Ch 99Y (115.25 MHz)
1187.0000	1124.0000	AM	Nationwide	DME Ch 100Y (115.35 MHz)
1188.0000	1125.0000	AM	Nationwide	DME Ch 101Y (115.45 MHz)
1189.0000	1126.0000	AM	Nationwide	DME Ch 102Y (115.55 MHz)
		AM	Gloucestershire	DME
1190.0000	1127.0000	AM	Nationwide	DME Ch 103Y (115.65 MHz)
1191.0000	1128.0000	AM	Nationwide	DME Ch 104Y (115.75 MHz)

Base	Mobile	Mode	Location	User & Notes
1192.0000	1129.0000	AM	Nationwide	DME Ch 105Y (115.85 MHz)
1193.0000	1130.0000	AM	Nationwide	DME Ch 106Y (115.95 MHz)
1194.0000	1131.0000	AM	Nationwide	DME Ch 107Y (116.05 MHz)
1195.0000	1132.0000	AM	Nationwide	DME Ch 108Y (116.15 MHz)
1196.0000	1133.0000	AM	Nationwide	DME Ch 109Y (116.25 MHz)
		AM	Barkway	DME
1197.0000	1134.0000	AM	Nationwide	DME Ch 110Y (116.35 MHz)
1198.0000	1135.0000	AM	Nationwide	DME Ch 111Y (116.45 MHz)
1199.0000	1136.0000	AM	Nationwide	DME Ch 112Y (116.55 MHz)
1200.0000	1137.0000	AM	Nationwide	DME Ch 113Y (116.65 MHz)
1201.0000	1138.0000	AM	Nationwide	DME Ch 114Y (116.75 MHz)
		AM	Cambridge	DME
1202.0000	1139.0000	AM	Nationwide	DME Ch 115Y (116.85 MHz)
1203.0000	1140.0000	AM	Nationwide	DME Ch 116Y (116.95 MHz)
1204.0000	1141.0000	AM	Nationwide	DME Ch 117Y (117.05 MHz)
1205.0000	1142.0000	AM	Nationwide	DME Ch 118Y (117.15 MHz)
1206.0000	1143.0000	AM	Nationwide	DME Ch 119Y (117.25 MHz)
1207.0000	1144.0000	AM	Nationwide	DME Ch 120Y (117.35 MHz)
		AM	Sumburgh	DME
1208.0000	1145.0000	AM	Nationwide	DME Ch 121Y (117.45 MHz)
		AM	Brecon	DME
1209.0000	1146.0000	AM	Nationwide	DME Ch 122Y (117.55 MHz)
1210.0000	1147.0000	AM	Nationwide	DME Ch 123Y (117.65 MHz)
1211.0000	1148.0000	AM	Nationwide	DME Ch 124Y (117.75 MHz)
1212.0000	1149.0000	AM	Nationwide	DME Ch 125Y (117.85 MHz)
1213.0000	1150.0000	AM	Nationwide	DME Ch 126Y (117.95 MHz)

The UK Scanning Directory
Scanner Log Sheet

Sheet............ of ....

Frequency (MHz)		Mode	Location	Callsign	Comments
Simplex	Duplex				

Thank you for your contribution to the UK Scanning Directory and to the hobby in general. All requests for anonymity and unsigned submissions will be treated in the strictest confidence. If you need more space, please feel free to use blank sheets of paper or photocopy this one.

RSLs RESTRICTED SERVICE LICENCES